W9-CGK-261

JUN. 4.2006

Date Due

BRODART, CO. Cat. No. 23-233-003 Printed in U.S.A.

Wildlife-Habitat Relationships

$28.50 University of Wisconsin Press Received Nov. 1992

Michael L. Morrison, Bruce G. Marcot,
and R. William Mannan

Wildlife-Habitat Relationships

Concepts and Applications

The University of Wisconsin Press

LIBRARY
UAS—JUNEAU

The University of Wisconsin Press
114 North Murray Street
Madison, Wisconsin 53715

3 Henrietta Street
London WC2E 8LU, England

Copyright © 1992
The Board of Regents of the University of Wisconsin System
All rights reserved

5 4 3 2 1

Printed in the United States of America

Figure 2.7 is from *An Introduction to Population Ecology* by G. E. Hutchinson, Yale University Press. Copyright © 1978 by Yale University.

Library of Congress Cataloging-in-Publiation Data
Morrison, Michael L.
Wildlife-habitat relationships: concepts and applications/
Michael L. Morrison, Bruce G. Marcot, R. William Mannan.
364pp. cm.
Includes bibliographical references and index.
ISBN 0-299-13200-5
1. Habitat (Ecology) 2. Animal ecology. Marcot, Bruce G.
II. Mannan, R. William. III. Title.
QH541.M59 1992 574.5′22–dc20 91–37591

LIBRARY
UAS—JUNEAU

To E. Charles Meslow and Jack Ward Thomas,
who taught us that wildlife conservation truly succeeds
when practiced with honor, rapport, and rigor

Contents

WILLIAM A. EGAN LIBRARY
UNIVERSITY OF ALASKA SOUTHEAST
11120 GLACIER HIGHWAY
JUNEAU, AK 99801-8676

Figures and Illustrations

Figures

Illustrations

Tables

Preface

The purpose of this book is to provide more than just a general introduction to wildlife biology. Several introductions are available, beginning with Aldo Leopold's *Game Management*, supplemented by Raymond F. Dasmann's *Wildlife Biology*. More recent works include James A. Bailey, *Principles of Wildlife Management*; William L. Robinson and Eric G. Bolen, *Wildlife Ecology and Management*; Frederick F. Gilbert and Donald G. Dodds, *The Philosophy and Practice of Wildlife Management*; Stanley H. Anderson, *Managing Our Wildlife Resources*; and Malcolm L. Hunter, Jr., *Wildlife, Forests, and Forestry*. We assume that readers are familiar with basic principles of inventory and monitoring of animals and their habitats. The books *Inventory and Monitoring of Wildlife Habitat*, edited by A. Y. Cooperrider, R. J. Boyd, and H. R. Stuart, and *Wildlife Habitats in Managed Forests: The Blue Mountains of Oregon and Washington*, edited by Jack Ward Thomas, provide overviews of basic sampling methodologies categorized by habitat type and major taxonomic groups. These texts and others have as their major objective the enlightenment of college undergraduates and nonprofessionals who aspire to become professional wildlife biologists or who are interested in habitat management concepts and issues.

Here we attempt to advance from the point at which these texts leave off. Our objective is to provide more in-depth information on the concepts associated with wildlife-habitat relationships. Our approach combines basic field zoology and natural history, ecological theory, and quantitative tools. From our perspective, a synthesis of these topics lies at the heart of understanding all natural ecological communities.

This book is intended for advanced undergraduate and graduate students with a background in general wildlife biology and principles of ecology, at both organismic and population levels. We hope it also provides interpretations of ecological theory useful for the practicing wildlife biologist. A course in statistics through analysis of variance, including correlation, regression, and inferential statistics, is recommended but not required to proceed through most chapters. The chapter on multivariate analysis is written so that the general principles underlying the use of such techniques can be readily grasped with or without a formal course on the subject.

We attempt to draw on the best examples of topics we raise, regardless of the species of animal involved or its geographical location. We do concentrate, however, on terrestrial vertebrates, especially birds and mammals, from temperate latitudes, since most examples pertinent to our book come from studies of these animals. We have tried to strike a balance between offering wide-ranging examples and providing adequate detail about each topic.

We emphasize the need for critical evaluation of methodologies and their applications to wildlife research. Wildlife biologists often base management decisions on low sample sizes, biased sampling techniques, and/or inappropriate analyses and interpretations, by design or default or through lack of understanding of rigorous scientific analysis. Although we agree that, at times, some information is better than none at all, we must at least be aware of how these problems increase the risk associated with management decisions. Technical tools available to scientists are outpacing our attention to rigorous pursuit of the scientific method. High-speed computers and "canned" statistical packages, for example, have made multivariate statistics available to those with limited statistical training. To aid the student and the professional, we have tried to explain fundamental concepts of wildlife habitat theory and assessments so that the use of advanced technical tools is more acceptable, more often sought, and more appropriately applied. Ultimately, the success of conservation efforts depends on gathering, analyzing, and interpreting reliable information on species composition, communities, and habitats. We hope this volume encourages such rigor.

The book was conceived in discussions the authors had at the 1984 symposium "Wildlife 2000: Modeling Habitat Relationships of Terrestrial Vertebrates." Each of us at one time has designed and taught graduate-level courses concerning wildlife-habitat relationships. The need for a text such as this one is clear; no such work currently exists. The "Wildlife 2000" symposium and the book that emerged from it brought together experts from various fields but could not present a unified review of the state of the art; more important, we thought it essential that the critical topic of wildlife and their habitats be evaluated under one cover. While it is well known that animals are tightly bound to specific configurations and types of habitats, no single, comprehensive review is available. Although we cannot survey all available literature, we trust this book will fill the gap that exists at this writing.

We designed the book to cover both theoretical and applied aspects of wildlife-habitat relationships with an emphasis on the theoretical framework under which such relationships are studied. We begin with a look at the historical and legislative background under which emphasis on wildlife-habitat relationships evolved. Next we examine habitat from an ecological and evolutionary perspective, habitat fragmentation, ways in which habitat can be measured and the data then analyzed, and ways the behavior of animals fits into analysis of habitat relationships. We close by discussing the development of predictive models and multivariate analysis. Throughout, we describe the directions we think the study of wildlife and their habitats should take.

Acknowledgments

We thank David Capen and Stanley Temple for their thorough reviews of the manuscript. We also thank Allen Fitchen of the University of Wisconsin Press for helping to guide the project to completion. Lori Merkle of the University of California, Berkeley, helped immensely in so many ways with correspondence, compilation, and other tasks; her work is greatly appreciated. We also thank Jerald Morse for helping with producing the figures.

Science proceeds toward understanding and predicting the natural world and progresses best through collective ventures. Discussions with colleagues since the inception of this project helped to gel many of our concepts. We especially appreciate dialogues with William Block, Larry Harris, Richard Holthausen, Winifred Kessler, Lloyd Kiff, E. Charles Meslow, Reed Noss, David Patton, Martin Raphael, Len Ruggiero, Hal Salwasser, Jack Ward Thomas, Jared Verner, David Wilcove, and a host of wildlife biologists and ecologists practicing conservation in the field. The actual face and shape of the land provide the ultimate—and, essentially, the only authentic—measures of wildlife conservation efforts. We hope this volume helps in the mission to correctly assess the results of such efforts.

Wildlife-Habitat Relationships

1 The Study of Habitat: A Historical Perspective

> Some birds live on mountains or in forests, as the hoopoe and the brenthus. . . . Some birds live on the sea-shore, as the wagtail. . . . Web-footed birds without exception live near the sea or rivers or pools, as they naturally resort to places adapted to their structure.
>
> —Aristotle, *Historia animalium* [344 B.C.] 1862

Introduction

An animal's habitat is, in the most general sense, the place where it lives. All animals, except humans, can live in an area only if basic resources such as food, water, and cover are present and if the animals have adaptations that allow them to cope with the climatic extremes and the competitors and predators they encounter. Humans can live even if these requirements are not met, because we can modify environments to suit our needs or desires and because we have access to resources such as food or building materials from all over the world. For these reasons, humans occupy nearly all parts of the earth, but other species of animals are restricted to particular kinds of places.

The distribution of animal species among environments and the forces that cause these distributions frequently have been subjects of human interest, but for different reasons at different times. The primary purpose of this introductory chapter is to review some of the reasons why people study the habitats of animals and to outline how these reasons have changed over time. We also introduce the major concepts that will be addressed in this book.

Curiosity about Natural History

Throughout recorded history humans, motivated by their own curiosity, have observed and written about the habits of animals. The writings of naturalists were, for centuries, the only recorded sources of information about animal-habitat relationships. Aristotle was the first and undoubtedly the best of the very early naturalists. He observed animals and wrote about a wide variety of subjects, including breeding behavior, diets, migration, and hibernation. Aristotle also noted where animals lived and occasionally speculated about the reasons why.

> A number of fish also are found in sea-estuaries; such as the saupe, the gilthead, the red mullet, and, in point of fact, the greater part of the gregarious fishes. . . . Fish penetrate into the Euxine [estuary] for two reasons, and firstly for food. For the feeding is more abundant and better in quality owing to the amount of fresh river-water that discharges into the sea. . . . Furthermore, fish penetrate into this sea for the purpose of breeding; for there are recesses there favorable for spawning, and the fresh and exceptionally sweet water has an invigorating effect upon the spawn. (Aristotle [344 B.C.] 1862)

Interest in natural history waned after Aristotle's death. Politics and world conquest received the focus of attention during the growth of the Roman Empire, and interest in religion and metaphysics suppressed creative observation of the natural world during the rise of Christendom (Beebe 1988). As a result, little new information was documented about animals and their habitats for nearly 1700 years after Aristotle's death. Yet, as Klopfer and Ganzhorn (1985) noted, painters in the medieval and pre-Renaissance periods still showed an appreciation for the association of specific animals with particular features of the environment. "Fanciful renderings aside, peacocks do not appear in drawings of moors nor moorhens in wheatfields" (Klopfer and Ganzhorn 1985:436). Thus, keen observers noticed relationships between animals and their habitats during the "Dark Ages," but few of their observations were recorded.

The study of natural history was renewed in the seventeenth and eighteenth centuries. Most naturalists during this period, like John Ray (seventeenth century) and Carl Linnaeus (eighteenth century), were interested primarily in naming and classifying organisms in the natural world (Eiseley 1961). Explorers made numerous expeditions into unexplored or unmapped lands during this period, often with the intent of locating new trade routes or identifying new resources. Naturalists usually accompanied the expeditions and collected or recorded information about the plants and animals they observed. Many new facts about the existence and distribution of animals worldwide were collected during this time, and the advance in knowledge generated considerable curiosity about the natural world.

During the nineteenth century naturalists continued to describe the distribution of new plants and animals, but they also began to formulate ideas about how the natural world functioned. Charles Darwin was the most prominent of these naturalists. His observations of the distributions of similar species was one set of facts among many that he marshaled to support his theory of evolution by natural selection (Darwin 1859). The work of Darwin is highlighted here not only because he recorded many new facts about animals but also because the theory of evolution by natural selection forms the framework and foundation of the field of ecology.

Curiosity about Ecological Relationships

In the early 1900s curiosity about how animals interacted with their environments provided the impetus for numerous investigations into what are now called ecological relationships. Interest in these relationships initially led to detailed descriptions of the distribution of animals along environmental gradients or among plant communities. Merriam (1890), for example, identified the changes that occurred in plant and animal species on an elevational gradient, and Adams (1908) studied changes in bird species that accompanied plant succession. Early biologists postulated that climatic conditions and availability of food and sites to breed were the primary causes of the distributions of animals they observed (see Grinnell 1917).

Later biologists, however, soon recognized that the distributions of some animals could not be explained solely on the basis of climate and essential resources. David Lack (1933) was apparently the first to propose that some animals (in this case, birds) recognize conspicuous features of appropriate environments and use these features to actively select a place to live. Areas without these features, according to Lack, generally would not be inhabited, even though they might contain all the necessary resources for survival. Lack's ideas gave birth to the concept of habitat selection and stimulated considerable research on animal-habitat relationships during the succeeding fifty years.

Svardson (1949) developed a general conceptual model of habitat selection and Hilden (1965) later expressed similar ideas. Their models characterized habitat selection as a two-stage process in which individuals first use general features of the landscape to broadly select among different environments and then respond to more subtle habitat characteristics to choose a specific place to live. Harris (1952) attempted to experimentally test the idea of habitat selection by presenting individual prairie deer mice (*Peromyscus maniculatus bairdi*) and woodland deer mice (*P. m. gracilis*) with the choice of artificial woods or artificial fields. He found that test animals preferred the artificial habitat that most closely resembled the natural environment of their own subspecies and concluded that the mice were reacting to visual cues provided by the artificial vegetation. Wecker (1963) later attempted to determine whether early experience (that is, learning) played a role in habitat recognition and selection in deer mice. He found that the behaviors associated with habitat selection were primarily controlled by what Mayr (1974) called closed genetic programs: that is, they were not greatly affected by early experience. Similar experiments conducted by Klopfer (1963), however, revealed that habitat selection by chipping sparrows (*Spizella passerina*) could be modified to some extent by early experience.

Svardson (1949) also suggested that factors other than habitat cues, such as the presence of conspecifics and interspecific competitors, could influence whether an animal stayed in a particular place. Habitat selection was, therefore, recognized as a complicated process involving several levels of discrimination and a number of potentially interacting factors. G. Evelyn Hutchinson formally articulated the multivariate nature of the causes of animal distribution in his presentation of the concept of the *n*-dimensional niche (Hutchinson 1957). From the 1960s to the present development of the concept of the niche, particularly the assessment of the effects of interspecific competition on the distribution of animals, has stimulated many detailed studies of how animals use their habitats. These studies have shown clearly that, for some species, habitat selection is influenced by conspecifics (Butler 1980), interspecific competitors (Werner and Hall 1979), and predators (Werner et al. 1983), as well as features of the environment that are directly or indirectly related to resources needed for survival and reproduction (see Chapter 2 for more details about niche and habitat selection).

Hunting Animals for Food or Sport

The earliest humans relied, in part, on killing animals for survival, and they undoubtedly recognized and exploited the patterns of association between the animals they hunted and the kinds of places where these animals were most abundant. Similarly, men who later made their living by trapping and hunting or could afford the luxury of hunting for sport knew where to find animals and could likely have speculated accurately about the habitat features that influenced the abundance of game species. Marco Polo reported, for example, that in the Mongol Empire in Asia, Kublai Khan (A.D. 1215–1294) increased the numbers of quail and partridges available to him for falconry by planting patches of food, distributing grain during the winter, and controlling cover (Leopold 1933). This advanced system of habitat management suggests a general understanding of the habitat requirements of the target game birds, but it is unlikely that the information was obtained through organized studies of habitat use. Also, the men who hunted and trapped for subsistence or sport rarely recorded their knowledge about habitats for posterity.

Not until people began to attempt to systematically apply biology to the management of game as a "crop" in the early 1900s did they realize that "science had accumulated more knowledge of how to distinguish one species from another than of the habits, requirements, and interrelationships of living populations" (Leopold 1933:20). The absence of information about the habitat requirements of most animals and the desire to increase game populations by manipulating the environment stimulated detailed investigations of the habitats and life histories of game species. H. L. Stoddard's work on bobwhite quail (*Colinus virginianus*), published in 1931, exemplifies early efforts of this kind.

Studies similar to Stoddard's have been conducted on most game animals in North America between 1930 and today (for example, Bellrose 1976; Wallmo 1981; Thomas and Toweill 1982), but many of these studies only summarized general habitat associations and did not identify critical habitat components. During the last two decades the number of hunters has increased dramatically while undeveloped land available for managing wild animal populations has decreased. Therefore, the need to manage populations more intensively is great, and detailed knowledge of habitat requirements is essential for this task. Studies of the habitat requirements of game animals continue to be conducted, as one can easily see by reviewing recent scientific journals on wildlife management.

Concern for Animal Species Threatened by Human Activities

Human activities have dramatically disturbed natural environments in North America (and elsewhere in the world) during the 1900s. These disturbances have primarily been associated with the rapid increase in the size of the human population and the exploitation of natural resources for human use. Concern about the negative effects of human activities on animal populations and other aspects of the natural environment eventually led to the passage of a number of laws in the United States designed to reduce environmental disturbance. Some of these laws also stimulated research on the habitat requirements of animal species (table 1.1).

Table 1.1. Important legislation in the United States that stimulated study, preservation, or management of animal habitat

Title	Date	Action
Migratory Bird Conservation Act	1929	Provided for the establishment of wildlife refuges
Migratory Bird Hunting Stamp Act	1934	Required a federal migratory bird hunting license; funds used to purchase lands for refuges
Fish and Wildlife Coordination Act	1934	Authorized conservation measures in federal water projects and required consultation with the U.S. Fish and Wildlife Service and states concerning any water project
Pittman-Robertson Federal Aid in Wildlife Restoration Act	1937	Provided for an excise tax on sporting arms and ammunition to finance research on a federal and state cooperative basis
Multiple Use–Sustained Yield Act	1960	Stipulated that national forests would be managed for outdoor recreation, range, timber, watershed, and wildlife and fish
Classification and Multiple Use Act	1969	Mandated multiple use management on lands administered by the Bureau of Land Management
Wilderness Act	1964	Gave Congress authority to identify and set aside wilderness areas
National Environmental Policy Act (NEPA)	1969	Stipulated that environmental impact statements would be prepared for any federal project that affected the quality of the human environment; wildlife habitat considered part of that environment
Endangered Species Conservation Act	1973	Provided protection for species and the habitat of species threatened with extinction
Sikes Act Extension	1974	Directed secretaries of Agriculture and Interior to cooperate with state game and fish agencies to develop plans for conservation of wildlife, fish, and game

Table 1.1. (*Continued*)

Title	Date	Action
Forest and Rangelands Renewable Resources Planning Act (FRPA)	1974	Called for units of the National Forest System to prepare land management plans for the protection and development of national forests
Federal Land Policy and Management Act	1976	Mandated land use plans on lands administered by the Bureau of Land Management
National Forest Management Act (NFMA)	1976	Stipulated that the plans called for in the FRPA would comply with NEPA and that management would maintain viable populations of existing native vertebrates on national forests

Source: Based in part on Gilbert and Dodds 1987:17.

Among the laws that have produced the most activity in the study of animal habitats are the National Environmental Policy Act of 1969, the Endangered Species Conservation Act of 1973, the Federal Land Policy and Management Act of 1976 and the National Forest Management Act of 1976 (Bean 1977). Legislators designed these laws, in part, to ensure that wildlife and other natural resources were considered in the planning and execution of human activities on public lands. Knowledge of the habitat requirements of animal species obviously is required before the effects of an environmental disturbance can be evaluated, before a refuge can be designed for an endangered species, or before animal habitats can be maintained on lands managed under a multiple-use philosophy. Biologists responded to the need for information about habitat requirements by studying, often for the first time, numerous species of "nongame" animals and by developing models to help predict the effects of environmental changes on animal populations (Verner, Morrison, and Ralph 1986).

The increase in environmental awareness during the 1970s also led to the development of nongame management programs within many state fish and game agencies. Inadequate funding currently limits most of these programs (funding is usually less than 5 percent of the expenditures for management of game animals), but the programs often emphasize identifying and managing habitat. State nongame programs, therefore, have added substantially to existing information about animal-habitat relationships.

Ethical Concerns

Another impetus for studying habitat relates to an ethical concern for the future of wildlife and natural communities. This concern is, in part, a humanistic one, insofar as the health of natural systems affects our use and enjoyment of natural resources in the broadest sense. From a utilitarian viewpoint, the world also is our habitat, and its health directly relates to our own. The ethical concern, however, transcends humanism in that wildlife and natural communities are intrinsic to the world in which we have evolved and now live. Writers of legal as well as ethical literature have argued that nonhuman species have, in some sense, their own natural right to exist and grow. The study of wildlife and habitat in this context may deepen our appreciation for and ethical responsibility to other species and natural systems.

Why should we be concerned about species and habitats that offer no immediate economic or recreational benefits? Several rather standard philosophical arguments offer complementary or even conflicting rationales. One viewpoint argues for conserving species and their environments because we may someday learn how to exploit them for medical or other benefits. Another viewpoint argues for preserving species for the unknown (and unknowable) interests of future generations; we cannot speak for the desires of our as-yet-unborn progeny who will inherit the results of our management decisions.

Generally, a traditional conflict has pitted ethical humanism against humane moralism. Ethical humanism, as championed by Guthrie, Kant, Locke, More, and Aquinas, argues that animals are not "worthy" of equal consideration; animals are not in some sense "up to" human levels, in that they do not share self-consciousness and personal interests. In effect, this argument allows us to subjugate wildlife and their habitats. Kant argued as much. He advanced his idea on a so-called deontological theme (from the Latin *deos*, or "duty"). That is, rights—specifically, human rights—allow us to view animals as having less value because they are less rational (or are arational); we humans have the duty to manage species and the freedom to subjugate them.

On the other hand, humane moralism, as championed by Betham (of the animal liberation movement) and Singer, argues that not all humans have equal values and that some animals deserve the focus of ethical consideration. According to this argument, humans are moral agents. Animals and, by extension, their habitats require consideration equal to that given humans, even if they do not ultimately receive equal treatment.

There also is a third ethical stance, one which may form the central impetus for studying and conserving wildlife and their habitat: an ecological ethic. The ecological ethic, as proposed by Callicott, has been most eloquently advanced by Leopold in his *Sand County Almanac*, although elements of his philosophy can be traced to Henri Bergson, Teilhard de Chardin, and John Dewey. The focus of ethical consideration in this view is on both the individual organism and the community in which it resides. Concern for the community essentially constitutes Leopold's ecological ethic, a holistic ethic which is concerned with the relationships of animals with each other and with their environment.

Leopold wrote of soil, water, plants, animals, oceans, and mountains, calling each a "natural entity." In his view, animals' functional roles in the community, not solely their utility for humans, provide a measure of their value. By extension, then, in order to act morally, we must maintain our individual human integrity, our social integrity, and the integrity of the biotic community.

Following such an ecological ethic, a concern for the present and future conditions of wildlife and their habitats motivates the writing of this book. The sad history of massive resource depletion, including extinctions of plant and animal species and large-scale alteration of terrestrial and aquatic environments, must, in our view, strengthen a commitment to further understanding wildlife and their habitats. Understanding is the necessary overture to truly living by an ecological ethic.

Concepts Addressed

This book covers both the theoretical and applied aspects of wildlife-habitat relationships with an emphasis on the theoretical framework under which researchers should study such relationships. An appropriate way to begin a preview of the concepts covered in subsequent chapters is to define the term *habitat*. A review of even a few papers concerned with the subject of habitat will show that the term is used in a variety of ways. Frequently *habitat* is used to describe an area supporting a particular type of vegetation. This use probably grew from the phrase *habitat type,* coined by Daubenmire to refer to "land units having approximately the same capacity to produce vegetation" (1976:125).

We, however, view habitat as a concept that is related to a particular species, and sometimes even to a particular population, of plant or animal. *Habitat,* then, is an area with the combination of resources (like food, cover, water) and environmental conditions (temperature, precipitation, presence or absence of predators and competitors) that promotes occupancy by individuals of a given species (or population) and allows those individuals to survive and reproduce. Habitat of high quality can be defined as those areas that afford conditions necessary for relatively successful survival and reproduction over relatively long periods when compared with other similar environments. (We recognize, though, that the habitats of some animals are ephemeral by nature, such as early seral stages or pools of water in the desert after heavy rains.) Conversely, marginal habitat supports individuals, but their rates of survival and reproduction are relatively low, and/or the area is usually suitable for occupancy for relatively short, or intermittent, periods. Thus, quality of a habitat is related to the rates of survival and reproduction of the individuals that live there (Van Horne 1983), to the vitality of their offspring, and to the length of time the site remains suitable for occupancy.

Understanding why a particular population or species occupies a given area, why it occupies only specific areas in a region, or why it occupies only specific continents often requires knowledge of the organism's ecological relationships, its evolutionary history, the climatic history of the area, and even the history of the movement of landmasses. We provide in Chapter 2 an overview of the forces, factors, and processes that determine why animals are found where they are and how they came to be there. The information presented emphasizes that both past and present conditions play a significant role in defining the habitat of an animal. In short, we provide in Chapter 2 the conceptual framework we feel is necessary before the study of habitat can proceed successfully.

It is well known that human use of natural environments—for example, for agriculture, urban development, or timber production—often renders them unsuitable for many of the animals that otherwise would live there. It is not as well known, however, that patches of natural environment that remain undisturbed but are surrounded by human-altered landscapes also may be unsuitable for some animals. Unsuitability of such patches is frequently related to their small size and isolation. Habitat fragmentation, the reduction and isolation of patches of natural environment, is widespread and can negatively influence animal populations by reducing the size of patches below a required minimum, by changing temperature and moisture regimes within patches, by reducing rates of recolonization, and by exposing animals to increased rates of predation, competition, and parasitism.

We must understand the dynamics and habitat requirements of populations sensitive to habitat fragmentation if we are to succeed in designing and establishing effective habitat reserves. We review in Chapter 3 some of the basic tenets of landscape ecology and discuss the dynamics of patches in a fragmented environment and of populations that occupy such environments (that is, metapopulations). We also provide guidelines for answering important questions about managing fragmented environments for wildlife: How big should habitat patches be? How should habitat patches be distributed over the landscape? Should habitat patches be connected with corridors?

One of the first steps one must take before developing conservation plans for environments that are being disturbed (for example, fragmented) is to identify those species or groups of species upon which the plan will be founded. (We assume here that at least some management decisions will be based on the habitat requirements and population dynamics of the animals that occupy the environment in question.) The next step is to ascertain the conditions that promote survival and reproduction in those species, that is, to identify their habitats. These steps appear simple, but in fact, they usually require considerable thoughtfulness and may involve rather complicated procedures. We provide in Chapter 4 a review and an analysis of the approaches that researchers have taken in the study of wildlife habitat. We address how to select species for study, what elements of the environment to measure, and how, when, and where to measure these elements. In Chapter 5 we discuss the theory and practices associated with the measurement of one of the more important aspects of habitat use, foraging behavior.

If one understands why a species or population occupies a given area and has developed techniques for measuring the factors that influence occupancy, then it may be possible to predict the distribution and abundance of that species or population in other areas or at other times. In Chapter 6 we review models used to predict wildlife-habitat relationships and examine how scientific uncertainty affects the use of these models. We also introduce knowledge-based and decision-aiding models and discuss model validation.

The idea that numerous factors can influence whether an animal occupies a given area, a concept formally articulated by Hutchinson (1957), lends itself to mathematical analysis with multivariate statistical techniques (see Green 1971). In Chapter 7 we review the use of these techniques in understanding and conceptualizing wildlife-habitat relationships.

The Future

Just as concepts and knowledge of wildlife and their habitats have changed over time, so we anticipate that many concepts we present in this book will also evolve. Among the areas receiving much current study are landscape ecology, habitat fragmentation, habitat prediction models, and statistical analyses of species distributions and habitat correlations. With a changing and growing knowledge base, we hope that the readers—current or future conservationists in the broadest sense—remain tied to an ecological land ethic and continue the pursuit of providing vital, productive habitats for humans and wildlife alike.

Literature Cited

Adams, C. C. 1908. The ecological succession of birds. *Auk* 25:109–53.

Aristotle. [344 B.C.] 1862. *Historia animalium*. London: H. G. Bohn.

Bean, M. J. 1977. *The evolution of national wildlife law*. Report to the Council on Environmental Quality. U.S. Government Document, stock no. 041–011–00033–5.

Beebe, W., ed. 1988. *The book of naturalists*. Princeton: Princeton Univ. Press.

Bellrose, F. C. 1976. *Ducks, geese, and swans of North America*. Harrisburg, Pa.: Stackpole Books.

Butler, R. G. 1980. Population size, social behavior, and dispersal in house mice: A quantitative investigation. *Animal Behavior* 28:78–85.

Darwin, C. 1859. *The origin of species*. New York: Penguin Books.

Daubenmire, R. 1976. The use of vegetation in assessing the productivity of forest lands. *Botanical Review* 42:115–43.

Eiseley, L. 1961. *Darwin's century*. Garden City, N.Y.: Doubleday, Anchor Books.

Gilbert, F. F., and D. G. Dodds. 1987. The philosophy and practice of wildlife management. Malabar, Fla.: Robert E. Krieger.

Green, R. H. 1971. A multivariate statistical approach to the Hutchinsonian niche: Bivalve molluscs in central Canada. *Ecology* 52:543–56.

Grinnell, J. 1917. Field tests of theories concerning distributional control. American Naturalist 51:115–28.

Harris, V. T. 1952. *An experimental study of habitat selection by prairie and forest races of the deer mouse, Peromyscus maniculatus*. Contributions from the Laboratory of Vertebrate Biology no. 56. University of Michigan, Ann Arbor.

Hilden, O. 1965. Habitat selection in birds. *Annales Zoologici Fennici* 2:53–75.

Hutchinson, G. E. 1957. Concluding remarks. *Cold Spring Harbor Symposium on Quantitative Biology* 22:415–27.

Klopfer, P. H. 1963. Behavioral aspects of habitat selection: The role of early experience. *Wilson Bulletin* 75:15–22.

Klopfer, P. H., and J. U. Ganzhorn. 1985. Habitat selection: Behavioral aspects. In *Habitat selection in birds*, ed. M. L. Cody, 435–53. New York: Academic Press.

Lack, D. 1933. Habitat selection in birds with special reference to the effects of afforestation on the Breckland avifauna. *Journal of Animal Ecology* 2:239–62.

Leopold, A. 1933. *Game management*. New York: Charles Scribner's Sons.

Mayr, E. 1974. Behavior programs and evolutionary strategies. *American Scientist* 62:650–59.

Merriam, C. H. 1890. Results of a biological survey of the San Francisco Mountains region and desert of the Little Colorado River in Arizona. USDA Bureau of Biology, *Survey of American Fauna* 3:1–132.

Stoddard, H. L. 1931. *The bobwhite quail: Its habits, preservation, and increase*. New York: Charles Scribner's Sons.

Svardson, G. 1949. Competition and habitat selection in birds. *Oikos* 1:157–74.
Thomas, J.W., and D.E. Toweill, eds. 1982. *Elk of North America*. Harrisburg, Pa.: Stackpole Books.
Van Horne, B. 1983. Density as a misleading indicator of habitat quality. *Journal of Wildlife Management* 47:893–901.
Verner, J., M.L. Morrison, and C.J. Ralph, eds. 1986. *Wildlife 2000: Modeling habitat relationships of terrestrial vertebrates*. Madison: Univ. of Wisconsin Press.
Wallmo, O.C., ed. 1981. *Mule and black-tailed deer of North America*. Lincoln: Univ. of Nebraska Press.
Wecker, S.C. 1963. The role of early experience in habitat selection by the prairie deer mouse, *Peromyscus maniculatus bairdi*. *Ecological Monographs* 33: 307–25.
Werner, E.E., J.F. Gilliam, D.J. Hall, and G.G. Mittelbach. 1983. An experimental test of the effects of predation risk on habitat use in fish. *Ecology* 64: 1540–48.
Werner, E.E., and D.J. Hall. 1979. Foraging efficiency and habitat switching in competing sunfishes. *Ecology* 60:256–64.

2 Habitat from an Ecological and Evolutionary Perspective

Introduction: Conceptual Framework

The reader must understand the ecological framework upon which this book was built in order to approach each chapter in the correct frame of mind. Inherent in our discussions is the notion that the present distribution and abundance of animals is a result of adaptation, mediated by the process of natural and other forms of selection, to biotic and abiotic factors. It is well known that climatic conditions greatly influence the external morphology and physiology of animals. Many ecological "rules" have been built on these morphological and physiological conditions: Allen's Rule states that certain extremities of animals are relatively shorter in the cooler parts of a species' range than in the warmer parts (see fig. 2.1); Bergmann's Rule asserts that geographic races of a species possessing smaller body size are found in the warmer parts of the range and races of larger body size in cooler parts; and Gloger's Rule says that dark pigments increase in warm and humid habitats. These are three of the most notable rules and are usually taught in undergraduate ecology courses (see Kendeigh 1961: 9–10, for a review of these and other rules). These rules do not apply uniformly to all species, nor are they accepted without controversy. Ricklefs (1973:134–36) asserted that many of the variations in body size attributed to environmental factors by various "rules" may actually result from competitor- or predator-induced relationships. The rules raise, however, a critical point: the distribution, activity, and abundance of animals are not only mediated by current abiotic and biotic conditions but also set against a background of adjustments—adaptations—made by the animals to forces that may now be absent or at least ebbing.

Figure 2.1. Rabbits of the genus *Lepus* from North America, demonstrating Allen's Rule. *Upper left to lower right:* arctic hare (*L. arcticus*), a resident of tundra habitats; snowshoe hare (*L. americanus*), of boreal coniferous forests; black-tailed jackrabbit (*L. californicus*), widespread in the western United States; antelope jackrabbit (*L. alleni*), of the Sonoran Desert of the southwestern United States and northwestern Mexico. (After paintings by Louis Agassiz Fuertes, as adapted by Ricklefs 1973, fig. 12.5.)

Thus, the field of biogeography plays a critical role in the analysis of patterns of habitat use. We must realize that development of an understanding of why a population is found in a specific area requires knowledge of the organism's ecological relationships: why it is associated with various edaphic, temperature, and biotic regimes. To understand why it is found only in certain areas of the continent requires knowledge of climatic history, which may have resulted in the isolation of scattered, relict communities. Finally, analysis of the evolutionary history of the population itself and of the geologic history of the landmasses is probably necessary to understand why the population is found only on certain continents (see Cox and Moore 1985:vii).

We are also interested in the influence of competitive interactions on the distribution and abundance of animals. Further, some believe that these competitive interactions may play important roles in shaping morphology and behavior, which could, then, be manifested in distribution and abundance. Because many, if not most, of the competitive interactions between species likely took place in the past, all we observe today is the result of these interactions. Thus, failure to observe intense interactions between two species does not disprove competition. Competition thus becomes unfalsifiable, the "ghost of competition past" (Connell 1980). We do not hope to solve this dilemma here. Rather, we intend to use such controversies to show that determination of factors responsible for the animals present in an area are complicated, interactive, and usually not altogether apparent.

This chapter, then, examines the conceptual framework for viewing wildlife-habitat problems followed throughout the book. We argue strongly that unless theorists of habitat selection and interspecific interactions consider the evolutionary perspective when developing wildlife-habitat relationships, further progress in this field will be limited. We believe this is the primary reason why our models have repeatedly failed to account for much of the variability in habitat use we see in nature.

Evolutionary Perspective

Occupation of an area and thus interspecific interactions are based upon the geologic events that structured the area originally. Most students have been introduced to the concept of the geologic timetable: that chart that relates the development of the Earth to the evolution of plants and animals. Volcanoes, earthquakes, changing climatic patterns (such as droughts), and the like constantly remind us today that many geologic forces are still at work. Although many of these processes are not as intense today as they were in the past, a new evolutionary force now impacts animal communities: human development. Here we are most concerned with events shaping our recent past, the Pleistocene to the present (that is, the Recent epoch). The use of habitat and the distribution and abundance of a species must be considered in light of these geologic events. It is surprising, therefore, that very few studies have incorporated such considerations into analysis of wildlife-habitat use. Our intent, then, is to view habitat use in light of gross geologic events. It will become clear that many species were preadapted to the area-habitat they now occupy (see Price 1984). One must understand the underlying reason for a species' presence in an area before developing a full appreciation of its habitat requirements and interspecific interactions.

Pleistocene

The Pleistocene epoch, which began two to three million years ago, is thought to have ended only some ten thousand years before the present. In a sense, our current Recent epoch is similar to an interglacial period of the Pleistocene (Cox and Moore 1985:205–11). Regardless of how we label the epoch, or the exact number of years involved, the advance and retreat of the continental ice sheet and associated glaciers obviously had a dramatic impact on the ground covered and the associated environmental conditions. There were apparently four major advances of the continental ice sheet in North America: the Nebraskan, Kansan, Illinoian, and Wisconsin. Between these periods of glaciation were three interglacial periods: the Aftonian, Yarmouth, and Sangamon. These advances and retreats were, of course, extremely slow by human standards. Thus, although habitats were destroyed, animals had time to seek new areas to occupy. The obvious physical destruction of habitat, however, was not the only impact on the environment. Such an enormous sheet of ice does not just suddenly appear: it was caused, of course, by a change in the Earth's temperature and precipitation patterns. Scientists have only speculated about the causes of these changes, with proposals ranging from the terrestrial-based (for example, volcanic activity caused changes in the Earth's atmosphere) to the extraterrestrial (alteration of the Earth's orbit caused changes) (Cox and Moore 1985:212–13). Further, areas close to but not covered by the ice also underwent dramatic physical and environmental changes because of their proximity to the ice fields. Thus, the Earth slowly experienced a dramatic change in environmental conditions. It appears that most of the species currently occupying the Earth are the survivors of the abiotic and biotic influences of the Pleistocene, especially the last several advance-retreat cycles. Most modern species of birds first appeared in the late Pliocene and were not numerous until the Pleistocene. At least 75 percent of the species found in the middle and upper Pleistocene deposits, however, represent living forms (see Selander 1965).

Although the areas we now call Canada and Alaska were for the most part covered in ice, numerous ice-free pockets, or refugia, did exist. Along the coast and near-shore islands, the ocean apparently had a mediating affect on temperatures (Heusser 1977). The result was a chain of forested refugia that extended south below the ice-covered regions (Heusser 1965; see fig. 2.2). Palynological evidence—deposits of pollen in sediment layers—has allowed a good reconstruction of the plants on these refugia and has provided a record of plant development during interglacial periods. In the Pacific Northwest, for example, vegetation on ice-free parts of the Olympic Peninsula through much of the middle and late Wisconsin glacial period probably resembled the edge of the Pacific Coast Forest in

Figure 2.2. Postulated glacial-age refugia of the aridlands and certain other regions of North America. The northern portion of the area indicated as the Sonoran Refugium probably was occupied by the biota of the Great Basin, and the southern part, by that of the Sinaloan Shrublands. (From Hubbard 1973, fig. 2.)

Alaska as it appears today, where Sitka spruce (*Picea sitchensis*) and mountain hemlock (*Tsuga mertensiana*) make contact with subalpine grassland or meadow. Interglacial (interstade) vegetation appears to approach that of the present time on the peninsula (Heusser 1965, 1977). Thus, scientists have been able to largely reconstruct the composition of forests during the Pleistocene; this is critical information if we are to reconstruct the animal communities present at the time (fig. 2.3).

Refugia were also present below the southern reach of the ice sheets. Because of the changes in environmental conditions, the snow level in the

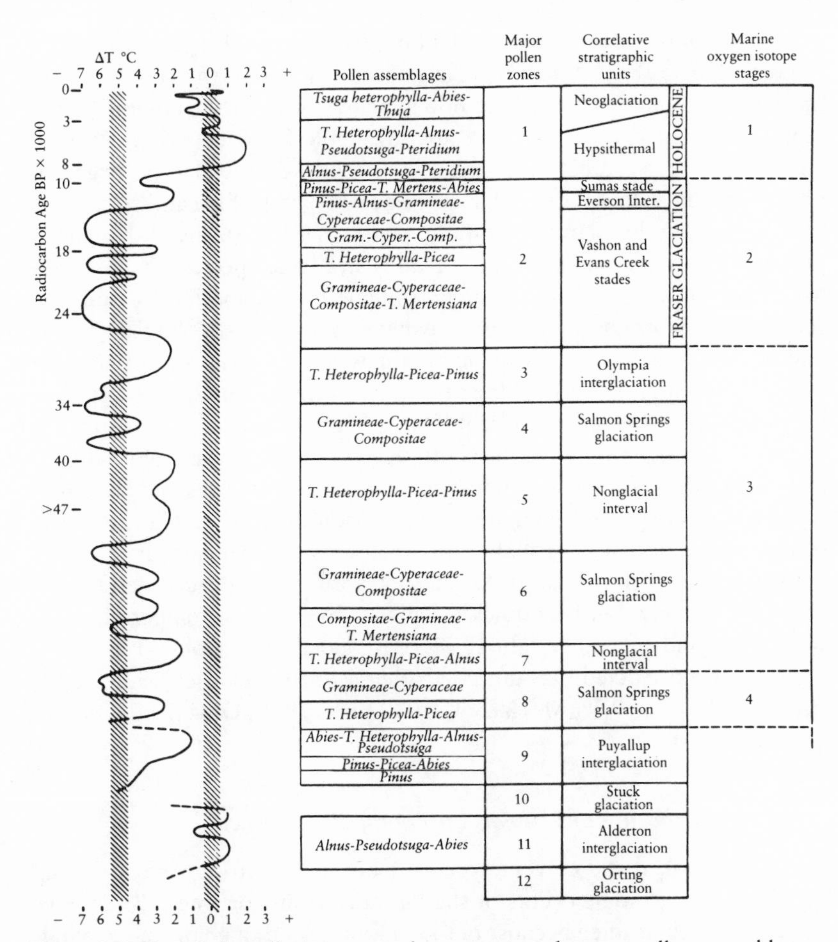

Figure 2.3. Correlation of reconstructed temperature changes, pollen assemblages, major pollen zones, and stratigraphic units on the Pacific slope with marine oxygen isotope stages. Temperature changes are amounts above or below the present mean for July; vertical guides indicate sea level and subalpine forest-alpine tundra transition. Note that the record below 40,000 B.P. is not on the same scale as the portion above this time plane. (From Heusser 1977, fig. 16.)

mountain chains of the continental United States was much lower than what we see today. During times of glaciation in the Cascade Mountains, for example, July temperatures were apparently close to freezing, and precipitation—probably as snow—and cloudiness were extensive (Heusser 1977). In the southwestern United States, Wisconsin glaciation depressed the biotic zones by 900–1200 m, so woodlands and forests occurred over much of the ice-free region (Hubbard 1973; see also Johnson 1977; Brakenridge 1978). During late Pleistocene glaciation, the snow level in the Rocky Mountains was 250 to 600 m below the present snowline (Richmond 1965), and the total land area available for occupation by most animal species was much less than it is currently. As the glaciers began to retreat northward and upward, bare or nearly bare soil became accessible for colonization by plants. Naturally, plants that had survived the glacial periods on refugia were the most readily available species for colonization. In turn, new habitats became available for occupation by animals. Again, species inhabiting refugia may have been the first colonists. As the late Wisconsin glacial period ended in the western interior of Canada, for example, an early invasion of boreal forest, dominated by spruce but lacking pine, spread from the adjacent United States onto deglaciated surfaces and till over stagnant ice. In the south the spruce forest was eventually replaced by prairie, about six thousand years before the present. The rapid replacement of spruce-dominated boreal forest by grasslands in the early postglacial period was probably a response to warmer and drier growing seasons (Ritchie 1976). In Minnesota, spruce forest succeeded tundra about 10,000 B.P. but was quickly replaced by pine and temperate hardwoods (Amundson and Wright 1979). Postglacial succession of vegetation has been reconstructed for numerous other areas (see Hansen 1938, 1953; Ritchie and Hare 1971; Mack et al. 1978; Birks 1980; Oliver, Adams, and Zasoski 1985).

Distribution of Animals during the Pleistocene

Some have linked the current type, number, and distribution of animal species to the geologic events of the Pleistocene just outlined. Birds have been intensively studied because of their mobility, although of course wings would not be required to keep pace with advancing and retreating glaciers. Rather, birds are of special interest because they could potentially move between refugia, flying over unsuitable habitat.

Mengel (1964) presented an interesting scenario to account for the occupation of North America by wood warblers (Emberizidae: Parulinae). By his reasoning, the retreat of glacial ice opened the way for occupation of vast areas by generalist species able to adjust to varying environmental conditions. The "pool" of potential colonists is unknown but is probably related to the types of species and their abundance on refugia and more southern habitats. Using a precursor of the black-throated green warbler (*Dendroica virens*) as an example, Mengel postulated that this pro-*virens* was able to occupy vast reaches of western North America as the environment warmed (see fig. 2.4). Most warblers apparently evolved

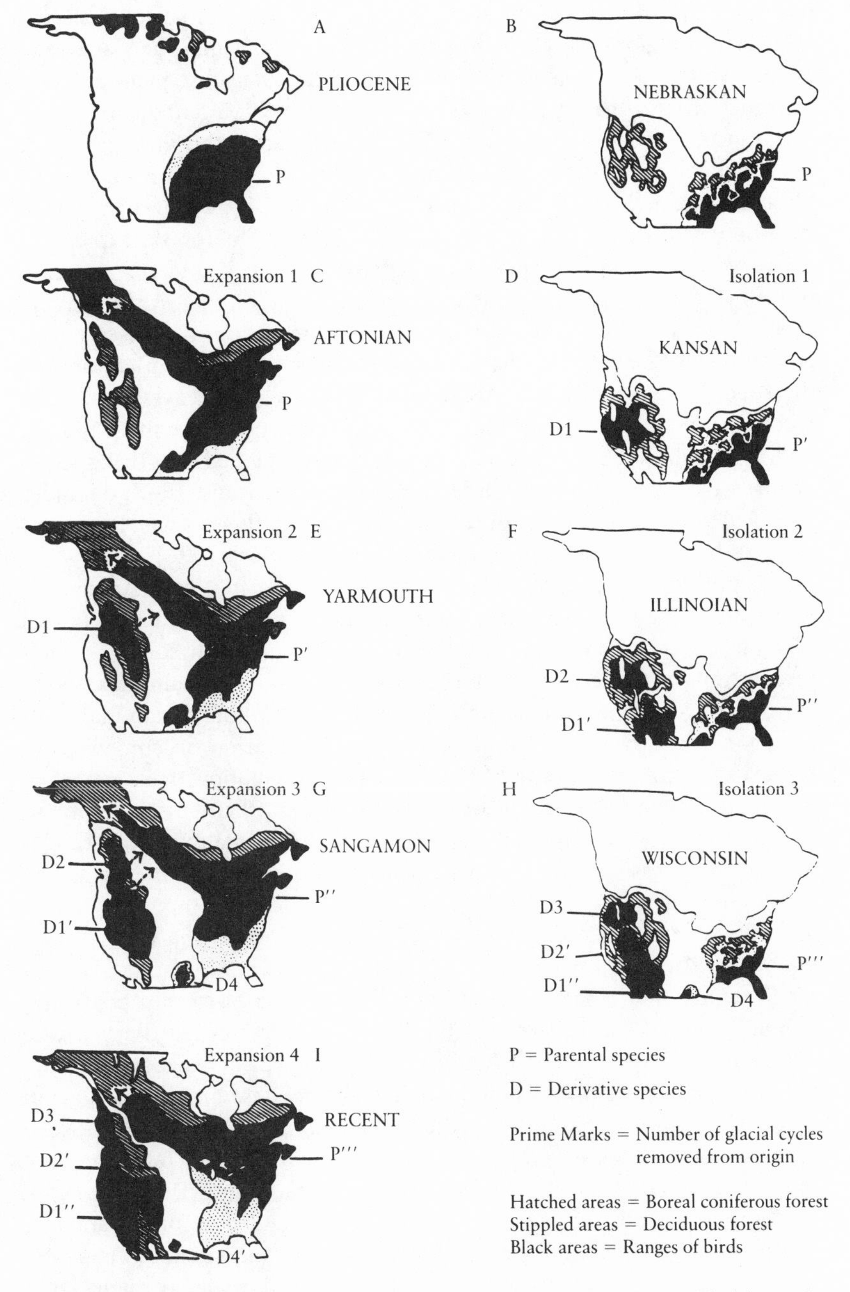

Figure 2.4. A model sequence showing the effects of glacial flow and ebb on the adaptation and evolution of a hypothetical ancestral wood warbler and its descendants. The details of glacial boundaries are approximate. (From Mengel 1964, fig. 4.)

in more tropical regions of North, Central, and South America, colonizing western regions of North America by passing through the Caribbean Islands and the southeastern United States. With the subsequent advance of the ice sheets, pro-*virens* was isolated to the north on refugia and to the south in the warmer regions of the Southwest. During this glacial period, these isolated populations either adapted to their changing surroundings or became extinct. With the warming and concomitant retreat of the glaciers, the populations were able to expand their range as the forests expanded; many were now distinct species (see Mengel 1970). Pro-*virens* was not, of course, the only species present on these isolated refugia. Although any speculation about the type and intensity of competition on these refugia is tenuous, it is safe to say that interactions must have taken place among the species present. Because most birds are small and leave poor fossil records, most of our knowledge concerning such processes must remain highly speculative. Similar scenarios have been developed for other species of birds (see Mengel 1970; Hubbard 1969, 1973; Rand 1948; Selander 1965). Although the general premise of Mengel (1964) is well accepted, many of his specifics are not (see Hubbard 1973; Flack 1976).

The present distribution of mammals, like birds, in North America is largely the result of Pleistocene events. During glacial advances, boreal mammals were apparently widespread in lowlands well south of their present ranges. Concurrent with the movements of these mammals northward during interglacial periods were movements of boreal mammals into montane areas. Here, because of the effect of elevation on climate, cool refugia were present. Many of these montane populations have persisted in boreal refugia far south of the northern range of their closest relatives, and the zonation of mammalian distributions on some mountain ranges in the southwestern United States apparently resulted from Pleistocene faunal movements (Vaughan 1972:318–19). In the mountains of Arizona and New Mexico, for example, discrete populations of the water shrew (*Sorex palustris*) are separated by as little as 300 km of unsuitable habitat (Vaughan 1972:319). The current distributions of certain birds and mammals appear to be the results of similar processes. In the Cascade Mountains of western North America, for example, a hybrid zone existing between species within the mammalian genus *Tamiasciurus* and the avian genus *Dendroica* apparently resulted from the relaxation of Pleistocene separation which allowed the species to overlap (and hybridize) in the zone of overlap between their preferred types of habitat (Smith 1981).

Migration is also apparently tied to the ebb and flow of ice and the colonization of new habitats. It is unclear, however, if we should consider migration as an adaptation of basically tropical birds to newly unglaciated habitats or as an attempt (adaptation) of temperate-zone birds to escape the harsh northern winter (see Stiles 1980). Although Stiles agreed that speciation was undoubtedly connected with the advances and retreats of the Pleistocene glaciers, it does not necessarily follow that such animal groups had their origins in North America. Stiles described a scenario in which a single wintering population breeds over a broad expanse of North America. Differentiation and speciation might occur in the breeding range with little effect on the winter distribution, such that eventually several closely related, morphologically similar species might winter together. This, in fact, occurs in Central America. A fundamental reason for migration may be related to exploitation of the large summer bloom of productivity at northern latitudes. This would in turn allow larger clutch sizes and higher productivity. Thus, migrant members of a population would contribute more offspring to successive generations than their sedentary counterparts; in time the genetic constitution to migrate would become fixed (Stiles 1980).

Patterns of migration and seasonal aspects of habitat use have definite ramifications for our attempts to determine why animals occupy the areas they do. It is not often recognized that migratory birds breeding in North America spend most of the year on wintering grounds: five to seven months on the wintering areas, plus several months spent in migration, is not uncommon (Keast 1980; Stiles 1980). The sheer number of fall migrants must have a tremendous effect on the patterns of resource use of permanent-resident species. Thus, one might expect selection against any tendency of permanent residents to invade the "migrant niche." This applies both to North and Central American resident species. Thus, exploitation of resources and competition for these resources likely played important roles in the establishment of migration and migrational areas and routes and are expressed in morphology and habitat use (see Cox 1968; Keast and Morton 1980). It is important to remember, however, that the forces shaping these decisions were likely based on numerous other abiotic and biotic events, many of which no longer actively influence the species. We are left with an effect without any apparent cause.

In his review of habitat selection in birds, Cody (1985) thought that the selection of different winter and summer habitats posed a set of "compromises" for a bird. (These ideas can, of course, be extended to any animal that migrates.) The two separate areas imposed morphological, behavioral, and physiological constraints on the abilities of the bird to efficiently use either the winter or the summer habitat. Quoting Fretwell's important monograph (1972), Cody went on to note that survival in wintering habitats may affect choice of breeding habitats and may even dominate adaptive morphology; adaptations to breeding habitat may therefore not be optimal. This scenario relates nicely to our previous discussion of the evolution of separate winter and summer grounds and the processes by which and locations where species evolved (Stiles 1980).

Finch bills, for example, seem more strongly influenced by their granivorous winter diet than by their mostly insectivorous diet during breeding (Cody 1985). Salomonson and Balda (1977) found that Townsend's solitaires (*Myadestes townsendi*) are flycatching insectivores during breeding but are frugivorous when they winter in Arizona. In the Sierra Nevada, California, many small-gleaning and hover-gleaning foliage insectivores become flakers and probers of the bark of tree limbs and trunks during winter in response to changes in resource abundance and location (Morrison et al. 1985). Given these and other examples, it is unlikely that these species could possess a morphology equally well adapted for such different modes of resource acquisition (see Cody 1985). Migration also greatly affects wing morphology and body size. Morrison (1983) found that, in the western United States, two populations of Townsend's warblers (*Dendroica townsendi*) apparently existed: one a shorter-winged group that wintered in the United States, the other a longer-winged group that migrated to Mexico and Central America (see also Grinnell 1905). Both groups had separate breeding areas. Cody (1985) thought that such morphological adaptations might well compromise the habitat- and foraging-site selection of these species during breeding. In his studies of North American *Dendroica* warblers, Greenberg (1979) showed that while the largest warbler species bred in conifers and woodlands, they were highly mobile and gregarious opportunists on their wintering grounds in Mexico and Central America.

The winter versus summer dichotomy is further complicated by the choice of habitat while in transit. In the southwestern United States, many woodland and forest warblers crossing, resting, and feeding in the Mojave and Sonoran deserts in fall and spring are certainly in seemingly atypical habitats. They must have behavioral and/or morphological adaptations, however, that allow them to successfully exploit these habitats. Here we have another set of compromises (Cody 1985).

Post-Pleistocene Events

Mengel (1970) postulated that the Great Plains served as an isolating mechanism between populations in eastern and western North America. Regardless of the accuracy of Mengel's thesis, it is certainly true that the Great Plains have undergone dramatic changes in resources and environmental conditions since the coming of Europeans to North America. Fire regimes have been changed, irrigation on a massive scale has been developed, native plants have been largely replaced by exotics, exotic species of animals have been introduced, native populations have been greatly reduced or extirpated, and on and on. While we will leave the philosophical and social ramifications of these changes to others, it is evident that these impacts have directly altered the composition of plant and animal communities. Further, the subsequent interactions between these altered communities have had marked indirect effects. Knopf (1986) presented an impressive account of the changes that have taken place in Colorado since colonial times. By his account, almost 90 percent of the avian species now present in Colorado were not present in 1900. The development of riparian (floodplain) forests associated with human development of the Great Plains primarily caused these changes. Forests on the Great Plains have provided corridors for the movement of forest birds across grasslands that previously served as a barrier to dispersal.

Human-induced changes in plants and animals of the Great Plains have altered the natural process of colonization and speciation. For example, several species of birds—once apparently separated by the Great Plains —now have overlapping ranges caused, apparently, by human-induced changes in vegetation such as agricultural and residential development. The familiar blue jay (*Cyanocitta cristata*) of eastern North America and the Steller's jay (*C. stelleri*) of the West now overlap and hybridize in Colorado (Williams and Wheat 1971). Hybridization within various avian taxa in the Great Plains has attracted much attention (see West 1962; Rising 1970; Emlen, Rising, and Thompson 1975).

Such changes are not restricted to the Great Plains, of course. In the eastern United States, habitat changes caused by human development has allowed the blue-winged warbler (*Vermivora pinus*) to expand its range and population size in many areas. This expansion has come at the expense of the similar golden-winged warbler (*V. chrysoptera*), which the blue-winged warbler hybridizes with and replaces (fig. 2.5; see Gill 1980). Agricultural clearing of native pine forests apparently caused hybridization between subspecies of the common grackle (*Quisculus quiscula*) in Louisiana (Yang and Selander 1968).

Hybridization represents, of course, a recombination of genetic material of species or subspecies. By creating opportunities for hybridization, modern society is tampering with the evolutionary process in promoting the radiation of ecologically isolated, congeneric species. Just as extinction of species causes a loss of genetic diversity, so does the loss of distinct, diverging genotypes through human-induced hybridization. Our main concern here is the effect such genetic changes has on habitat use and resource exploitation.

It is not our intent to fully review human-induced habitat changes and their impacts on animals. We should mention, however, the widespread

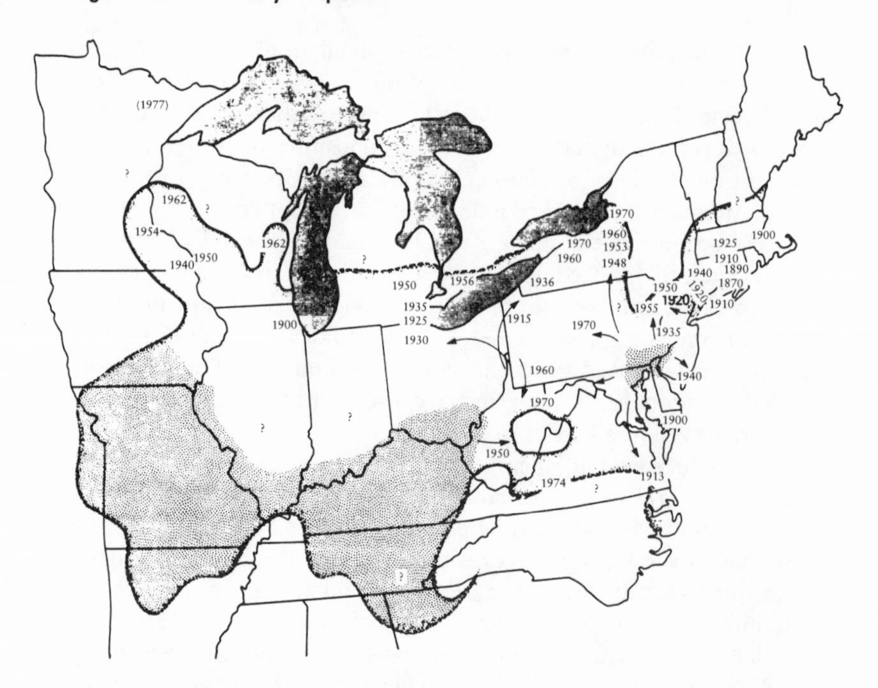

Figure 2.5. Distribution and spread of the blue-winged warbler during the last century. Dates on the map indicate when blue-wings first became established at a locality. The stippled area is the approximate range in the mid-1800s. The range boundaries are approximations, and the arrows indicating patterns of spread are hypotheses only. (From Gill 1980, fig. 3.)

changes taking place outside North America on the wintering grounds of many of our migratorial species (mainly birds, of course). The cutting of large stands of forest trees, the use of pesticides, and subsistence and sports hunting all impact populations to varying degrees. Regardless of the amount and quality of habitat available in North America to a population for breeding purposes, they have little meaning if wintering grounds are absent.

Our overall point is simple, but the details are complicated. One must study the relationships between an animal and its environment in the context of past events that initially shaped the adaptation of the animal to its surroundings. Many of the relationships currently expressed may not be a result of forces now impacting the animal: Past competitors may be extinct or at least extirpated from the current range of the species; a population may be influenced primarily by catastrophic events that occur only infrequently (like "bottlenecks" of Wiens 1977; see fig. 2.6); and the range occupied by a species may be restricted even though seemingly suitable

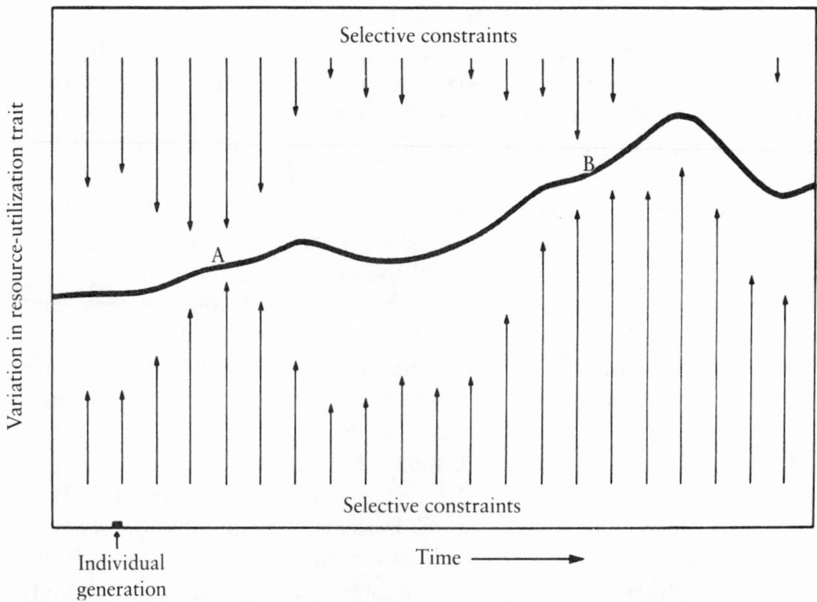

Figure 2.6. Changes in the mean phenotypic expression of some resource-utilization trait (such as foraging behavior, bill size) in a population (solid line) and in the population variance about this mean (shaded area) are influenced by the conditions of changing selective constraints or intensity associated with a variable environment. *A* and *B* indicate times of ecological crunches, during which selection is intense. The duration of an individual generation is indicated on the time axis. (From Wiens 1977, fig. 1.)

habitat is available elsewhere. Referring to the first point, the range could be restricted because of the influence of a now-extinct species, but the extant species may not be mobile enough to occupy other areas. Here, no amount of animal-vegetation correlation will reveal why the range is so restricted. It is a simple point, but recognizing and analyzing it is complicated. Unfortunately, many of these problems cannot be directly quantified through field studies. Recognition of them, along with their incorporation in study design and subsequent analyses and conclusions, however, can at least remove many of the nagging uncertainties involved in management decisions. Subsequent discussions and analyses of study designs and methods presented in this book will incorporate these questions.

The Concept of Habitat Selection

Within the evolutionary framework just developed lies the entire spectrum of theories that attempt to account for the selection and use of an area by a species. Underlying these theories is the conceptual idea that animals somehow perceive the correct configuration of habitat required for their survival. Such a concept has important overtones for our discussions of the measurement of habitat; that is, we must be able to measure what the animal "sees" if we hope to develop meaningful descriptions (for example, models) of habitat use.

Early ecologists, especially students of plant growth and distribution, were well aware that the distribution of plants was closely tied to soil type. Plants were found to be useful indicators of the type and fertility of soil present (Clements 1920; see also Morrison 1986). Clements and early animal ecologists were also aware that plants could be used as at least loose indicators of animal presence. Because animals are mobile and because plants supply only part of their habitat requirements, ecologists knew that plants would be of limited usefulness in describing the distribution and ecology of wildlife (see Chapter 4). Certainly we are all aware of the classic life zone concept of Merriam (1898), whereby different sets of animal species are distributed in accordance with the distribution of major vegetation types (such as small mammals in the Canadian life zone). What we are concerned with here is the distribution and abundance of animals within these life zones or vegetative communities. To examine this subject, we should first discuss how an animal perceives its environment.

One of the earliest papers on habitat selection was a classic work on birds by Hilden (1965). Hilden based his ideas on a contrast between proximate and ultimate factors. Proximate factors caused the release of innate behaviors that resulted in a certain "settling pattern." This pattern of habitat selection was, of course, related to the long-term reproductive success and survival of the species. Hilden further postulated that various "positive" and "negative" factors added to some threshold level—the "stimulus summation"—resulted in selection of an area; some stimuli were more important to the animal than others. Hilden knew of the evolutionary problem involved in determining reasons for patterns of habitat use noted in a species. Although he concluded that interspecific competition was a driving force in habitat selection, he recognized that such forces probably operated in the past. Hilden did not, however, draw any distinction between summer and winter (breeding versus nonbreeding) habitat requirements, nor did he give any emphasis to noncompetitive factors (like weather) and their possible influence on patterns of habitat selection. Nevertheless, Hilden's paper formed a firm foundation upon which most subsequent analyses of habitat selection were built.

Another classic study devoted to evaluating the factors primarily responsible for selection of habitats was an experimental analysis of deer mice (*Peromyscus maniculatus*) by Wecker (1964). Wecker attempted to determine the relative strengths that innate behavior and learning played in habitat selection. Using an outdoor pen 30 m (100 feet) long, he compared groups of mice with different genetic makeups and pretest experience in different habitats. He concluded that patterns of habitat selection are genetically determined and that learning at an early age will enforce these patterns. Furthermore, experience in atypical habitats would not override the genetic template determining the habitats selected. Wecker's results indicated that certain "key environmental stimuli" elicited the pattern of habitat selection noted. Wecker extended his results to develop a scenario that could reflect different steps in an evolutionary sequence leading from behavior primarily dependent on learning to that dependent primarily on innate control: habitat restriction through social factors and homing; recognition of the field environment through learning; learning capacity reduced to cues associated with the field habitat; imprinting to the field environment through exposure very early in life; and innate determination of the habitat response. Thus, preadaptation must play an important role in determining the pattern of habitat selection evident in a species.

Concomitant with studies of the overall behavioral pattern associated with habitat selection were studies exploring the niches of various species (for a review of the development of niche theory, see Whittaker and Levins 1975). G. Evelyn Hutchinson (1957, 1978) caused a substantial revolution in our thinking of the niche, and thus habitat selection, when he published his ideas on the multivariate nature of the niche (the "*n*-dimensional hypervolume"; see fig. 2.7). This work caused ecologists to realize that animals operated in a much more complicated world than previous graphs and charts indicated. Further, the interrelated and interactive nature of the various niche axes began to become apparent. This is not to say that the pioneering work of Grinnell (1917) was not an essential part of the development of current thought. Our point is that, following Hutchinson's work, for the first time ecologists everywhere began to search for new, multivariable methods of exploring the niche and means of habitat selection.

One of the first studies to statistically evaluate the "Hutchinsonian niche" was a work on birds by James (1971). Using multivariate procedures (see Chapter 7), she quantified the habitat relationships of breeding birds in Arkansas. Further, she used these data to address the more general question of how birds selected their habitat. She used the term *niche gestalt* to describe the basic life-form of the vegetation that characterized the habitat of a species. This approach differed from previous, community-based, studies. As outlined by James, inherent in the term *gestalt* are the concepts that each species has a characteristic perceptual world, that it responds to its perceptual world as an organized whole, and that it has a predeter-

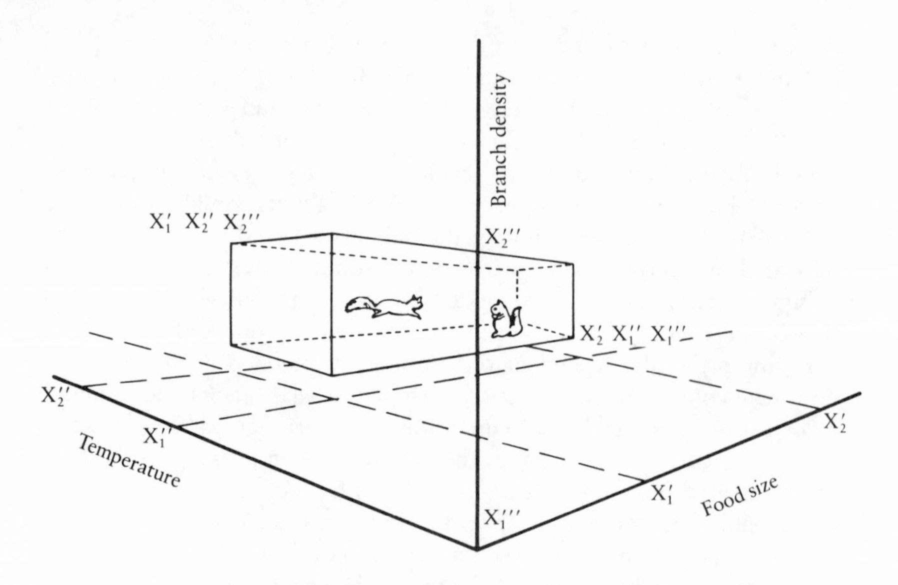

Figure 2.7. A three-dimensional orthogonal fundamental niche; x' might define food size, in this case the mean diameter of various seeds or fruits, notably acorns; x'' might define temperature tolerance, and x''', some measure of the density of branches, between certain diameters, in unit volume of physical space. (From Hutchinson 1978, fig. 99.)

mined set of specific search images. These characteristics are assumed to be at least partially genetically determined, but they are also modified by experience. Our problem is that we cannot see through the eyes of an animal; we cannot directly know what the animal perceives or what influences these perceptions have on its behavior. James was well aware of this problem, noting that her approach "can be defended only if one assumes that predictable relationships exist between the occurrence of a bird and of its characteristic vegetational requirements" (James 1971:217). A plethora of papers followed James's study, involving birds (see Anderson and Shugart 1974; Whitmore 1975; Conner and Adkisson 1977; MacKenzie and Sealy 1981; Anderson, Ohmart, and Rice 1983), small mammals (Dueser and Shugart 1978; Kitchings and Levy 1981; Murúa and Gonzàlez 1982), and various other taxonomic groups (Reinhert 1984; Szaro 1988).

The increasing popularity of multivariate analyses to describe patterns of habitat use and detail community structure resulted in a 1980 symposium, "The Use of Multivariate Statistics in Studies of Wildlife Habitat" (see Capen 1981). This symposium, while not restricted to avian studies, was dominated by them. This, we believe, reflected the emphasis avian ecologists gave to such studies compared with people working on other taxonomic groups. The concentration on avian studies was still evident four years later, when "Wildlife 2000," a symposium on modeling wildlife-habitat relationships (see Verner, Morrison, and Ralph 1986), was again clearly dominated by bird papers (even though an effort was made to secure papers on nonavian species). Most of these studies used similar methods and analyses to describe the habitat used by the population based—either explicitly or implicitly—on James's rationale of the niche gestalt and correlative relationships between animals and habitat.

Experimental manipulations of population densities have shown that, in many cases, the density of one species can affect the density of another. The classic study of Paine (1966) on intertidal predator-prey relationships—while not concerning vertebrates—set the tone for studies that followed on vertebrate populations. Paine examined the hypothesis that local diversity of animal species is related to the number of predators and their efficiency in preventing single species from monopolizing some important resource that is in limited supply. Where predators capable of preventing monopolies are missing or experimentally removed, the systems become less diverse. Experimental manipulations of habitat and/or densities of vertebrate populations followed Paine's conceptual framework. In contrast with the dominance of bird studies in multivariate analyses of habitat use, numerous small mammal and reptile studies have focused on manipulation of population numbers; this is due to the high mobility of birds relative to these other taxa.

In a series of field and laboratory experiments using the plethodontid salamanders *Desmognathus fuscus* and *D. monticola,* Keen (1982) tested the hypothesis that interspecific interference competition plays no role in microhabitat utilization and activity patterns. Single-species and mixed-species groups were tested to determine if the presence or absence of potential competitors influenced either species' selection of substrate texture or moisture, utilization of cover objects, and other aspects of habitat use and activity. Keen concluded that the two species possessed very different patterns of habitat use and activity, yet interference competition did play a role in spatial distribution. Tinkle (1982) was unable, however, to detect changes in habitat selection, survivorship, or population density following removal experiments of a desert rodent community. Using three species of heteromyid rodents, Lemen and Freeman (1986) conducted removal experiments to again determine the role of interference competition in determining patterns of habitat use observed in the field. They found that removal of the larger species (*Dipodomys merriami*) had a positive effect on a smaller species (*Perognathus longimembris*) that had a similar diet. They concluded that interference competition was a present, but weak, force in determining habitat utilization by these species.

Manipulations of habitat—by clear-cutting, herbicide application, and the like—and food—especially the addition of supplemental food—have also been used to analyze species-species relationships (Sullivan and Sullivan 1982; Grubb 1987). Results vary, of course, based on the animals involved, the experiments attempted, and various temporal and spatial aspects of each particular study. These studies indicate, however, that an animal not only must perceive the "correct" habitat but also must perceive the presence of potential competitors and predators. Experimentation with avian vocalizations have shown that certain species may evaluate the number and type of birds present by listening to the number and intensity of songs when choosing an area in which to settle (Krebs 1977; but see Morrison and Hardy 1983).

The question of what influences habitat selection is thus complicated by real and potential competitive interactions. Natural or even human-induced changes in animal populations may remove potential competitors from one area while creating a problem for a species in another area. For example, human-caused changes in habitat have apparently allowed populations of brown-headed cowbirds (*Molothrus ater*) to increase, thus increasing the rate of nest parasitism on other birds by this species (Brittingham and Temple 1983).

Given the numerous factors complicating our determination of how an animal perceives its environment, is it possible to develop a general model of habitat perception? Hutto (1985) thought that habitat selection could best be viewed as a hierarchical decision-making process. Although he restricted his discussion to nonbreeding migratory birds (because of the data set he had available to him), we believe his ideas—with modification—can be extended to most vertebrates (see fig. 2.8). At the broadest level, habitat selection is primarily innate: evolutionary forces have led to the establishment of migratory routes, wintering areas, and the like that constrain the animal to certain geographic boundaries. As Hutto noted, certain areas simply lie outside the utilized area of a species. Within the restricted area, however, the animal can explore alternatives and make choices based on the costs and benefits associated with the use of each habitat, such as food availability. Finally, the animal must choose specific sites within the habitat (that is, microhabitat). This final, finest level can be influenced by a host of factors, including mate availability and presence of competition.

Earlier, Johnson (1980) advanced the idea of selection order that described the hierarchical nature of habitat selection. By his scheme, one selection process will be of higher order than another if it is conditional

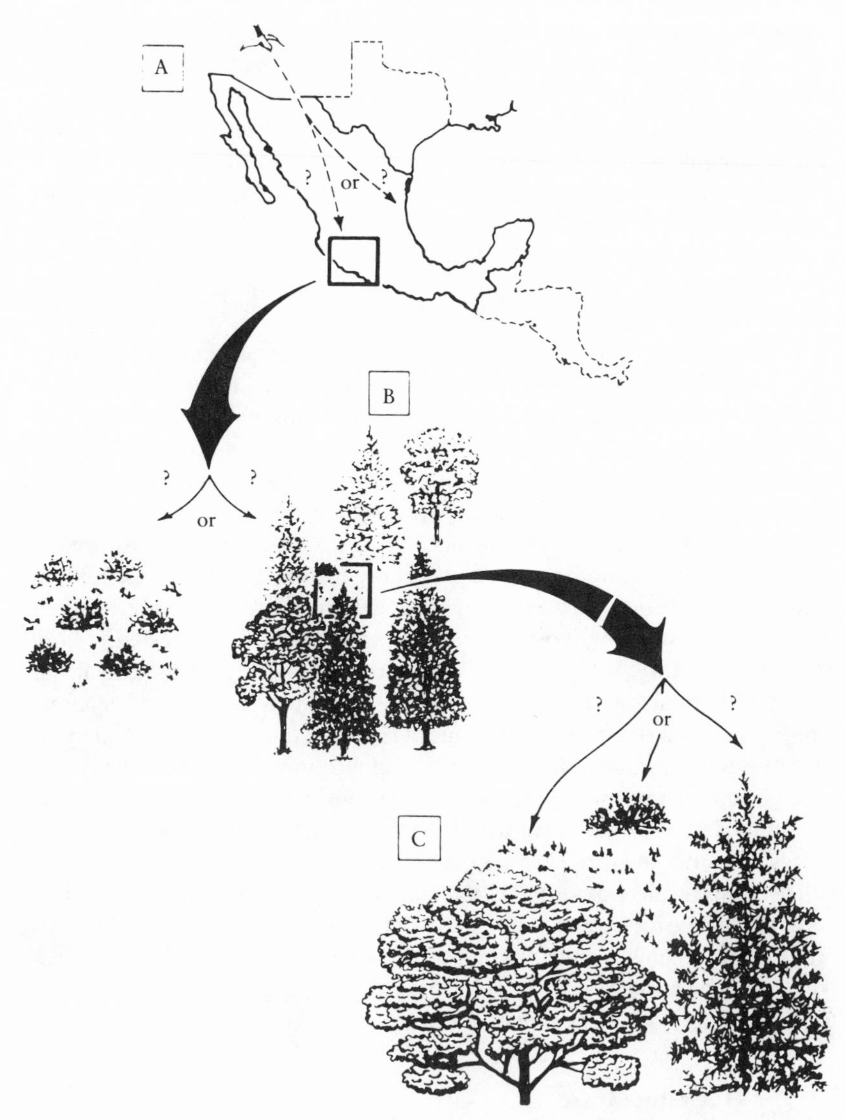

Figure 2.8. A diagrammatic illustration of the hierarchical decision-making process involved in "choice" of nonbreeding habitats by a migratory bird. The process at level *A* is probably inflexible and innate, while the processes at levels *B* and *C* may very well involve exploratory assessment of intrinsic costs and benefits associated with each option. (From Hutto 1985, fig. 7.)

upon the latter. First-order selection was defined as the selection of physical or geographic range of a species. Within that range, second-order selection determines the home range of an animal or social group. Third-order selection pertains to the use of various habitat components within the home range. Finally, if third-order selection determines a feeding site, the actual procurement of food items from those available at that site can be termed fourth-order selection. This hierarchical selection has a unifying nature: habitat studies and feeding studies are not qualitatively distinct; they are simply of different orders. Johnson's hierarchical concept of selection had been implicitly recognized by previous authors.

It should be readily apparent that development of a conceptual framework for the analysis of how animals select their habitat is extremely complicated. Throughout this book, but especially in Chapters 4 and 5, we discuss various methods for tackling these problems. It should also be evident that intensive replication and rigorous experimentation will be required in many cases. Taken separately, many of the factors influencing animal populations may not negate, say, models that seek to predict general animal-habitat relationships. These factors cannot be taken separately, however, for the animal must cope with what it finds in an area or move on. This is why most laboratory experiments are criticized for not patterning reality. While such investigations are an important part of our study of habitat selection—they allow us to control factors and separate minor from major effects—they cannot replicate the interrelated and interactive environment the animal faces, an environment that often defies any prediction (just follow your local weather report!).

When designing studies of wildlife-habitat relationships, one must, in all cases, consider the evolutionary processes—both biotic and abiotic—that likely shaped the way animals interact with their environment. Further, one must consider these processes when trying to explain why animals respond to change—both natural and human-induced—in the way they do. In essence, then, one should examine the "big picture": studying the growth of individual trees may not explain the presence of the forest.

Literature Cited

Amundson, D.C., and H.E. Wright, Jr. 1979. Forest changes in Minnesota at the end of the Pleistocene. *Ecological Monographs* 49:1–16.

Anderson, B.W., R.D. Ohmart, and J. Rice. 1983. Avian and vegetation community structure and their seasonal relationships in the lower Colorado River valley. *Condor* 85:392–405.

Anderson, S.H., and H.H. Shugart, Jr. 1974. Habitat selection of breeding birds in an east Tennessee deciduous forest. *Ecology* 55:828–37.

Birks, H. J. B. 1980. The present flora and vegetation of the moraines of the Klutlan Glacier, Yukon Territory, Canada: A study in plant succession. *Quaternary Research* 14:60–86.

Brittingham, M. C., and S. A. Temple. 1983. Have cowbirds caused forest songbirds to decline? *BioScience* 33:31–35.

Brakenridge, G. R. 1978. Evidence for a cold, dry full-glacial climate in the American Southwest. *Quaternary Research* 9:22–40.

Capen, D. E., ed. 1981. *The use of multivariate statistics in studies of wildlife habitat*. USDA Forest Service General Technical Report RM–87.

Clements, F. C. 1920. *Plant indicators*. Washington, D.C.: Carnegie Institution of Washington.

Cody, M. L. 1985. An introduction to habitat selection in birds. In *Habitat selection in birds*, ed. M. L. Cody, 3–56. New York: Academic Press.

Connell, J. H. 1980. Diversity and the coevolution of competitors; or, The ghost of competition past. *Oikos* 35:131–38.

Conner, R. N., and C. S. Adkisson. 1977. Principal component analysis of woodpecker nesting habitat. *Wilson Bulletin* 89:122–29.

Cox, C. B., and P. D. Moore. 1985. *Biogeography: An ecological and evolutionary approach*. 4th edition. Oxford: Blackwell Scientific Publications.

Cox, G. W. 1968. The role of competition in the evolution of migration. *Evolution* 22:180–92.

Dueser, R. D., and H. H. Shugart, Jr. 1978. Microhabitats in a forest-floor small mammal fauna. *Ecology* 59:89–98.

Emlen, S. T., J. D. Rising, and W. L. Thompson. 1975. A behavioral and morphological study of sympatry in the indigo and lazuli buntings of the Great Plains. *Wilson Bulletin* 87:145–79.

Flack, J. A. D. 1976. Bird populations of aspen forests in western North America. *Ornithological Monographs* no. 19.

Fretwell, S. 1972. *Populations in a seasonal environment*. Princeton: Princeton Univ. Press.

Gill, F. B. 1980. Historical aspects of hybridization between blue-winged and golden-winged warblers. *Auk* 97:1–18.

Greenberg, R. 1979. Body size, breeding habitat, and winter exploitation systems in *Dendroica*. *Auk* 96:756–66.

Grinnell, J. 1905. Status of Townsend's warbler in California. *Condor* 7:52–53.

Grinnell, J. 1917. The niche-relationships of the California thrasher. *Auk* 34: 427–33.

Grubb, T. C., Jr. 1987. Changes in the flocking behaviour of wintering English titmice with time, weather and supplementary food. *Animal Behaviour* 35: 794–806.

Hansen, H. P. 1938. Postglacial forest succession and climate in the Puget Sound region. *Ecology* 19:528–42.

Hansen, H. P. 1953. Postglacial forests in the Yukon Territory and Alaska. *American Journal of Science* 251:505–42.

Heusser, C. J. 1965. A Pleistocene phytogeological sketch of the Pacific Northwest and Alaska. In *The Quaternary of the United States*, ed. H. E. Wright, Jr., and D. E. Frey, 469–83. Princeton: Princeton Univ. Press.

Heusser, C. J. 1977. Quaternary palynology of the Pacific slope of Washington. *Quaternary Research* 8:282–306.

Hilden, O. 1965. Habitat selection in birds. *Annales Zoologici Fennici* 2:53–75.
Hubbard, J. P. 1969. The relationships and evolution of the *Dendroica coronata* complex. *Auk* 86:393–432.
Hubbard, J. P. 1973. Avian evolution in the aridlands of North America. *Living Bird* 12:155–96.
Hutchinson, G. E. 1957. Concluding remarks. *Cold Spring Harbor Symposium on Quantitative Biology* 22:415–27.
Hutchinson, G. E. 1978. *An introduction to population ecology*. New Haven: Yale Univ. Press.
Hutto, R. L. 1985. Habitat selection by nonbreeding, migratory land birds. In *Habitat selection in birds*, ed. M. L. Cody, 455-76. New York: Academic Press.
James, F. C. 1971. Ordinations of habitat relationships among breeding birds. *Wilson Bulletin* 83:215–36.
Johnson, D. H. 1980. The comparison of usage and availability measurements for evaluating resource preference. *Ecology* 61:65–71.
Johnson, D. L. 1977. The late Quaternary climate of coastal California: Evidence for an ice age refugium. *Quaternary Research* 8:154–79.
Keast, A. 1980. Migratory Parulidae: What can species co-occurrence in the north reveal about ecological plasticity and wintering patterns? In *Migrant birds in the Neotropics: Ecology, behavior, distribution, and conservation*, ed. A. Keast and E. S. Morton, 457–76. Washington, D.C.: Smithsonian Institution Press.
Keast, A., and E. S. Morton, eds. 1980. *Migrant birds in the Neotropics: Ecology, behavior, distribution, and conservation*. Washington, D.C.: Smithsonian Institution Press.
Keen, W. H. 1982. Habitat selection and interspecific competition in two species of plethodontid salamanders. *Ecology* 63:94–102.
Kendeigh, S. C. 1961. *Animal ecology*. Englewood Cliffs, N.J.: Prentice-Hall.
Kitchings, J. T., and D. J. Levy. 1981. Habitat patterns in a small mammal community. *Journal of Mammalogy* 62:814–20.
Knopf, F. L. 1986. Changing landscapes and the cosmopolitism of the eastern Colorado avifauna. *Wildlife Society Bulletin* 14:132–42.
Krebs, J. R. 1977. The significance of song repertoires: The Beau Geste hypothesis. *Animal Behaviour* 25:475–78.
Lemen, C. A., and P. W. Freeman. 1986. Interference competition in a heteromyid community in the Great Basin of Nevada, USA. *Oikos* 46:390–96.
Mack, R. N., N. W. Rutter, V. M. Bryant, Jr., and S. Valastro. 1978. Reexamination of postglacial vegetation history in northern Idaho: Hager Pond, Bonner Co. *Quaternary Research* 10:241–55.
MacKenzie, D. I., and S. G. Sealy. 1981. Nest site selection in eastern and western kingbirds: A multivariate approach. *Condor* 83:310–21.
Mengel, R. M. 1964. The probable history of species formation in some northern wood warblers (Parulidae). *Living Bird* 3:9–43.
Mengel, R. M. 1970. The North American Central Plains as an isolating agent in bird speciation. In *Pleistocene and Recent environments of the central Great Plains*, ed. W. Dort, Jr., and J. K. Jones, Jr., 279–340. Dept. of Geology Special Publication no. 3, Univ. of Kansas, Lawrence.
Merriam, C. H. 1898. *Life zones and crop zones*. USDA Biological Services Bulletin no. 10.

Morrison, M. L. 1983. Analysis of geographic variation in the Townsend's warbler. *Condor* 85:385–91.

Morrison, M. L. 1986. Bird populations as indicators of environmental change. In *Current ornithology*. Vol. 3, ed. R. F. Johnston, 429–51. New York: Plenum Press.

Morrison, M. L., and J. W. Hardy. 1983. Vocalizations of the black-throated gray warbler. *Wilson Bulletin* 95:640–43.

Morrison, M. L., I. C. Timossi, K. A. With, and P. N. Manley. 1985. Use of tree species by forest birds during winter and summer. *Journal of Wildlife Management* 49:1098–102.

Murúa, R., and L. A. Gonzàlez. 1982. Microhabitat selection in two Chilean cricetid rodents. *Oecologia* 52:12–15.

Oliver, C. D., A. B. Adams, and R. J. Zasoski. 1985. Disturbance patterns and forest development in a recently deglaciated valley in the northwestern Cascade Range of Washington, U.S.A. *Canadian Journal of Forest Research* 15: 221–32.

Paine, R. T. 1966. Food web complexity and species diversity. *American Naturalist* 100:65–75.

Price, P. W. 1984. Communities of specialists: Vacant niches in ecological and evolutionary time. In *Ecological communities: Conceptual issues and the evidence*, ed. D. R. Strong, Jr., D. Simberloff, L. B. Abele, and A. B. Thistle, 510–23. Princeton: Princeton Univ. Press.

Rand, A. L. 1948. Glaciation, an isolating factor in speciation. *Evolution* 2:314–21.

Reinhert, H. K. 1984. Habitat variation within sympatric snake populations. *Ecology* 65:1673–82.

Richmond, G. M. 1965. Glaciation of the Rocky Mountains. In *The Quaternary of the United States*, ed. H. E. Wright, Jr., and D. E. Frey, 217–30. Princeton: Princeton Univ. Press.

Ricklefs, R. E. 1973. *Ecology*. Newton, Mass.: Chiron Press.

Rising, J. D. 1970. Morphological variation and evolution in some North American orioles. *Systematic Zoology* 19:315–51.

Ritchie, J. C. 1976. The late-Quaternary vegetational history of the western interior of Canada. *Canadian Journal of Botany* 54:1793–818.

Ritchie, J. C., and F. K. Hare. 1971. Late-Quaternary vegetation and climate near the Arctic tree line of northwestern North America. *Quaternary Research* 1:331–42.

Salomonson, M. G., and R. P. Balda. 1977. Winter territoriality of Townsend's solitaires (*Myadestes townsendi*) in a piñon-juniper-ponderosa pine ecotone. *Condor* 79:148–61.

Selander, R. K. 1965. Avian speciation in the Quaternary. In *The Quaternary of the United States*, ed. H. E. Wright, Jr., and D. G. Frey, 527–42. Princeton: Princeton Univ. Press.

Smith, C. C. 1981. The indivisible niche of *Tamiasciurus*: An example of nonpartitioning of resources. *Ecological Monographs* 51:343–63.

Stiles, F. G. 1980. Evolutionary implications of habitat relations between permanent and winter resident landbirds in Costa Rica. In *Migrant birds in the Neotropics: Ecology, behavior, distribution, and conservation*, ed. A. Keast and E. S. Morton, 421–35. Washington, D.C.: Smithsonian Institution Press.

Sullivan, T. P., and D. S. Sullivan. 1982. Population dynamics and regulation of Douglas squirrel (*Tamiasciurus douglasii*) with supplemental food. *Oecologia* 53:264–70.

Szaro, R. C., ed. 1988. *Management of amphibians, reptiles, and small mammals in North America.* USDA Forest Service General Technical Report RM–166.

Tinkle, D. W. 1982. Results of experimental density manipulation in an Arizona lizard community. *Ecology* 63:57–65.

Vaughan, T. A. 1972. *Mammalogy.* Philadelphia: W. B. Saunders.

Verner, J., M. L. Morrison, and C. J. Ralph, eds. 1986. *Wildlife 2000: Modeling habitat relationships of terrestrial vertebrates.* Madison: Univ. of Wisconsin Press.

Wecker, S. C. 1964. Habitat selection. *Scientific American* 211:109–16.

West, D. A. 1962. Hybridization in grosbeaks (*Pheucticus*) of the Great Plains. *Auk* 79:399–424.

Whitmore, R. C. 1975. Habitat ordination of passerine birds of the Virgin River valley, southwestern Utah. *Wilson Bulletin* 87:65–74.

Whittaker, R. H., and S. A. Levins, eds. 1975. *Niche: Theory and application.* Benchmark Papers in Ecology no. 3. Stroudsburg, Pa.: Dowden, Hutchinson, and Ross.

Wiens, J. A. 1977. On competition and variable environments. *American Scientist* 65:590–97.

Williams, O., and P. Wheat. 1971. Hybrid jays in Colorado. *Wilson Bulletin* 83: 343–46.

Yang, S. Y., and R. K. Selander. 1968. Hybridization in the grackle *Quiscalus quiscula* in Louisiana. *Systematic Zoology* 17:107–43.

3 Habitat Fragmentation

> The commendable plans are those that develop systems of unity, that effect a resolution of all man-made and natural forces and thus create a new landscape of order and repose. —Simonds (1961:22)

> We cannot command nature except by obeying her. —Francis Bacon (ca. 1610)

Introduction

The ultimate aim of planning for habitats is the conservation of associated wildlife species. In the end, three main ingredients determine whether species are conserved: availability of habitats, behavior of individual animals, and dynamics of populations. Availability of habitat determines what resources and environments are accessible. Behavior establishes how animals select their resources and interact with their environments. And population dynamics dictate rates of population change and how habitats are occupied.

To maintain viable populations of wildlife species, sufficient resources and adequate environmental conditions must provide for reproduction, foraging, resting, cover, and dispersal of animals at a variety of scales across space and time. Sufficient amounts, types, and arrangements of resources must provide for the needs of reproductive individuals on daily, seasonal, and yearly bases. Habitat also must be well distributed over a broad geographic area to allow breeding individuals to interact within and among populations spatially and among generations temporally.

One aspect of habitat distribution affecting persistence of breeding individuals and populations is habitat fragmentation, or the increase in isolation and decrease in size of resource patches. We use the term *patch* here to refer to an area which has more or less homogeneous environmental conditions. Fragmentation of environments threatens population viability especially of species that select "interior conditions," such as forest interiors, and for which the presence of edges is detrimental. On the other hand, a limited degree of environmental fragmentation can benefit species that thrive on edges or that require more than one kind of environment.

No single factor has been a greater cause of declines in wildlife populations than loss of habitat. And no one aspect of changes in habitat conditions has been more insidious and difficult to understand than that of fragmentation. Fragmentation affects quality of habitat of a given species in subtle ways. It changes types and quality of the food base. It changes microclimates by altering temperature and moisture regimes. It changes availability of cover and brings species together which normally have little contact, and thus may increase rates of nest parasitism, competition, and predation. It also can increase contact with and exploitation by humans.

Studying and predicting effects of fragmentation is one important topic that has emerged in recent years from the young science of landscape ecology (Forman and Godron 1986). To interpret and predict both deleterious and advantageous effects of fragmenting contiguous environments, it is necessary to understand related topics of population dynamics and environmental disturbance as studied in landscape ecology.

We begin this chapter by defining key terms and reviewing some of the fundamental tenets of landscape ecology. Next we discuss landscape dynamics of patches, including descriptions of patch patterns, the role of corridors between patches, types of landscape processes, and population dynamics at the landscape scale. We also discuss modeling and measuring fragmentation, managing patch patterns, and monitoring population responses. Planning human presence and resource use at the landscape scale is discussed, including implications of current theory for assessing potential effects of fragmentation. We provide a case example of how one federal agency is managing to reduce fragmentation. Finally, we close by discussing future study needs.

Definitions and the Science of Landscape Ecology

Definitions and History

Landscape ecology is a new science drawn from old disciplines. As an amalgamation of landscape architecture, zoogeography, plant geography, and synecology, its roots reach back to these European sciences of the nineteenth century and before. It has been only in the latter half of the twentieth century, however, that these elements have been combined in a unique constellation of disciplines. Landscape ecologists largely arose from plant geographers, zoogeographers, and population ecologists.

The basic unit of study in this field, the landscape, can be defined as "a part of the space on the earth's surface, consisting of a complex of systems, formed by the activity of rock, water, air, plants, animals and man and that by its physiognomy forms a recognizable entity" (Zonneveld 1979). Note in this definition the presence of several systems that influence the presence of landscape characteristics. Also, the landscape is a recognizable, distinct entity, much like the concept of a vegetation stand or an ecological community. Landscapes, like vegetation stands, can be mapped and classified.

Landscape ecology is the study of the response of species or communities to patterns across more than one patch. Many of the basic tenets of landscape ecology derive from island biogeography, zoogeography, and phytogeography: the sciences of the distributions and movements of animals and plants across islands and broad geographic areas.

Much of landscape ecology has focused on dynamics of plants and animals in patches, especially on islands and similarly isolated environments. Traditional island theory tells us that, on islands, population dynamics and community structures are determined by several main factors (fig. 3.1): size of the island (the smaller the island, the greater the local extinction rates); distance from the colonizing species pool (the further away, the greater the extinction rates); and attributes of the species, including dispersal ability, demographics (survivorship, recruitment), and habitat use specialization.

In a continental setting, environmental fragmentation represents the degree to which patches are increasingly isolated from other similar patches. The dynamics of environmental fragmentation probably operate differently than dynamics of oceanic islands and might affect species differently, depending on their habitat affiliations, dispersal capabilities, and other characteristics.

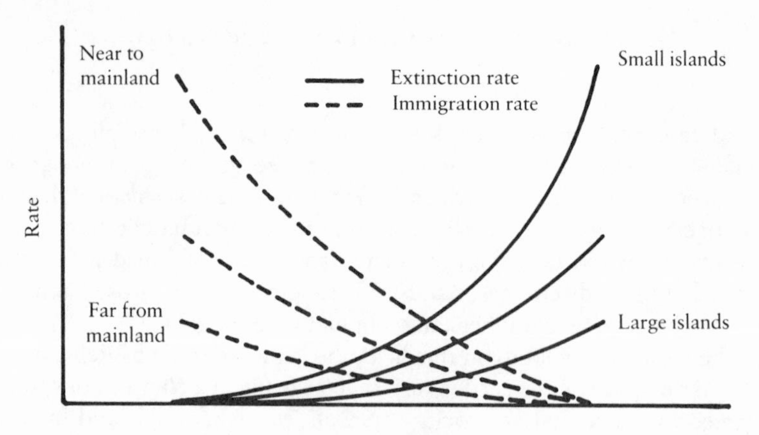

Figure 3.1. Equilibrium of species richness (number of species) on islands is affected by colonization and extinction rates. (As adapted by Samson 1980:246, from MacArthur and Wilson 1967, fig. 8.)

Why Consider Fragmentation?

Why is fragmentation of environments a concern? As Harris (1988) observed, until recently wildlife managers had traditionally viewed edges and ecotones as desirable (Leopold 1933; Thomas, Maser, and Rodiek 1979; Yoakum and Dasmann 1971). In their classic textbook on wildlife management, Yoakum and Dasmann (1971) recommended that managers should strive to increase the amount of edge, because wildlife flourish when habitats meet. (Note here that their use of *habitat* refers to vegetation or environments in general, rather than an area with specific conditions used and selected by a specific species, as we have defined the term in this book.) A high floral and faunal richness has typically been associated with edge environments (Hogstad 1967; Jakucs 1972; Johnston 1947; Sammalisto 1957; Thomas, Maser, and Rodiek 1979). In 1938 Lay declared that woodland openings and clearings are valuable for attracting wildlife. In recent decades, however, many once-contiguous forests have been severed, and adverse effects on fauna are being manifested, such as by increased predation on avian nests (Yahner, Morrell, and Rachael 1989) or by overbrowsing by white-tailed deer (*Odocoileus virginianus*) (Alverson, Waller, and Solheim 1988). The traditionally touted advantages, desirability, and prescriptions for creating edges and forest openings are under scrutiny (Harris 1988; Reese and Ratti 1988; Yahner 1988).

A number of wildlife and plant species have been shown to be associated with conditions existing in the interior of relatively large patches of mature vegetation or to be adversely affected by the proximity of early seral stage vegetation and associated edges. The composition and relative abundance of associated species within biotic communities vary among landscapes with various amounts and spatial distributions of habitats. Thus, fragmentation affects species richness of the communities, population trends of some species, and overall biological diversity of the ecosystem. Whether the effects are unfavorable or beneficial depends on the desired species composition and abundances in the area. This dependence has strong implications for the management of landscapes.

Wilcove (1987) identified four ways that fragmentation can lead to local extinction: A species can be initially excluded from the protected patches. Patches can fail to provide habitat because of decrease in size or loss of internal heterogeneity. Fragmentation creates smaller, more isolated populations that are at greater risk from catastrophes, demographic variability, genetic deterioration, or social dysfunction. Fragmentation can disrupt important ecological relationships; this can cause secondary extinctions from loss of key species and adverse influences of external alien environments and edge environments.

Forest fragmentation also is a concern in tropical forests. Klein (1989) found that communities of nearly all species of dung and carrion beetle were more abundant in intact forest than in forest fragments or in clearcuts in Manuas, Brazil. Forest fragments had fewer species, smaller populations, and smaller sized beetles than did contiguous forests. Also, the beetles seldom moved from intact forest into the fragments, even though the fragments were isolated by less than 350 m and only for two to six years. In 1 ha forest fragments, dung decayed more slowly because of the sparser beetle communities. Thus, through its effect of reducing the dung decomposition agents, forest fragmentation indirectly contributed to differences in nutrient cycling rates and other related ecosystem processes, possibly to long-term site productivity.

Similar concerns for adverse effects on long-term forest productivity were hypothesized by Marcot (in press) for north temperate conifer forest. Fragmentation of temperate forests in the western United States might be reducing or excluding populations of mycophagous (fungi-eating) small rodents, which are key dispersal agents for distributing a number of species of hypogeous (below-ground) fungus, especially mychorrizae (Maser, Trappe, and Ure 1978). Mychorrizae and related fungi are essential symbionts of western conifers, especially of Douglas-fir (*Pseudotsuga menziesii*) but also of pines (*Pinus* spp.), true firs (*Abies* spp.), hemlocks (*Tsuga* spp.), oaks (*Quercus* spp.), and alders (*Alnus* spp.). The fungus grows with the tree's root hairs and assists in nutrient uptake. The small rodents that act as key dispersal agents require old forest stands or large down wood

(Franklin et al. 1981; Maser, Trappe, and Li 1984). Forest fragmentation could be reducing the availability of these old forest stands or vegetation elements, with the possible indirect effects of reducing long-term productivity of nearby forests and forest plantations.

Excessive fragmentation of environments—including aquatic, estuarine, grassland, and forest—has been identified as a key concern in some wildlife planning strategies. One prime example is the set of regulations (36 CFR 219) implementing the National Forest Management Act (NFMA) of 1976. Under these regulations, national forest lands administered by the USDA Forest Service must develop land management plans. The regulations provide several guidelines pertinent to planning habitats in landscapes, including mandates to preserve long-term productivity. They also call for maintaining population viability by ensuring the continued, well-distributed existence of all native and desired nonnative species. The NFMA regulations also mandate protecting biological diversity of all species and communities of plants and animals on national forest lands.

Similarly, the Endangered Species Act, with recent revisions, mandates protection of threatened and endangered plant and animal species. Such species typically are very habitat-specific with narrow geographic and environmental distributions and use specific and often scarce or declining habitats. They are also typically rare, and some with large bodies and large home ranges use a variety of environments and resources across landscapes for meeting individual, breeding, and population needs for persistence. Additional federal legislation, such as the Migratory Bird Conservation Act and regulations on air and water quality, have implications for providing habitats at landscape and geographic scales.

Thus, environmental fragmentation is a habitat planning and conservation issue in that increasing fragmentation can affect population persistence and diversity of species and communities. This is true especially with older forest environments, including old-growth forests. Fragmentation can isolate individuals, breeding units, and subpopulations of patch-interior species. Such isolation increases the risk of local extinctions (Soule 1986) because of increased variance of population size from chance survival and reproduction events; fluctuations in the environment and quality and quantity of resources; and increased susceptibility of loss of smaller, more isolated patches to catastrophic events such as windstorms and fires. Local extinctions decrease biological diversity, measured by the number of species, the presence of rare or habitat-specific species, and the kinds of vegetation structures, environments, and resources available in an area.

Recently, in many countries, public interest in habitat planning has risen because of increased concern over related issues. Such issues include reduction of biological diversity, increases in environmental fragmentation and its perceived adverse effects on diversity, and loss of old-growth forests and other scarce and declining habitats (see Norse et al. 1986). Thus, from the perspectives of social and public interests, as well as from the scientific perspective, environmental fragmentation and management of habitats across landscapes are legitimate areas of concern and study.

Landscape Dynamics

In this section, we explore the basics of landscape dynamics, including disturbance of patch dynamics, descriptions of patch patterns, the role of corridors between patches, landscape processes, and population dynamics at the scale of landscape. Although this section is not intended as a thorough review of the theory and concepts of landscape ecology, key topics have been selected to aid understanding of and accounting for effects of landscapes on wildlife populations.

Dynamics of Patches

Occurrence and distribution of patches in a landscape are not static. They shift over time under systematic forces such as succession and erosion and under catastrophic forces such as storms and short-term disturbance by humans. A body of empirical and theoretical literature has addressed the dynamics of vegetation patches across landscapes and the role of various disturbance agents (see Pickett and White 1985).

Catastrophic disturbances generally introduce early successional patches into a landscape and reduce existing, older successional patches to remnants. This is the typical process by which mature forests become fragmented into remnant blocks. Catastrophic disturbance also affects patch size, generally increasing the size of early successional patches and reducing the size of late successional patches. Some disturbances, however, may affect within-patch characteristics more than between-patch characteristics. Heterogeneity within vegetation patches might be increased by opening canopies, creating tree falls, and increasing light and heat transmittal to lower canopy levels and the ground surface. Disturbance changes regimes of energy and nutrient flow in a landscape, although it may either enhance or depress particular flows depending on the type of disturbance.

Shugart (1984) proposed a way to classify disturbances based on the relationship between landscape area and the area disturbed (fig. 3.2). His system characterizes disturbances as ranging from small tree falls affecting less than 1000 m^2 in individual forests of 100 ha or so, to wildfires affecting up to 100 ha on recreational forests of over 10,000 ha, to hurricanes affecting 1,000,000 ha or more on islands of equal area. Shugart proposed

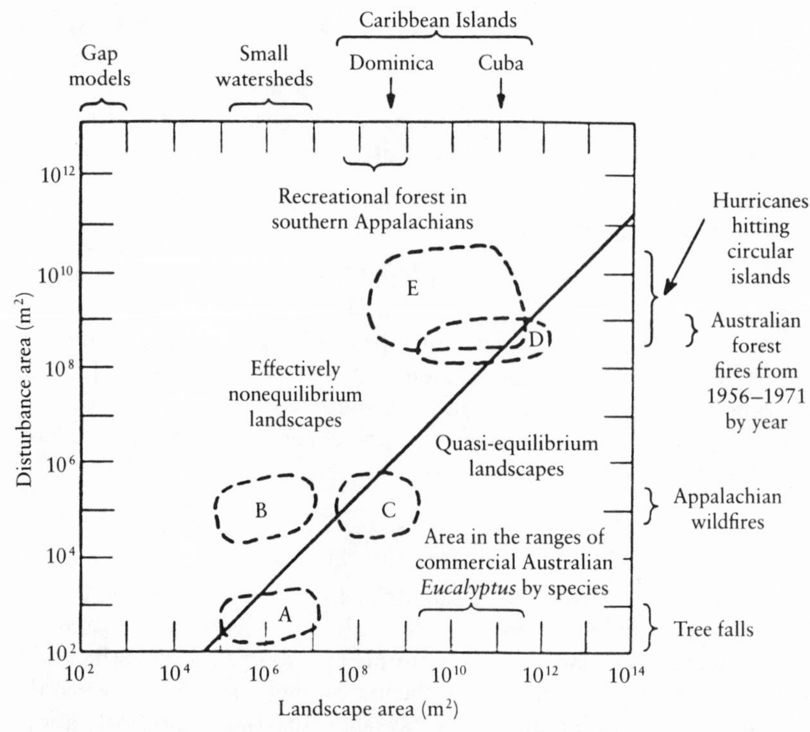

Figure 3.2. Classification of disturbance regimes based on area disturbed and total landscape area. (As adapted by Shugart 1984:165, from Shugart and West 1981.)

that a state of equilibrium of conditions in a landscape undergoing disturbance entails something on the order of a landscape area approximately fifty times the size of the disturbance. Thus, a small woodlot can absorb tree fall disturbances, but the area of a national forest or national park is needed to absorb random wildfires to remain equilibratory landscapes. In landscapes not in equilibrium, where disturbances vary in size, it may be necessary for the manager to alter the scale of disturbance or to increase the area under management in order to provide for equilibrium disturbance conditions.

Another way to characterize disturbances is by the degree of change of the environment. Four general types of disturbance can be depicted in this way (fig. 3.3). What may be called Type I disturbances are environmental catastrophes such as volcanoes and extensive wildfires having major impact on a large geographic area. Type II disturbances are more localized and include the effects of storms, insects, and disease. Type III disturbances are widespread but of relatively low impact per unit time and include the chronic or systematic effects of predators and competitors and gradual

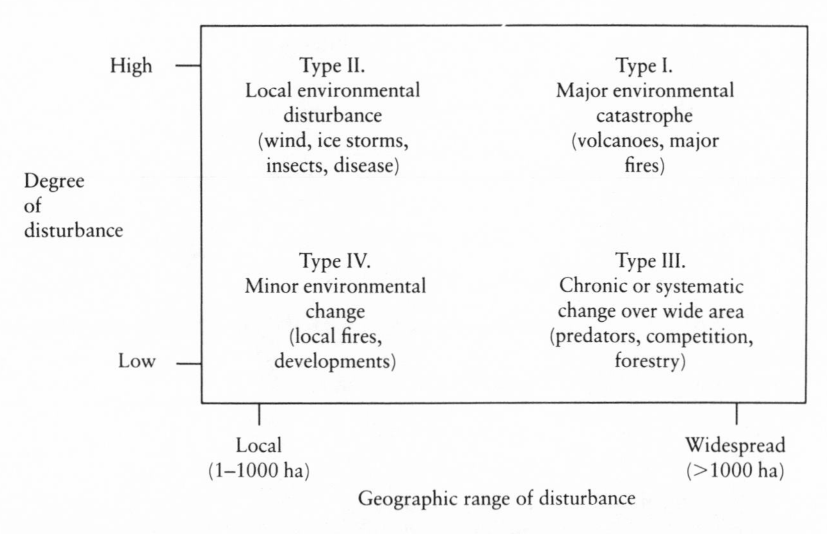

Figure 3.3. Four types of disturbances shown by degree of disturbance and geographic area affected

shifts in land use over a wide area. Type IV disturbances are localized and generally low in impact and include minor environmental changes caused by local fires and human developments.

Much of the concern about effects of forest fragmentation derives from local reduction and isolation of habitats from Type II and Type IV disturbances. Secondary effects of fragmentation, however, such as increases in predatory or competitor species, occur on a more widespread scale and are difficult to predict and monitor locally. Increasingly, resource planners wish to track cumulative effects of Type IV disturbances. Such cumulative effects may become Type III disturbances over a broader area. These more widespread effects should be assessed as part of any fragmentation monitoring program or research study.

Environmental conditions characterizing edges of remnant patches—frequent and direct sunlight and marked fluctuations in temperature and moisture—often penetrate into a patch. Thus, the ecologically effective patch size, that is, the amount of truly interior environment, is smaller than the physical patch size. Similarly, patch shape is affected by fragmentation. Disturbance and fragmentation may create edges, islands, rings, peninsulas, and other patch shapes not conducive to patch-interior species but beneficial to edge-successional and early successional species. In general, the number and configuration of patches and their changes over time vary under different disturbance conditions (see Ranney, Bruner, and Levenson 1981).

We can also characterize environmental fragmentation according to the

Table 3.1. Implications of different scales of habitat dispersion for various attributes of fragmented landscapes

	Dispersion	
Attribute	Geographic	Structural
Size (m^2)	Large, > 1000	Small, < 10
Isolation	Usually medium to large	Usually small
Boundary gradient	Steep	Shallow
Impact of extrinsic disturbance	Confined to edge and up to a few hundred meters in	Throughout
Vulnerability to functional disruption	Medium to small	Medium to large
Scale of organism affected	Large generalist to medium specialist	Medium specialist to small specialist
Advantages for conservation	Usually has intact interior	Usually of greater total extent

From "Scale and the Spatial Concept of Fragmentation," in *Conservation Biology*, 4:2, pp. 197–202, 1990. Reprinted by permission of the Society of Conservation Biology, and Blackwell Scientific Publications, Inc..

spatial scale at which it occurs and by the body size and area use patterns of associated wildlife species. Lord and Norton (1990) viewed fragmentation as a discontinuity of any environmental gradient and thus independent of scale. They suggested that dispersion of patches be viewed on a wide range of scales, from structural to geographic (table 3.1).

Descriptions of Patch Patterns

Understanding the effects of disturbances and landscape processes on environmental fragmentation is the key to describing and predicting patch patterns. At the landscape scale of a few to several dozen patches, mathematical depictions of configuration of patches, travel corridors and pathways, and edge structures are useful means of characterizing patterns of patches across a landscape.

Forman and Godron (1986) reviewed numerous ways to depict patch patterns. These included representing patches as matrices and networks. With such a representation, mathematics of networks can be used to analyze how particular patch patterns provide for connectivity—potential movement of animals—among patches. Forman and Godron also summarized in a matrix six measures of patch characteristics, including patch shape, isolation, accessibility, interaction, and dispersion (table 3.2).

Patch patterns also have been described from the perspective of the edges present in a landscape. The ability of animals to move through edges of various types has been called porosity or permeability of edges. From this perspective, some models of animal movement in a landscape relate directly to patterns of patches as they affect movement highways, filters, and barriers. Such models are useful for describing species-specific conditions of isolated and connected patches (Buechner 1987a, b; Stamps, Buechner, and Krishnan 1987). Buechner (1987b) modeled movement of animals across insular parks and concluded that the direction and magnitude of movements can be affected by perimeter-to-area ratios of the patches, edge permeability, behavior and habitat preferences of the species, and relative sizes of dispersal sinks and source pools of moving animals.

Connectivity of patches and permeability of edges vary according to a species' body size, habitat specificity, and area of home range. What acts as habitat entirely suitable in type and amount for sustaining a small-bodied, habitat-specific species such as the red tree vole (*Arborimus longicaudus*), a specialist on Douglas-fir, also might act, at a much wider scale, merely as a dispersal stepping-stone for a larger-bodied, less habitat-specific species such as the mountain lion (*Felis concolor*).

That different species respond differently to a given landscape reflects the notion of environmental grain size. Animals with different resource orientations, body sizes, and home range areas perceive and use patches and resources at different scales. A specific set of patches might appear coarse-grained to a narrow-ranging species and fine-grained to a wide-ranging species. Thus, assessments of whether a particular patch configuration and degree of fragmentation is beneficial or deleterious largely depend on the characteristics of the species using the landscape.

Effects of Patch Size

The size of patches or amounts of a given type of environment present in a landscape directly affect colonization by individuals, persistence of individuals and breeding units, and numbers of species in the area. The size of forest patches in eastern deciduous forests has been related to the probability of occurrence of songbird and other bird species (Whitcomb et al. 1981; see fig. 3.4). Such probability curves are called incidence functions. Incidence functions describe the likelihood of a species being found within a patch of habitat of specific area. Incidence functions can also be interpreted as the proportion of patches of a given size having the species present.

Table 3.2. Measures of patch characteristics in a matrix

Shape of a Patch

$$D_i = P/2(A\pi)^{1/2} \tag{1}$$

where D_i is a shape index of patch i, P is the patch perimeter, and A is the patch area.

Isolation of a Patch

$$r_i = 1/n \sum_{j=1}^{n} d_{ij} \tag{2}$$

where r_i is an index of the isolation of patch i, n is the number of neighboring patches considered, and d_{ij} is the distance between patch i and any neighboring patch j.

Isolation of Patches

$$D = \sum(v_x^2 + v_y^2) \tag{3}$$

where D is an index of the isolation of all patches present. Patches are located on a grid with x and y coordinates. The average location and the variance for all patches are calculated for the y coordinate. $v_x^2 + v_y^2$ are the variances on the x and y coordinates, respectively.

Accessibility of a Patch

$$a_i = \sum_{j=1}^{n} d_{ij} \tag{4}$$

where a_i is an index of the accessibility of patch i; d_{ij} is the distance along a linkage between patch i and any of the n neighboring patches j.

Interaction Among Patches

$$I_i = \sum_{j=1}^{n} A_j/d_j^2 \tag{5}$$

where I_i is the degree of interaction of patch i with n neighboring patches; A_j is the area of any neighboring patch j; and d_j is the distance between the edges of patch i and any patch j.

Dispersion of Patches

$$R_c = 2d_c(\phi/\pi) \tag{6}$$

where R_c is an index of dispersion; d_c is the average distance from a patch (its center or centroid) to its nearest neighboring patch; and ϕ is the average density of patches. $R_c = 1$ with randomly distributed patches; $R_c < 1$ for aggregated patches; and $1 < R_c \leq 2.149$ for regularly distributed patches. Thus, R_c is a measure of aggregation.

Source: R. T. T. Forman and M. Godron, *Landscape Ecology*. Copyright © 1986 by John Wiley & Sons. See Forman and Godron 1986:188–89 for specific references for equation 5.

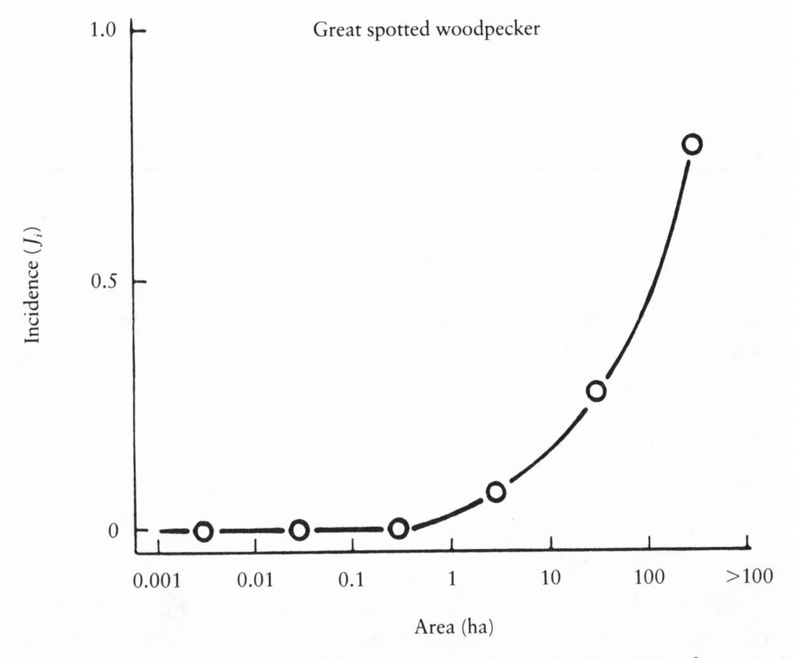

Figure 3.4. Incidence function of the great spotted woodpecker (*Picoides major*). (From Moore and Hooper 1975, in Wilcove et al. 1986:243.)

Incidence functions also are used to depict the number of different species present in an area—an island or a patch—of a given size. Such relationships are called species richness–area relationships and are mathematically depicted as

$$S = CA^z$$

where S is the number of species present, A is the area of the island or patch, C is a scaling constant that varies by taxon and location, and z is the rate at which the number of species increases with increasing area. From a number of studies of species richness on oceanic archipelagoes, z varies from approximately 0.24, for breeding land and freshwater birds in the West Indies and land vertebrates in islands of Lake Michigan, to 0.49, for breeding land and freshwater birds in islands of the Gulf of Guinea (MacArthur and Wilson 1967). Preston (1962a, b) calculated a theoretical value of 0.263. When plotted on a log-log relationship, where $\log S = \log C + z \log A$, the function appears as a straight line, and z becomes the slope of the line. (Also see Coleman 1981 for a theoretical discussion of other species-area distributions.) In consequence of the z factor, as a general rule of thumb, twice the number of species seem to require ten times the area (Darlington 1957; see also Harris 1984).

Howe, Howe, and Ford (1981) reported that area of small rainforest remnants in New South Wales, Australia, was the best single predictor of bird species richness. Their species-area relationship suggested a z value of 0.50. Patch isolation, disturbance by livestock, and distance from water, however, all tended to reduce the number of resident species found. Lomolino (1984) reported that species-area and species-isolation relationships of terrestrial mammals on nineteen archipelagoes were consistent with basic predictions from the equilibrium theory of island biogeography. Also, he noted that the species-area relationship weakened (the z value declined) with more isolated islands and with less vagile species (see also Lomolino 1982). Later in this chapter we will discuss whether such generalizations derived from studies of oceanic islands also apply to isolated patches in continental settings.

Effects of Other Factors

The likelihood of the presence of particular species, or the levels of total species richness, is affected by more than just area (Usher 1985). Other key factors include the presence of essential resources, such as food, water, and nest-building materials; heterogeneity of environments; and occurrence of competitors, predators, and diseases.

Freemark and Merriam (1986) reported that diversity of forest avifauna among twenty-one forest fragments in an agricultural landscape near Ottawa, Canada, depended on both size and heterogeneity of forests but that large fragment size was critical for the presence of forest-interior and resident-related species. Askins (1984; also Askins, Philbrick, and Sugeno 1987) reported that diversity and abundance of forest-interior birds in southeastern Connecticut correlated with both area of forest fragments and regional forest area and that regional distribution of forest was the more significant factor. Sites more isolated from other forests had fewer forest-interior species.

Lynch and Whigham (1984) recounted that local abundances of bird species breeding in interior upland forests in coastal Maryland were significantly influenced by forest area, isolation, structure, and floristics. They found that forest fragmentation effects are complex and species-specific. Although fragmentation generally had negative effects on forest-interior species, forest structure and floristics were more important than patch size and isolation for many species.

Interchange with partially isolated subpopulations or demes in adjacent geographic areas could affect persistence of a species in a given area. Such "metapopulation" dynamics are discussed below.

The presence of other species can greatly influence the basic species-area effect. For example, orb spiders (Araneae) reach extremely high densities on subtropical islands lacking predators (Schoener and Toft 1983).

Soule, Wilcox, and Holtby (1979) concluded that extinction rates of large mammals in nineteen nature reserves in East Africa were inversely related to reserve size but that even the largest reserves failed to provide sufficient habitat for maintaining viability of associated populations over a few centuries. Kushlan (1979) cautioned that size alone is an inadequate criterion for establishing nature reserves. He cited the case of Everglades National Park in Florida, which is losing species despite its relatively large size. The park is not a self-contained ecosystem and is highly dependent on nutrients and water received from outside the park boundaries.

In a study of chaparral-requiring birds in patches of chaparral of various sizes in southern California, Soule and his colleagues (1988) reported that four biotic and biogeographic variables accounted for 90 percent of the variation in bird species richness. Greater bird species richness correlated with younger age of the patch, larger area in chaparral, larger total area of the canyon study sites, and presence of coyotes (*Canis latrans*) and absence of gray foxes (*Urocyon cinereoargenteus*). The researchers hypothesized that the presence of coyotes suppressed the predatory actions of foxes, while the coyotes themselves did not prey on the birds.

The Role of Corridors

Much has been written and debated about the utility of providing corridors between patches. Corridors have been recommended for providing travel lanes for animals between refugia (MacClintock, Whitcomb, and Whitcomb 1977; Harris and Gallagher 1989). Corridors are viewed as a major means to link otherwise fragmented and isolated patches.

The concept of corridors has been used at a variety of scales, from linking national forests and national parks within and between physiographic provinces to linking specific forest patches within a subbasin area. At the broadest scale, corridors might allow for interaction of breeding individuals among otherwise isolated populations.

As a means of linking patches of habitat used seasonally by some species, the concept of the migration corridor has been used to expand protected areas in the central cordillera of Costa Rica (Stiles and Clark 1989). The Zona Protectora La Selva corridor has been established to link the Reserva Biologica La Selva with Reserva Forestal Cordillera Volcanica Central and Parque Nacional Braulio Carrillo (Harris and Gallagher 1989). Species such as the three-wattled bellbird (*Procnias tricarunculata*) and the snowcap (*Microchera albocoronata*) use the transect for seasonal migration movements (Stiles and Clark 1989; fig. 3.5).

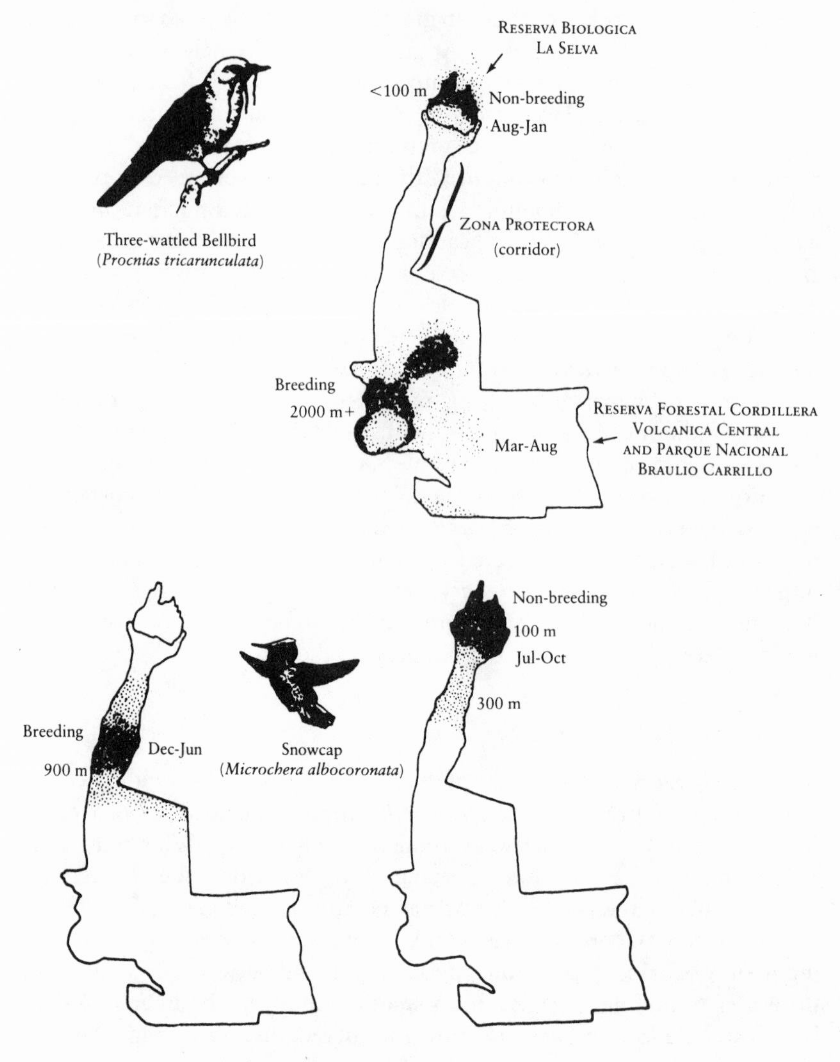

Figure 3.5. Example of a movement corridor established in the Costa Rican highlands. (From Stiles and Clark 1989:423.)

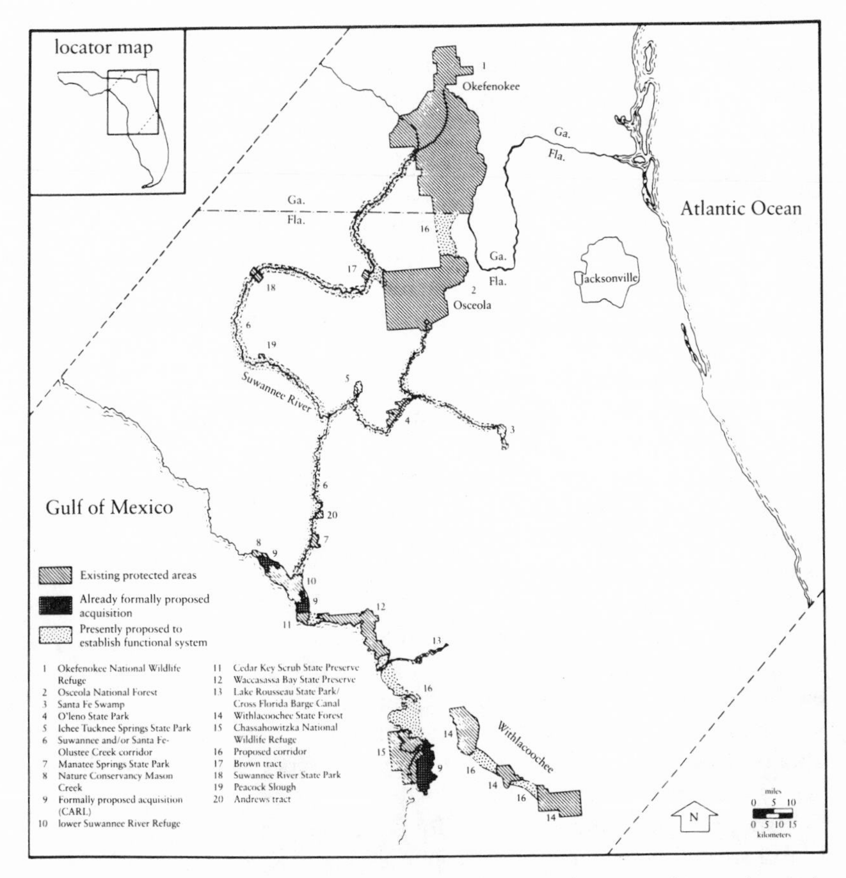

Figure 3.6. Corridors used to link a regional system of protected areas in north Florida and south Georgia. (From Noss and Harris 1986:306.)

Harris and Noss (1985; see also Noss and Harris 1986) proposed a regional system of protected areas in north Florida and south Georgia. A network of riparian and coastal corridors would link a number of existing protected areas—including national forests, national wildlife refuges, state parks, and other preserves (see fig. 3.6).

At a narrower scale, corridors would provide less vagile species cover from predators and harsh environmental conditions so they could move between areas for feeding, breeding, and resting. Such corridors would provide connections between patches large enough to be used as part or all of an animal's home range (daily, seasonal, or annual). As with the concept of patch patterns and environmental grain, corridors might function at different scales for different species. For example, stringer stands of mature forest cover at least four hundred feet wide positioned downslope to a topographic saddle might provide a travel lane and cover for elk, whereas down wood distributed throughout a single forest plantation might provide for movement of salamanders and lizards across the stand.

Much has been written on recommending corridors as part of habitat conservation plans, but much empirical data on use of corridors is circumstantial or absent. Intuitively, the notion of connecting patches at various scales with various structures of vegetation and other environmental elements is a good one, but more studies are needed to determine which types of corridors function as assumed for various species, habitats, and landscape contexts.

The idea of providing corridors has arisen because of the notion that when patches providing for individuals' needs are isolated, the likelihood of successful dispersal and colonization declines. What is the likelihood that a particular corridor with well-defined boundaries actually will be used by target animals? The likelihood of success probably varies by vagility and habitat-specificity of the species. Aquatic animals will most likely use stream and riparian corridors, whereas upland bird species will just as likely ignore such areas.

A generalized solution to linking patches within and across landscapes might be found in providing for a specific kind of matrix. That is, instead of establishing narrow passageways of vegetation with well-defined boundaries which may be at risk from windstorms and other catastrophic events, an alternative is to provide specific vegetation types and cover conditions across the planar landscape. Such a matrix approach does not lock habitat into specific routes and may allow for better resilience and recovery from loss of specific stands from catastrophic events.

Landscape Processes

In providing for adequate distributions of habitats within and across landscapes, we must understand the agents that change environmental conditions.

Landscape processes are abiotic as well as biotic. Abiotic processes affecting habitat patterns in landscapes include mass wasting, soil development, water and air quality, and the effect of parent material on vegetation development. In general, abiotic processes involve interchanges of materials—air, gases, particulates, soil, and water—within and among landscapes. Managing habitats at the landscape scale should account for how interchanges of materials from off-site affect on-site planning objectives and how on-site activities transport materials to other areas. Assessing cumulative effects of activities both on- and off-site is important when planning for specific habitat amounts and configurations.

Natural disturbances are also part of the abiotic processes of landscapes. Disturbances can be chronic, or slow. Examples include vegetation succession, climatic changes, and changes in fire frequencies and intensities. Disturbances can also be catastrophic, or fast, as with fire, storms, floods, and volcanoes.

Karr and Freemark (1985) offered a classification of factors to be evaluated in the study of disturbance (table 3.3). Factors to consider include type and regime of disturbance, type of biological system, and regional context. Karr and Freemark noted that biotic responses to disturbances vary from population extinctions to alterations in growth rates to changes in behavior and ecology of habitat selection.

Shugart and Seagle (1985) modeled patch occupancy by vertebrates in the presence of disturbance in a Tasmanian wet rainforest ecosystem and concluded that both diversity of the forest and its environmental constancy through time affect species richness in a landscape. The larger the landscape, the greater the buffer against environmental variation. Also, in their model, presence of competitive interactions among colonizers greatly decreased the number of potential species in a landscape.

Biotic processes include the roles of species invasions, as with nest parasites and competitors. Exploiting sites developed by humans, brown-headed cowbirds (*Molothrus ater*) have invaded the Sierra Nevada of California since 1930. They currently parasitize nests of twenty-two host species and may be threatening the continued survival of some species in the Sierra, especially the warbling vireo (*Vireo gilvus*) (Rothstein, Verner, and Stevens 1980).

Other biotic processes, such as epidemics—increases or invasions of pest species that depress population sizes or trends of other species—also affect distribution and abundance of a population at a landscape scale. Speciation and subspeciation are additional biotic processes that affect how a species responds to landscape conditions across broad geographic zones and over time spans of thousands of generations. As well, human disturbance, including exploitation, hunting, and introductions of exotic species, can be viewed as a biotic process.

Much of landscape ecology addresses how animals use heterogeneous environments. Studies of patterns of resource selection, foraging behaviors, and patterns and degrees of resource specialization all provide information on use of habitats. Dynamics of fauna relate to how animals move through the landscape. Thus, we also study types of movements, such as dispersal of juveniles, seasonal and annual migrations of adults, and other, less regu-

Table 3.3. Classification of factors to be evaluated in a study of landscape disturbance

Factor	Example	Biotic response result
Type of disturbance		
Physical factor	Drought impact on distribution of birds and frogs in tropical forest	Behavior and ecology of habitat selection
Biological factor	Yellow fever decimates howler monkey population	Evolution of disease resistance, extinction of host, or cyclic abundance changes in monkey
Interaction of physical and biological factors	Spring rain and bird distribution impact fishes in Everglades	Growth rates, survivorship, and reproduction of fish
Regime of disturbance		
Spatial dimension (scale)	Tree fall vs. hurricane distribution of large area of forest	Time to recolonization and assemblage of vertebrates will vary
Temporal dimension		
Frequency	Annual vs. irregular severe dry season	Physiological adaptation vs. behavioral, ecological, and evolutionary responses
Time of occurrence	Food and sedimentation during spawning period of fish	Destruction of cohort
Type of biological system		
Individual-species-population	Birds and lizards vary in mobility, longevity, and ability to turn off reproduction on land bridge island	Bird extinction rates high, whereas lizard extinction rates low on Barro Colorado Island, Panama
Assemblage/ecosystem environment type—biotic and abiotic	Fish more constrained by the chemistry of their aqueous environment than birds are by chemical toxicity of air	Different suite of physiological and behavioral adaptations, but air pollution from human society alters this balance

Table 3.3. (*Continued*)

Factor	Example	Biotic response result
Regional context		
Within area	Size of area and nature of the internal habitat mosaic	Survivorship, colonization, extinction pattern varies among islands
Between area and adjacent regions	Forest habitat islands in ocean vs. in agricultural area	Varying extinction and colonization dynamics among vertebrate groups; differing effects of colonists from adjacent patch of grass vs. water

Source: Karr and Freemark 1985:162.

lar movement patterns. Of interest in the design of landscapes are patterns, rates, and distances of species' movements, as well as breeding and social behaviors of associated species.

Population Dynamics at the Landscape Scale

Understanding and predicting the response of wildlife to patch configurations and fragmentation at the landscape scale require understanding population dynamics. In this section, we explore concepts of metapopulation dynamics. We discuss implications of the metapopulation concept for designing patch patterns at the landscape scale to sustain population viability. We then compare ecological concepts of oceanic islands with how populations use patches in a continental setting and highlight dynamics that the two settings share. In that way we can come to better understand how populations might respond to patterns of habitat patches and conditions at the landscape level. We then discuss effects of patch juxtaposition patterns, including that of environmental fragmentation. Finally, we discuss how population size and trends might reach equilibrium among patches.

Dynamics of Metapopulations

A metapopulation is a species whose range comprises geographically distinct patches interconnected through patterns of gene flow, extinction, and recolonization (Lande and Barrowclough 1987). Metapopulations generally occur when environmental conditions and species characteristics provide for less than a complete interchange of individuals and genetic re-

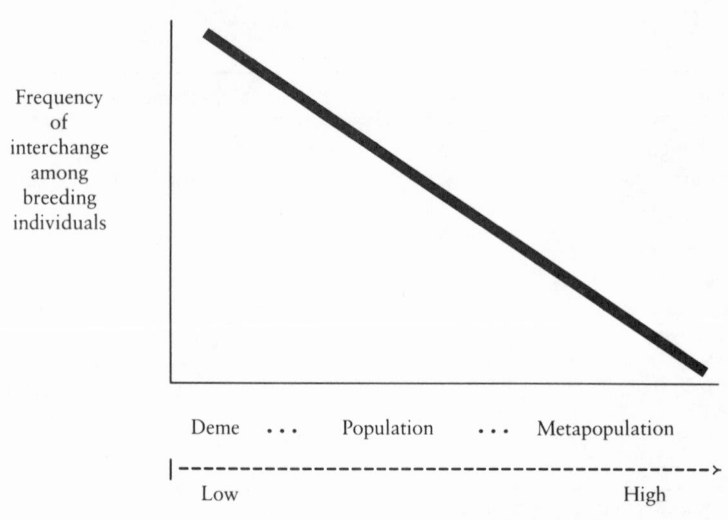

Figure 3.7. Gradient of interchange within metapopulations of varying internal structure.

sources among subpopulations. It occurs typically where habitat exists in a heterogeneous condition across a region, province, or landscape, serving to partially isolate breeding individuals.

Current literature refers to metapopulations in a variety of contexts. Metapopulations are typically conceived as pockets or subpopulations that interchange genetic material at a significantly higher rate within the subpopulation than with other subpopulations. As such, metapopulations can exhibit a gradient of internal structure ranging from nearly entirely interbreeding (panmictic), to loosely interbreeding with local pockets, to consisting of mostly isolated subpopulations that rarely interbreed (fig. 3.7). In this regard, the genetic structure of metapopulations can consist of a single, nearly panmictic genome; local demes isolated by distance from other demes; or mostly isolated subpopulations sharing genetic material with other subpopulations only infrequently through chance dispersal events. By definition, an entirely panmictic population has no metapopulation structure per se, because it acts as a single genetic and demographic entity. The difference between a metapopulation structure with more or less disjunct subpopulations and a clinal population structure with intergrading subpopulations, however, is a matter of degree of interchange among the subpopulation units.

An example of metapopulation distributions consisting of mostly isolated subpopulations is the occurrence of montane bird species across

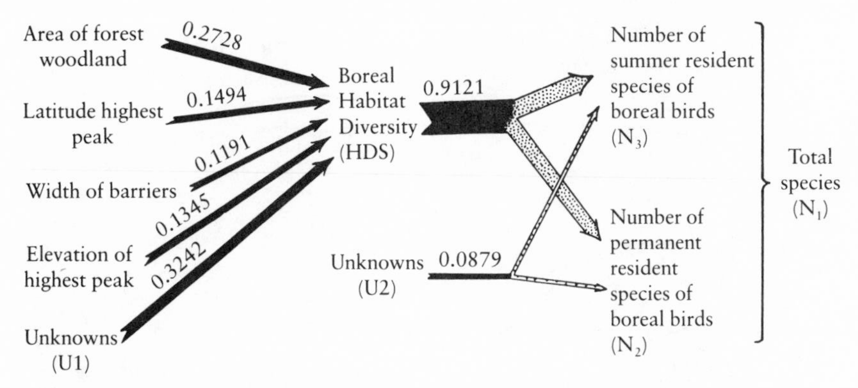

Figure 3.8. Sources of diversity of boreal bird communities in the southwestern United States. (From Johnson 1975:554.)

mountain ranges in the Great Basin (Johnson 1975, 1978). Johnson explored the sources of diversity of the boreal bird communities (fig. 3.8). He reported that the following known agents, listed in order of decreasing correlation, affected numbers of boreal bird species: area of forest woodland, latitude of the highest peak, elevation of the highest peak, and width of barriers (degree of isolation). Thus, the overall metapopulation distributions of the species were affected by local characteristics influencing local extinctions, as well as by isolation influencing colonization.

Typically, scientists perceive metapopulations as groups of individuals occupying locally restricted habitat areas. Over time, individuals in local patches or groups of patches might become extinct from chance variations in survivorship and recruitment or from catastrophic or systematic declines in the resource base (fig. 3.9). The site might eventually become recolonized by dispersing or "floating" individuals from adjacent patches. In this way, the pattern of patch occupancy over a broad geographic area might resemble a bank of blinking lights. Some lights remain lit for a longer period of time because they represent optimal conditions, areas with larger amounts of habitat, or patches central to the geographic range. Other lights might be mostly dark and blink on infrequently and only for short durations because they represent suboptimal conditions, areas with lesser amounts of habitat, or patches peripheral to the general geographic range.

Metapopulation structures may arise from a variety of factors, including subtraction of a previously contiguous population from habitat fragmentation or loss; invasion of hitherto unoccupied sites, as in expansion of a species' range due to newly available habitat or absence of key competitors, predators, or diseases; occurrence and colonization of sink habitat used in good reproductive years; heterogeneous distributions of resources,

Figure 3.9 (a–g). Schematic representation of landscape dynamics of patch colonization and occupancy.

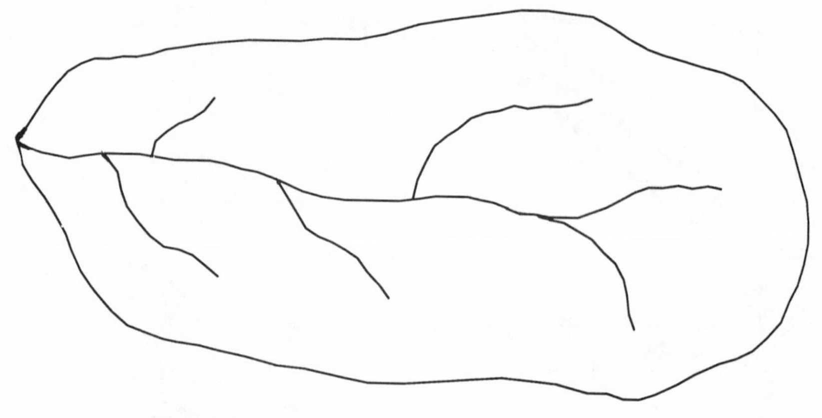

3.9a. Watershed boundary and riverine system

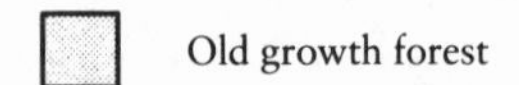

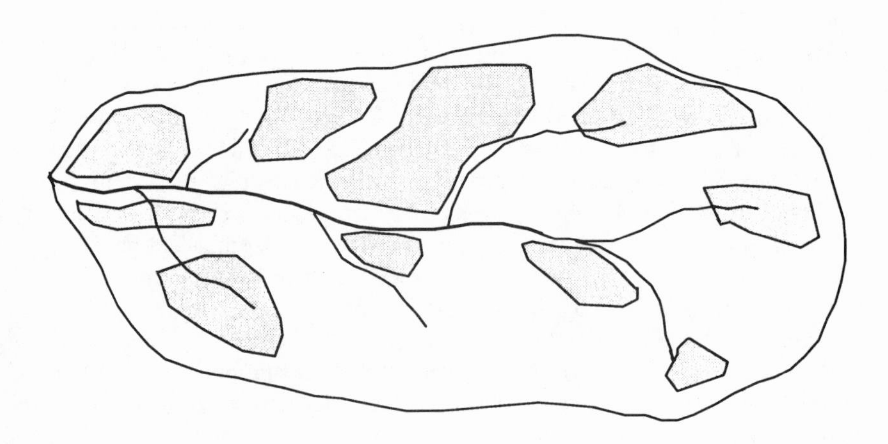

10 patches

3.9b. Occurrence of ten patches of old-growth forest

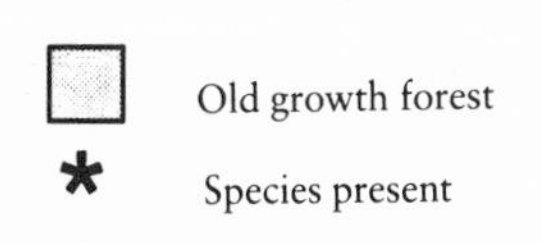

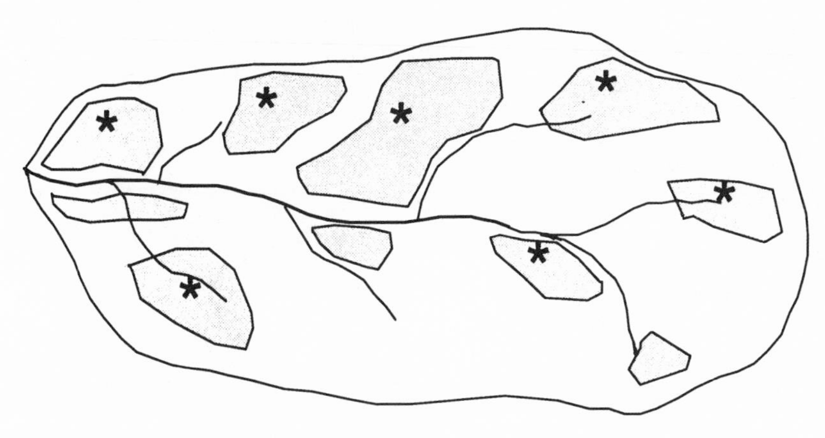

10 patches 7 occupied (70%)

3.9c. Occurrence of old-growth obligate vertebrate in seven of the ten patches

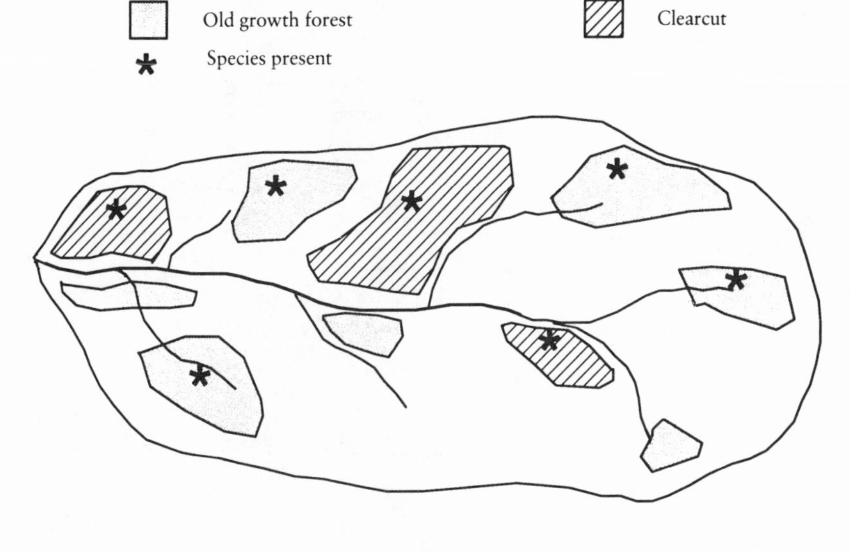

10 patches 7 occupied (70%)

3.9d. Selection of three patches for clear-cut timber harvesting

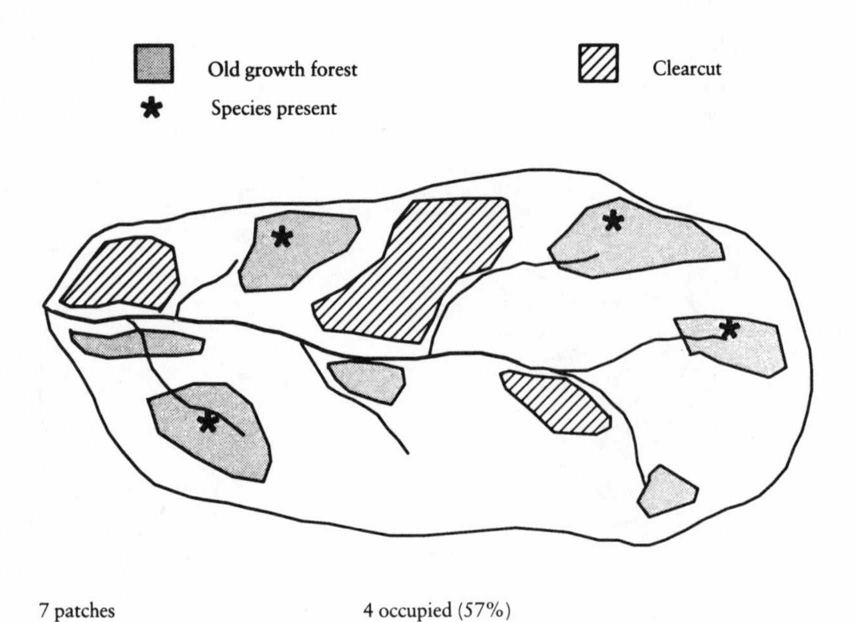

3.9e. Immediate result of harvest disturbance: loss of the species in three patches

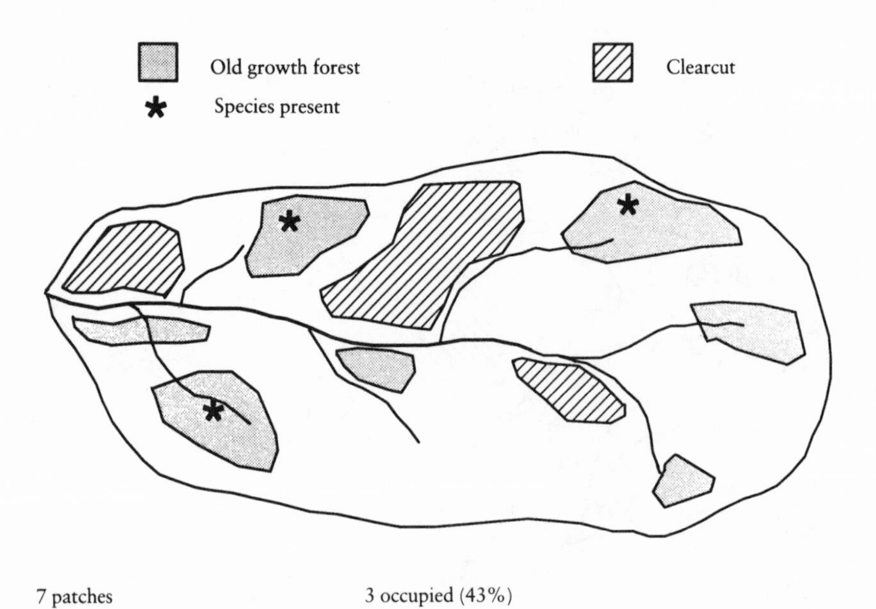

3.9f. Later loss of the species in a distant, isolated, smaller forest patch (faunal relaxation)

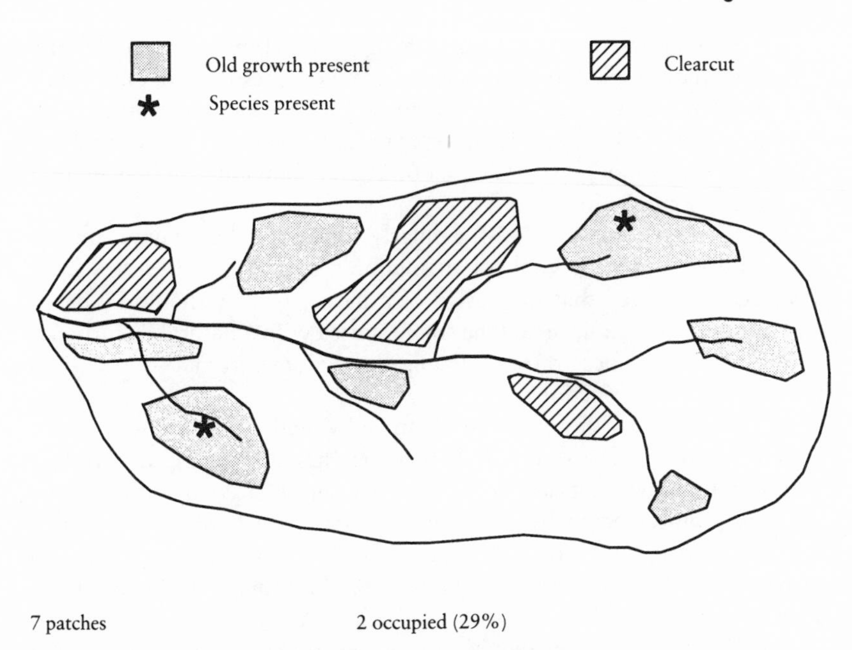

3.9g. Still later loss of the species in a larger forest patch recently isolated

such as tropical fruits, over space and/or time; heterogeneous or patchy distributions of habitat, such as lithic substrates or aquatic environments; and introductions of populations by humans.

Dynamics of metapopulations are complex. They entail spatial, temporal, and numerical dynamics, as biological and resource conditions change over time and across the species' range. From a management perspective, potential risks to extinction from metapopulation structures are difficult to analyze. The essential question for planning habitat to conserve viability of a metapopulation concerns estimating the probability that a well-distributed population will be maintained through time. The distribution of patches across a landscape, especially as it affects accessibility among individuals and subpopulations, is vital for helping to ensure a well-distributed population.

What are the implications of these dynamics for planning and managing patches of habitat? One possible guideline is to arrange patches of suboptimal (sink) habitats peripherally but close to optimal (source) habitats. This would allow for recolonization of sink habitats during periods of good resources. Often, however, we do not know what constitutes suboptimal and optimal conditions. In these cases, there are two possible courses of action. We can look to professional judgment and conduct a Delphi-type survey of expert opinion from which to craft initial patch planning guidelines. The guidelines should then be implemented and the response of individuals and populations monitored (see Chapter 4). Or we can base initial planning guidelines on habitat parameters in prediction models, such as habitat suitability index models (see Chapter 6). Again, individual and population responses should be monitored.

Pickett and Thompson (1978) similarly concluded that maintenance of colonization sources is vital for maintaining species in nature reserves. They recommended that the design of a nature reserve should be based on "minimum dynamic area," the smallest area with a natural disturbance regime which balances rates of emmigration from external sources with rates of internal extinctions.

At the scale of a home range of an individual or reproductive unit, another guideline is to provide habitat patches for feeding and resting within daily movement distances. Also, one should account for increased resource requirements when planning for foraging habitats near breeding habitats or structures; this would help plan for adequate patch sizes and proximities of different habitat types required for different needs at various stages of life.

The dynamics of patch colonization suggest that not all suitable patches will be occupied at any one time. Therefore, from a management perspective, one should not forgo conserving potentially suitable patches if they are not occupied in any one year. Rather, one should monitor patches over time to determine occupancy patterns before changing management direction that would severely reduce the quality of the patches. Seemingly vacant patches may still be vital for dispersal and distribution. Monitoring should be used to determine the frequency and type of use that such patches receive.

The concept of metapopulation dynamics may be useful for establishing criteria for listing a species as threatened, endangered, rare, or sensitive, as dynamics of local extinctions and colonization increase risks to viability. Conversely, the concept may provide guidelines for delisting a species as well. Developing delisting criteria would entail targeting a realistic recovery goal for an adequate population distribution, allowing for desired degrees of isolation and interactions of subpopulations and a realistic degree of patch occupancy.

A range of patch conditions or qualities may be needed to maintain well-distributed populations. One should consider areas used seasonally for migration, dispersal, and resting in order to assist in the interchange of individuals within and among subpopulations. In general it may be best to provide a range of habitat conditions for maintaining genetic diversity throughout ecotypes, subpopulations, and races of the metapopulation. Providing only optimal conditions—environments with the highest or most consistent occupancy—may not provide for long-term speciation and diversification of the genetic stock.

In general, understanding metapopulation dynamics is the key to providing the correct amount, quality, and distribution of habitat at various scales. It should be accounted for in guiding habitat rehabilitation and land acquisition, such as for seasonal use or migration (see fig. 3.5). Larger, contiguous patches are more desirable than smaller, isolated patches. In providing linkages between patches, however, there may be few options for design.

A metapopulation structure is not always inferior to an entirely panmictic population. It is not always clear if connecting subpopulations of a metapopulation is beneficial. Advantages conferred by a metapopulation structure include greater buffering against widespread disease or catastrophic events. Also, a greater variety of disjunct locations occupied may allow for greater evolutionary advantage. Local ecotypes may evolve to be more fit to local environmental conditions than if the genome was swamped from panmixia with all other subpopulations.

Oceanic Islands and Habitat Patches

Armed with a general understanding of metapopulation dynamics in a landscape context, we may next question the applicability of the theory of island biogeography to continental settings.

How do population dynamics on oceanic islands compare with habitat patches in a continental landscape setting? In some ways, population and metapopulation dynamics in continental habitat patches exhibit characteristics different from those on oceanic islands. On oceanic islands, for example, the surrounding "sea" is uninhabitable, and dispersal through such a region is an infrequent and chance event. In contrast, in continental landscapes, the intervening land may be suboptimal but usable for dispersal, resting, or seasonal or annual migration (Whitcomb 1977). Therefore, movement patterns among patches of habitats within a continental landscape may be more complex than those among islands.

Also, there are differences in the effects of patterns and the juxtaposition of patches of habitats. In continental landscapes, patch patterns directly affect occupancy rates and colonization dynamics and thus population persistence in an area. Whereas this is also true on oceanic islands, it is generally not considered in island equilibrium theory. Thus, depicting colonization and extinction processes may be more complex in forest landscapes than on islands.

There is also a difference in the effects of patch size. Although a small patch in a continental landscape may cease to function as habitat, much like a small, isolated oceanic island, it still may retain the same physiognomic structure. This could contribute to occurrence and persistence of a species in a landscape if adjacent or nearby conditions were suitable for the species in question. Patches in continental landscapes typically suffer invasion of edge effects into the patch interior. The smaller the patch, the greater is the fraction of the patch affected by edge. There is no direct equivalent to this effect on oceanic islands. Also, even if the patch is smaller than necessary to provide key requirements for the various stages of life of an associated species, the patch may still contribute valuable dispersal, feeding, cover, or resting conditions, depending on the landscape in which the patch occurs. An equivalent small oceanic island, on the other hand, would contribute nothing.

There may be differences in effects of competition between species on islands and that between species in continental landscapes. Studying birds in sixty-nine forest islands (shelterbelts) in eastern South Dakota, Martin (1981) found that absence of species was not due simply to habitat preferences or isolation. Competitive interactions among species influenced the numbers of individuals and species within the communities and the ecological structure of the communities, whereas the interaction of environmental conditions, chance, and competition determined distribution patterns of individual species among communities. In contrast, small mammal species on a Virginia barrier island showed no effect from competition on resource use by each species, although a high degree of habitat segregation was observed (Dueser and Porter 1986).

On the other side of the coin, oceanic islands and habitat patches in landscapes do share some common dynamics. Several key concepts are especially pertinent and useful when assessing and planning for habitat across a landscape.

Extinction phenomena is the first concept we will consider. The extinction rate of a species in both oceanic islands and in habitat patches in a landscape is a function of increasing isolation, such as that caused by fragmentation and decreasing area. On islands and in continental landscapes, extinction rates are generally higher with species specialized for particular resources or environments than with more generalist species.

Studying lizards on small Bahamian islands, Schoener and Schoener (1983) reported that time to extinction directly increased with island area. Above a certain size, colonization was rapid, but below the size colonizers rapidly became extinct. In continental habitat patches, incidence functions suggest similar relationships between patch size and species presence when the patch is sufficiently isolated from a species colonizing pool (Whitcomb et al. 1981; see fig. 3.4). Effects are less clear when the habitat patch is immersed in a landscape matrix of patches of varying environmental quality and interpatch distances.

A related concept in common to both oceanic and continental settings is occupancy rate. The occupancy rate of a species and species richness on islands, as in patches in landscape, are generally affected by more than just area. For example, Howe, Howe, and Ford (1981) reported that approximately 75 percent of the bird species in the extensive rainforest they studied in New South Wales also were detected in forest remnants. Species not recorded in the remnants were either rare in the extensive forest area or had large home ranges that were not supported by the isolated fragments. The rare species' absence from the fragments was probably due to the combined influence of the fragments' small size and isolation, and possibly to sampling effect. Because of low population sizes, chance extinction of rare species in the fragments was probably high, and rates of colonization into the fragments from the more extensive forest species pool was low. Also, the authors reported that they found several bird species typical of open country in the remnants and not in the extensive forests, emphasizing that remnants contain edge and early successional species.

Another pertinent concept is the founder principle. This states that a single set of immigrating individuals can begin a local subpopulation in a previously unoccupied area. Related to the founder principle is the rescue effect, which occurs when an immigrant fills an unoccupied, but previously occupied, patch (Brown and Kodric-Brown 1977). Understanding how segments of a population, such as nonbreeding floaters, can act as founders and rescuers is important in designing landscapes.

Lomolino (1984) reported that a rescue effect probably influenced the insular distribution of two species of vole (*Microtus pennsylvanicus* and *Blarina brevicauda*) in the Thousand Island region of the St. Lawrence River, New York. Distribution of the species on islands resulted from immigration selection—selection for the more vagile phenotypes in the mainland source population—and exclusion of predators on the islands.

When the environment is altered at a rate faster than the population can respond demographically, faunal relaxation, a decline in species richness or occupancy of islands or patches, might result. Faunal relaxation occurs in islands or in landscapes that have undergone recent isolation. Decline occurs because colonization rates have decreased but local extinction rates have not. Relaxation takes place with the sudden isolation of a rich environment.

Faunal relaxation may account for the decline or disappearance of seven bird species characteristic of mature forest in an oak-hemlock forest fragment in the Connecticut Arboretum between 1953 and 1976. In more recent years, however, the fragments have become less isolated as vegetation succession in extensive surrounding areas have produced conditions conducive to supporting mature forest bird species once again (Askins and Philbrick 1983). Faunal relaxation may be currently occurring in some of the national parks (Newmark 1986), although there is dispute over the evidence. It could occur in wildernesses over the next fifty years, as old forests become reduced on commercial timber production land, if patch linkages are not provided.

The result of faunal relaxation is either local extinction of a population or creation of relict populations. Relict populations seem to presently occur in odd and disjunct locations and may be holdovers from earlier environmental conditions. Occurrence and distribution of relict populations may reflect neither current environmental conditions nor dispersal capabilities of the species. Thus, it may be misleading to define habitat and landscape management requirements based on studies of current occurrence patterns and environmental associations of these relict populations.

In general, patterns of species presence across continental patches and landscapes, as across oceanic islands, result from differential colonization and extinction processes. These in turn are a function of such factors as the species' dispersal ability, life history, and demographic characteristics; variations in local weather and resource conditions; patch size and spacing; changes in corridors between patches and the types of environments used for dispersal; and the presence of competitors, predators, parasites, and diseases.

Population Equilibrium among Habitat Patches and Subpopulations

The point at which a species occupies patches of habitat within and among landscapes depends on dynamics at two scales: equilibrium of colonization with extinctions among localized patches in a particular landscape, and equilibrium of immigration with emigration and death among subpopulations of a metapopulation. These two scales overlap among species with various body sizes, use areas, and population structures. But the main principle is that of balance of local colonization and extinction rates. Population equilibrium thus is a function of dynamics of dispersal, seasonal migration, and other movements and represents a balance of demographic recruitment and loss.

Temple and Cary (1988) developed a computer simulation model of a landscape consisting of a hypothetical population of forest-interior birds whose fecundity was negatively related to their proximity to an edge in the landscape. The model demonstrated that if some species do exhibit a lower fecundity near edges, then their persistence in a landscape can be controlled

A. Elements of Habitat Fragmentation

Illustration A1a. Umner Strait, southeast Alaska: According to island biogeography, presence and persistence of wildlife species on true oceanic islands is governed by size of the island, distance from the mainland, and diversity of habitats within the island. In this case, a relatively small, homogeneous island distant from the mainland will have a low equilibrium number of wildlife species. (Compare with Illustration A1b.) (Photo by Bruce G. Marcot.)

Illustration A1b. Humboldt County, California: Example of a continental "island" or habitat patch of early successional vegetation resulting from timber harvest of mixed Douglas-fir (*Pseudotsuga menzeisi*)-hardwood forest. As vegetation in the patch undergoes secondary succession, wildlife associated with it will more resemble that of the surrounding forest. Wildlife communities can be expected to respond differently to this two-dimensional clearcut patch in a three-dimensional forest matrix, as compared with a three-dimensional forest patch retained in a two-dimensional clearcut matrix as shown in Illustration A2b. (Photo by Edward Toth, with permission.)

Illustration A2a. Kuiu Island, southeast Alaska: Unbroken, maritime, subboreal forest landscape of Sitka spruce (*Picea sitchensis*) and western hemlock (*Tsuga heterophylla*). The landscape is highly interconnected but offers little habitat for wildlife primarily associated with early successional vegetation. (Compare with Illustration A2b). (Photo by Bruce G. Marcot.)

Illustration A2b. Kuiu Island, southeast Alaska: Subboreal forest fragmented by timber harvest and road construction, showing a diversity of stand ages, edge contrasts, and an isolated forest island (lower left). Wildlife in such landscapes often are high in species richness, although the absence of travel corridors between the remnant old forests (top and bottom) might serve to restrict interchange of small, less vagile species. (Photo by Bruce G. Marcot.)

Illustration A3. Trinity County, California: Riparian gallery forests associated with major or minor water courses can act as habitat linkages for mustelids, many songbirds, and a host of other species, throughout a landscape. (Photo by Bruce G. Marcot.)

B. The Stand Scale: Examples of Effects of Human Influence on Forest Stand Structures

Illustration B1a. Trinity County, California: Results of simplification of a forest stand from even-age silviculture designed to maximize volume growth of timber. This second-growth stand consists solely of Douglas-fir (*Pseudotsuga menzeisii*) and lacks the diversity in structure and composition of vegetation and wildlife communities found in old-growth forests. (Photo by Bruce G. Marcot.)

Illustration B1b. Humboldt County, California: Recent changes in forestry operations on National Forest lands are emphasizing retaining vegetation elements that contribute to a diverse biota. These elements include hardwood trees, shrubs, down wood, and standing snags. Specific forestry operations designed to keep stands diverse vary by geographic region and forest type. This stand of Douglas-fir also contains Pacific madrone (*Arbutus menziesii*), California black oak (*Quercus kelloggii*), and other broad-leaf trees and shrubs that serve to diversify stand structures and wildlife communities. (Photo by Bruce G. Marcot.)

Illustration B2. Northern Black Forest, Federal Republic of Germany: Silvicultural management of *Pinus sylvestris* stands in Germany's Black Forest has resulted in a simple structure and low wildlife diversity. The veritable absence of snags, down wood, hardwoods, shrubs, and more than one layer of foliage provides poorly for a diverse fauna. (Photo by Bruce G. Marcot.)

C. The Landscape Scale: Examples of Effects of Human Land Use Activities on Edges and Patch Patterns

Illustration C1a. Kupreanof Island, southeast Alaska: Large clearcuts, fires, or other major disturbances that result in high-contrast edges between adjacent vegetation patches can be subject to a variety of secondary effects. These include invasion of the remaining forest by early successional plant species, wildlife predators and nest parasites. Greater variations in microclimatic conditions can be expected near such edges of forest stands than in the more equable forest interiors. Trees along such edges also are more susceptible to wind-throw from storms. (Photo by Bruce G. Marcot.)

Illustration C1b. Prince of Wales Island, Alaska: Example of strip-cut forestry designed to thin and open dense conifer stands without subjecting the forest to extreme conditions of high-contrast edges (compare with Illustration A2b). Although some blowdown of trees occurs along the strip edges, edge influence is minimized if the cuts are angled perpendicular to the prevailing direction of major storms and are kept narrower than one tree-height. (Photo by Bruce G. Marcot.)

Illustration C2. Himalayan Mountains, Utter Pradesh, India, near border of Tibet, China: Much of the *Cedrus deodara* and *Pinus rothsburgii* forests in the Himalayas has been cut and grazed, leaving highly erodable bare soils and causing a greater frequency of catastrophic floods downstream. (Photo by Bruce G. Marcot.)

by regulating the amount of edge present. With a simple population, that is, with zero or inadequate emigration from other regions where reproduction is better, interior-dependent populations in a severely fragmented landscape can become locally extirpated.

In a similar patch dynamics model, Fahrig and Merriam (1985) assessed survival of a population of white-footed mice (*Peromyscus leucopus*) among forest patches set in an agricultural landscape. The model predicted that mice in isolated woodlots would have lower growth rates and would be more prone to local extinctions than those in connected woodlots. These predictions were verified by empirical observations of wild populations.

Planning Human Presence and Habitats in Landscapes

Boundary, boundary, where shall it fall,
Can we believe, one size fits all?
If not we need new kinds of wisdom,
To make it fit the ecosystem.
—John Clark, "Ten Golden Rules of Marine Parks," *WII Newsletter* 5(1):17

The problem of designing landscapes for maintaining viability of wildlife populations involves designing for human presence as well. In this section, we discuss how humans have altered and used landscapes and, by using the ecological principles discussed earlier, how landscapes managed for human use can be designed for maintaining the viability of animal populations.

Designing Landscapes for Human Presence

Forman and Godron (1986) described a gradient of human-altered landscapes. Ranging from the least to the most disturbed, these are natural, managed, cultivated, suburban, urban, and megalopolis landscapes. At one extreme, there is little direct resource exploitation in natural landscapes of the national parks. At the other extreme, urban and megalopolis landscapes have been so heavily altered that there is little hope of providing habitats of substantial qualities and amounts for contributing to the viability of many naturally occurring vertebrate species. Thus, the real management challenge occurs in managed, cultivated, and suburban landscapes. For example, Knopf (1986) argued that riparian communities along the Platte River in eastern Colorado should be managed from the perspective of providing a dispersal corridor for bird species across the plains.

In the United States, a large portion of public lands occur as managed landscapes, according to Forman and Godron (1986). In many other parts of the world, much of the land available for providing habitat occurs in cultivated and suburban areas. Examples of cultivated landscapes include the highly terraced Himalayan foothills in Pakistan, India, Tibet, and Nepal, and much of the pastureland in New Zealand and Great Britain. Examples of suburban landscapes include villages in central and southern Africa and central Europe. The rest of this discussion explores designing landscapes for human presence in managed forest and grassland biomes.

The ultimate aim of designing viable managed landscapes is to sustain both resource use and species viability over generations. In a managed landscape, resources such as rangeland and timber are used, but conditions are actively regulated to help maintain desired wildlife species. *Maintain* means managing for the continued presence of a species, even if distribution and abundance patterns of species change from entirely natural conditions. In managed landscapes, some resources are used for economic, aesthetic, or cultural interests.

Landscape planners choose the composition and design of the spaces they create. The species of greenery and their amount and placement, the spatial configuration of patterns of human use, and other elements of landscape architecture all affect the ecological content of the landscape (Forman and Godron 1986). Some landscapes are designed to fit humans into an ecologically viable environment (see Zube 1986).

Forman and Godron (1986) pointed out that human perceptions and social custom can have great effects on design of landscapes. The quintessence of this principle can be found in the Black Forest of southwestern Germany, where for centuries conifer forests have been traditionally managed for optimal wood production in a manicured environment sterile of structural and wildlife species diversity. Only in recent years have adverse effects on long-term site production of wood spurred interest in designing more diverse forests there.

Some environmental psychologists and landscape planners have tried to understand the role of visual elements of a natural scene (see Leopold 1969). The work by Calvin, Dearinger, and Curtin (1972), Kaplan (1972, 1973, 1976), and Kaplan and Wendt (1972) disclosed that many suburban and urban people are more attracted to a fine-grained visual setting with trim, well-defined edges and little visual chaos. Such a preference translates to well-manicured forests, edged lawns, and woodlands with little understory. Successfully designing natural elements into managed, suburban, and urban landscapes requires understanding—and changing—these perceptions.

Forman and Godron (1986:75) offered that managing a natural landscape for both human presence and ecological objectives depends on keeping human activity inversely proportional to the sensitivity of landscape elements; protecting the areas of major flows in the landscape, such as animal movements; and maintaining natural disturbance regimes. Managing a remnant of a natural landscape focuses on these same three objectives, plus two more: minimizing isolation and minimizing human impacts from the surrounding matrix.

Designing Landscapes for Viable Populations

Given the goal for designing managed landscapes to help maintain viable wildlife populations, what concepts can we apply from landscape ecology for planning habitats in managed landscapes? First, we must precisely articulate the objective species and their habitats. Such management objectives would derive from and be affected by goals of existing landowners, prevailing policy or regulations, current land allocations and land use plans, existing environmental conditions and past management activities, and future capability of the land base.

For example, if the project is to acquire land for protecting habitats for highly sensitive vernal pools in northeastern California, then the major factor to consider might be capability of the land base. That is, how much area would be needed to contribute to a more or less self-contained biotic and hydrologic community to protect the habitat qualities?

If, however, the project is to derive management guidelines for linking national forests or parks with a corridor or network system in the southeastern United States (see figs. 3.5 and 3.6), then prevailing land use allocations and current environmental conditions would need high consideration, because these factors would directly determine the feasibility of land acquisition and the utility of the lands for providing adequate environmental conditions in the short term. Of course, it is vital in this example to define the wildlife species and their habitat requirements that are of interest. For example, if the objective is to provide corridors between forests and parks for dispersal of the Florida panther (*Felis concolor coryi*), then understanding that species' needs for reduced environmental fragmentation and its current distribution is paramount (Maehr 1990). Also, it is vital to understand in this case how the distribution of private lands and the goals of the landowners affect current and future environmental conditions, for these provide many of the potential vital links in the habitat patch network. In general, several guidelines apply.

Guidelines for Size and Edge of Habitat Patches

Providing adequate size of habitat patches and defining desirable limits of environmental fragmentation are important for maintaining the desired number of species within patches or the desired occupancy of patches by a species. One guiding concept for the size of habitat patches to provide may derive from the use of incidence functions. The study of eastern songbirds by Whitcomb and his colleagues (1981) suggested that the species most dependent on size of forest patches, the black-and-white warbler (*Mniotilta varia*), appeared in only half of four eastern hardwood forest study sites measuring at least 70 ha (175 acres) and in both study sites measuring at least 300 ha (750 acres). Other songbirds occurred more frequently in smaller patches.

Again, identifying the target species, as well as acceptable levels of occurrence, is central to developing specific guidelines. One objective, for instance, might be to manage for forest patches of adequate size so that more than half of the patches would provide for the most area-sensitive species. In the case of the study described above, the target species would be the black-and-white warbler, and its area requirement would be somewhere between 70 and 300 ha per forest patch. This species-specific objective is probably more desirable than an objective of providing only for richness in the number of species; with the latter, the more habitat- or area-sensitive species may be inadvertently excluded.

Soule and his colleagues (1988) provided a similar study of effects of area and vegetation conditions on chaparral-requiring bird species in southern California. Species-specific incidence functions suggested that most species reached at least two-thirds of their highest recorded percent occurrence in chaparral patches of at least 20 ha. This includes the more area-sensitive species such as the California thrasher (*Toxostoma redivivum*).

Providing adequate patch sizes also helps maintain patch-interior conditions. The effects of fragmentation of old-growth conifer forests in the Pacific Northwest provide one example (Franklin and Forman 1987). Forest fragments are subject to drying and invasion by early successional plant species along edges and at large openings. A rule of thumb is that such effects occur at least two tree heights or approximately 120 m (400 feet) into the forest stand from an edge, road, or large opening. Thus, a forest fragment less than or about 30 ha (80 acres) in size consists only of edge. Some environmental conditions, such as equable temperature and moisture regimes, are found only in interiors of forest stands. They are important for species such as spotted owls, which select roosts within old forests for their cool, equable temperatures (Barrows 1981). Thus, for protecting interiors of forests for wildlife species closely associated with old-growth temperate conifer forests of the western United States, a starting guideline would be to provide a patch size of at least 30 ha.

This is a gross generalization based on average condition, however. Edge effects on temperature, moisture, wind, and vegetation conditions also vary according to other factors. The following conditions will result in a greater degree of edge effect: edges transverse to the prevailing direction of windstorms; forests located at high elevations on slopes; adjacent vegetation with a high contrast of structure and species composition, such as an old-growth forest adjacent to an old field undergoing early stages of secondary succession (the greater the contrast of stand age and vegetation height, the greater the effect of edge); and the structure of the edge itself. If patches of snags, hardwood or broadleaf trees and shrubs, and other vegetation elements contributing to overall plant diversity are situated along edges, they may also encourage adverse effects of fire or insect pest spread or attract an undue density of predators or nest parasites.

Also, it is vital to recognize that in heavily fragmented landscapes, the last remaining patches of older or forested vegetation may play an important role. The patches may act as stepping-stones for dispersal of many species associated with the specific environmental conditions throughout the landscape. Removal of such patches because they fail to meet criteria for size and provision of interior conditions may result in the network of dispersal of wildlife being severed in the landscape. Also, small fragments of scarce or declining habitat might form the cores of a new habitat network under a landscape prescription that would emphasize their presence and role. Thus, although it is not necessarily desirable to create small patches, where they currently exist from past land management practices they might still play vital roles.

Guidelines for Juxtaposition of Patches

Designing managed forest landscapes also entails describing desired spatial arrangements and juxtapositions of patches. This has important implications for scheduling management activities at stand and subbasin scales. For example, Harris (1984) illustrated how a forest landscape can be managed for production of timber and provision of old forest conditions (fig. 3.10). In his scheme, a core of old-growth forest would be surrounded by long-rotation "spokes" of forest stands harvested on various schedules, at intervals of a century or more. In this manner, the general landscape would provide a substantial amount of dense, old forest cover beneficial for species associated with old growth. Mealey, Lipscomb, and Johnson (1982) provided a scheme for scheduling activities in a managed forest landscape so that various successional stages would be present at all times.

These simplified scenarios, however, best serve as starting points for developing real-world prescriptions for creating, restoring, or maintaining landscapes without undesirable fragmentation effects. Actual prescriptions would need to account for many additional variables. Here are just a few: landowner objectives for the landscape as a whole and for adjacent watersheds; capability of the land base to produce and maintain particular environmental conditions; possible increased risks of fire, insect out-

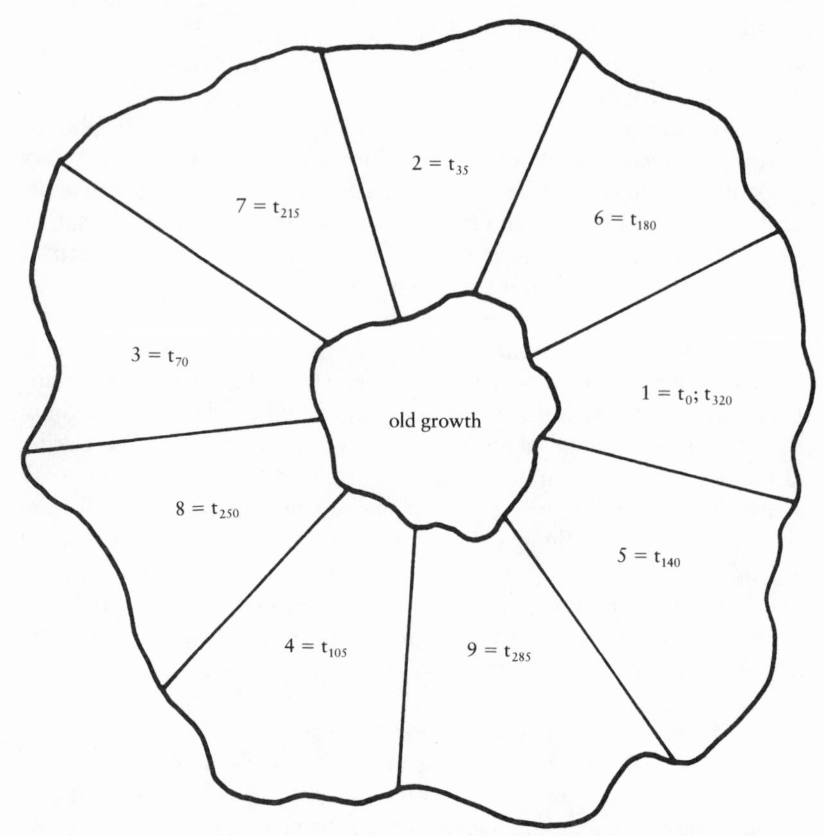

Figure 3.10. Schematic of a long-rotation, old-growth forest landscape. Alternate cutting of stands at various years (t) leads to maximum age difference between stands over a complete cutting cycle. (From L. D. Harris, *The Fragmented Forest*, The University of Chicago Press. © 1984 by the University of Chicago.)

breaks, storm damage to trees, floods, and erosion and their systematic and catastrophic effects that would result from creating specific patch patterns across the landscape; presence or need for developing transportation and access routes and the potential adverse effects on erosion and edge creation; absolute costs and costs in relation to anticipated benefits; and effects on the visual scene, including considerations for topography and public access and use areas such as campgrounds. New tools that may prove helpful in responding to some of these real-world considerations include a geographic information system armed with an up-to-date and accurate inventory of habitats and environmental conditions (see Chapter 6).

Adequate landscape design is objective- and location-specific. Thus, only general guiding concepts can be offered for juxtaposition. A landscape designer should ensure that vegetation structures are not placed adjacent to one another where they might create a high-contrast edge susceptible to windthrow or insect or predator invasion, or where penetration of one patch by adverse conditions of temperature, moisture, or early seral vegetation would occur. Depending on the wildlife species of interest and their specific environmental requirements, "feathering" vegetation structures along edges between patches might be more desirable than producing stark, high-contrast edges.

Second, the designer should ensure that patches of scarcer habitats are accessible throughout the landscape to associated species or to species most habitat-specific (stenotopic) and to those with lowest vagility. This might entail providing travel lanes or spacing habitat patches well within the species' dispersal distances. Patch juxtaposition patterns should be assessed for current conditions, conditions resulting immediately from land management activities, and conditions over time. Currently suitable juxtaposition patterns might become unsuitable as vegetation develops into new stages and structures or with the operation of other management activities.

Development of Working Hypotheses

Managing for populations at a landscape scale essentially involves establishing working management hypotheses about how populations will respond to environmental conditions at stand and landscape scales. Modeling population dynamics in landscapes and metapopulation dynamics at a broader geographic scale would help establish such working hypotheses. For example, incidence functions can be used to generate working hypotheses of expected rates of patch occupancy by a species. As well, equilibratory colonization and extinction rates among subpopulations can be used to craft hypotheses of distribution, size, and trends of metapopulations. Management hypotheses should be tested by monitoring environmental conditions and occupancy of patches over time.

Thus, more realistic expectations of patch occupancy and population distribution, size, and trends can be established for planning purposes. These would include specifying the proportion of patches expected to be occupied at any one time and over a specified interval. More realistic expectations would better inform decision-makers on planning issues and would help explain distributional patterns of species in fragmented or disjunct patches.

Implications for Planning for Habitat Diversity

Landscape design is a major facet of planning for habitat diversity in managed landscapes. Central to setting management objectives and designing landscape patterns is an accurate inventory of species and communities at basin and regional scales. Inventories should include depictions of historical distributions of native and introduced plant and animal species and communities. Terrestrial, riparian, aquatic, and other special edaphic communities should be included. Inventories should also depict current distribution and abundance of species and communities.

Planning for habitat diversity also entails developing classifications. Classifications of plant associations can be used for describing historical, existing, and potential future conditions. They should include classifications of seral, structural, and compositional stages of terrestrial, riparian, aquatic, and edaphic plant associations. Classifications of wildlife communities should also be developed and related to the classification of plant associations.

Inventories and classifications are tools used in planning habitat diversity. Development of standards and guidelines in habitat plans and explicit statements of desired future conditions for specific landscapes are also centrally required. Desired future conditions should describe specific distribution and abundance of native and introduced plant and animal species and communities. A desired future condition is a clear picture of how the landscape should look: the presence and distribution of patches. Desired future conditions are in turn derived from basin-specific management objectives. Such objectives describe desired distribution and abundance of plant and animal species and communities and desired patch patterns in a general way, such as with desired population levels or patch occupancy rates. The two tiers in designing landscapes—basin-specific objectives and desired future landscape conditions—are essential steps in planning diversity.

Development of Evaluation Procedures

Finally, as a landscape design is implemented as part of planning for habitat diversity in a managed landscape, several evaluation procedures need to be applied. One useful class of tools for assessing the effects of various landscape conditions is cumulative effects analysis. Cumulative effects analysis predicts effects on distribution and abundance of plant and animal species and communities from activities both on-site and off-site. In defining landscapes, one should consider the cumulative effects of all activities within a hydrologic unit such as a subbasin when prescribing activities on any one patch. Cumulative effects analysis helps describe future options available under each alternative and provides a richer information base from which to make management decisions.

Another evaluation procedure useful for designing landscapes is assessing disturbance processes and long-term productivity. Assessing disturbance processes entails describing past and future roles of various disturbance agents in shaping the composition, structure, and function of biotic communities. Ideally, the evaluation should project the type, frequency, extent, intensity, and location of disturbance agents. Types of disturbances to consider include fire, disease, wind, flood, and geologic actions such as volcanism and mass wasting. Assessing disturbances helps describe likely effects of management on long-term productivity, including effects on forest insect and disease problems; nutrient cycling, availability, and uptake; presence of dispersal agents; effects of pollutants and toxicants; and effects of climatic shifts.

Another key evaluation procedure is the development of a monitoring program. A monitoring program tests management hypotheses and theoretical relationships used to design landscapes. Monitoring is generally intended to be used as a feedback mechanism to alter the course of management if the biological system should react more adversely than anticipated (Walters 1986; Walters and Hilborn 1978; Holling 1984). Thus, a monitoring procedure should track "early warning signals," evidence that a plant or wildlife population is beginning to respond adversely to landscape conditions. Warning signals of this type should be "early" in order to provide time for remedial action should undesired population declines become evident.

Early warning signals for monitoring wildlife populations might include occurrence or increase in frequency of social dysfunction, such as the breakup of pair bonds in monogamous species or increases in the nonbreeding or the nonterritorial "floating" segment of the population. Monitoring early warning signals is useful for correcting reversible activities and is essential for halting irreversible activities. Examples of irreversible activities lending to monitoring include loss of ancient forests, excessive soil erosion, and declines in long-term productivity. Monitoring programs need to specify how results will enchance or reinforce planning direction and management standards and guidelines. Such criteria for reevaluation define a critical phase of the monitoring process and serve to close the loop between observation of effects and readjusting management direction.

Monitoring systems should signal when a species, distributed patchily as a metapopulation, is in decline. How one should monitor the size and trend of a metapopulation is problematic and entails species-specific solutions. Auteological studies should determine which factors affecting the dynamics of a metapopulation are most important for monitoring, including colonization events, occupancy of sources, sinks, or stepping-stones, and effects of fragmenting environments and the increasing amount of edge in a landscape.

Establishing Natural Preserves

Island biogeography and landscape ecology have been used to develop guidelines for establishing natural preserves. For example, Adamus and

Clough (1978) presented criteria for selected natural areas in Maine. Focusing on scarcity of species as one key criterion, they introduced a list of attributes of species that should inspire greater concern for providing natural areas. The list included species' site tenacity, seasonal mobility, area size needs, spatial distribution, endemism, tolerance to humans, reproductive capacity, and scarcity of habitat. The more restrictive these attributes are, the greater the desirability of establishing a natural area for such species becomes.

Case Study: Managing for Minimal Fragmentation

In the United States, the managed landscape on public lands is subject to great pressure from a variety of resource uses. Principal among these uses on national forest land is the harvesting of timber. In the Pacific Northwest, current federal legislation and regulations decree that forest stands shall be harvested in patch sizes of approximately sixty acres or less on national forest lands. On lands administered by other agencies, similar timber harvest patterns have been followed for a variety of reasons, including patchiness of land ownership patterns. The reason for such a dispersed-cutting approach seemed desirable when the landscapes consisted mostly of naturally occurring forests. Edge was created to enhance populations of game species, principally deer and elk.

As landscapes became more dissected with patch cuts, the pattern of remaining forested stands changed. Whereas once the cuts created welcome openings in a mostly forested landscape, they were now perceived as dominating the landscape, relegating residual forest patches to fragmented, isolated conditions. Edges, openings, and fragmentation of the once more-continuous forest cover are now viewed as threats to conserving species associated with old-growth forests and interior forest conditions.

At least one national forest, Willamette, in the Oregon Cascade Mountains, has instituted a test of a landscape management approach to reverse the effects of increasing forest fragmentation. The program has become known as minimum fragmentation and is based on the working management hypotheses of Franklin and Forman (1987). The overall objectives of landscape management on Willamette National Forest were to consider long-term productivity in the scheduling of management projects and timber harvests and to design landscapes with broader expanses of contiguous forest cover (Hemstrom 1989).

Components addressed in this landscape management approach include providing for snags, down wood, and riparian environments. Also considered are forest vegetation stand size and placement resulting from management activities. The procedure used to design and plan landscapes first establishes basin-scale objectives as guided by the standards set in the national forest land management plan. Then desired future conditions for specific subbasin areas are described.

Local district officials of the National Forest System then use the desired future conditions to design specific vegetation stand configurations for each subbasin or aggregate of subbasins. They follow four steps: accounting for existing land allocations described in the forest management plan; evaluating the current condition of the watershed; describing the desired conditions for management requirements (for example, considering habitat needs for management indicator species); and designing the landscape to meet desired future conditions.

The landscape is designed by specifying type, location, size, and timing of management activities, such as timber harvest and regeneration. A focus on particular landscape features, including visual resources, occurrence of natural forest stands, and winter range for big game, guides specific landscape designs. Activities are designed and scheduled to produce particular future conditions in subsequent decades, such as large, contiguous patches of forest in the next harvest cycle. The key to planning for such future conditions is planning subbasin conditions across the next harvest cycle on a much longer time scale than required by the previous two-to-five-year project action plans or the ten-to-fifteen-year national forest land management plans. Whether shorter-term planning procedures are conducive to longer-term landscape design remains to be tested by the next rounds of project and forest plans.

On the Blue River Ranger District of Willamette National Forest, all timber sales are considered with minimum fragmentation as the central objective guiding landscape design. Timber harvest, regeneration, and other forest management activities are placed and scheduled to reduce fragmentation effects of dispersed cuts and to create large and contiguous forest blocks in the future. Researchers are currently studying the effects of such landscape designs. Studies include investigations of population response by wildlife, including edge species such as deer and elk and nonedge, forest-interior species. Studies also are assessing effects on watershed processes, riparian vegetation, upslope processes, and soil movement. The expected payoffs from such a landscape design approach include increasing our understanding of effects of implementing the current forest land management plan and developing a geobased information system for monitoring landscape patterns.

Future Study Needs in Habitat Fragmentation and Landscape Design

What do we need to better apply principles of landscape ecology? First, we need additional empirical data on effects of forest fragmentation. Specifically, we need to understand which landscape parameters correlate with and directly affect species' abundance and distributions. We commonly need better understanding of species requirements.

We also need to better understand real-world dynamics of metapopulations as affected by landscape and regional distributions of environments. Such understanding has vital implications for planning patch sizes and patterns at landscape scales. We need better ways to predict effects of subdividing populations. This has implications for developing management requirements to provide for fragmentation-sensitive species.

Following Forman and Godron's suggestion, we also need research on linking human perceptions with ecological conditions and landscape design. Such research would straddle sociology, anthropology, economics, psychology, and ecology. Understanding how human perceptions are created and maintained—including perceptions of personal and cultural benefit—would "add an important dimension to landscape ecology, for understanding both a determinant of present patterns and an indicator of alternative futures" (Forman and Godroń 1986:72).

Perhaps what is most needed for applying scientific and theoretical principles to environmental fragmentation and landscape ecology is to foster a broader and clearer understanding of the objectives for land management. The history of conservation legislation is replete with legislative and litigative actions, political and antipolitical activism, and scientific and public debate designed to reallocate patterns of land use. In pluralistic societies, such revisiting of land use decisions should be allowed. To foster a rational and ecological land design to ensure long-term survival of species and long-term productivity of their ecosystems, however, a concurrence of objectives and a shared sense of purpose are essential first steps.

Literature Cited

Adamus, P. R., and G. C. Clough. 1978. Evaluating species for protection in natural areas. *Biological Conservation* 13:165–79.

Alverson, W. S., D. M. Waller, and S. L. Solheim. 1988. Forests too deer: Edge effects in northern Wisconsin. *Conservation Biology* 2:348–58.

Askins, R. A. 1984. Effect of regional forest configuration on the species richness and density of forest birds. Abstract of paper presented at the 1984 meeting of the American Ornithologists' Union, August, in Lawrence, Kans.

Askins, R. A., and M. Philbrick. 1983. Changes in the bird community as a forest fragment becomes less isolated. Abstract of a paper presented at the 1983 meeting of the American Ornithologists' Union, August, in New York.
Askins, R. A., M. J. Philbrick, and D. S. Sugeno. 1987. Relationship between the regional abundance of forest and the composition of forest bird communities. *Biological Conservation* 39:129–52.
Barrows, C. W. 1981. Roost selection by spotted owls: An adaptation to heat stress. *Condor* 83:302–9.
Brown, J. H., and A. Kodric-Brown. 1977. Turnover rates in insular biogeography: Effects of immigration on extinction. *Ecology* 58:445–49.
Buechner, M. 1987a. Conservation in insular parks: Simulation models of factors affecting the movement of animals across park boundaries. *Biological Conservation* 41:57–76.
Buechner, M. 1987b. A geometric model of vertebrate dispersal: Tests and implications. *Ecology* 68:310–18.
Calvin, J. S., J. A. Dearinger, and M. E. Curtin. 1972. An attempt at assessing preferences for natural landscapes. *Environment and Behavior* 4:447–70.
Coleman, B. D. 1981. On random placement and species-area relations. *Mathematical BioScience* 54:191–215.
Darlington, P. J. 1957. Zoogeography: The geographic distribution of animals. New York: John Wiley and Sons.
Dueser, R. D., and J. H. Porter. 1986. Habitat use by insular small mammals: Relative effects of competition and habitat structure. *Ecology* 67:195–201.
Fahrig, L., and G. Merriam. 1985. Habitat patch connectivity and population survival. *Ecology* 66:1762–68.
Forman, R. T. T., and M. Godron. 1986. *Landscape ecology*. New York: John Wiley and Sons.
Franklin, J. F., K. Cromack, W. Denison, A. McKee, C. Maser, J. Sedell, F. Swanson, and G. Juday. 1981. *Ecological characteristics of old-growth Douglas-fir forests*. USDA Forest Service General Technical Report PNW–118.
Franklin, J. F., and R. T. T. Forman. 1987. Creating landscape patterns by forest cutting: Ecological consequences and principles. *Landscape Ecology* 1: 5–18.
Freemark, K. E., and H. G. Merriam. 1986. Importance of area and habitat heterogeneity to bird assemblages in temperate forest fragments. *Biological Conservation* 36:115–41.
Harris, L. D. 1984. *The fragmented forest*. Chicago: Univ. of Chicago Press.
Harris, L. D. 1988. Edge effects and conservation of biotic diversity. *Conservation Biology* 2:330–32.
Harris, L. D., and P. B. Gallagher. 1989. New initiatives for wildlife conservation: The need for movement corridors. In *Preserving communities & corridors*, ed. G. Mackintosh, 11–34. Washington, D.C.: Defenders of Wildlife.
Harris, L. D., and R. F. Noss. 1985. Problems in categorizing the status of species: Endangerment with best of intentions. Paper presented at the sixteenth IUCN technical meeting, November 1984, Madrid, Spain.
Hemstrom, M. 1989. Paper presented at Old Growth symposium, March, USDA Forest Service, Portland, Oreg.
Hogstad, O. 1967. The edge effect on species and population density of some passerine birds. *Nytt Magasin for Zoologi* 15:40–43.

Holling, C. S. 1984. *Adaptive environmental assessment and management.* New York: John Wiley and Sons.

Howe, R. W., T. D. Howe, and H. A. Ford. 1981. Bird distributions on small rainforest remnants in New South Wales. *Australian Wildlife Research* 8: 637–51.

Jakucs, P. 1972. Dynamic relationship of the forests and swards: Quantitative and qualitative studies on the synecology, phytocenological, and structural relationship of the edge of forests. Budapest: Akademiai Kaido, Verlag Ungarischen Akad.

Johnson, N. K. 1975. Controls of number of bird species on montane islands in the Great Basin. *Evolution* 29:545–67.

Johnson, N. K. 1978. Patterns of avian geography and speciation in the intermountain region. In *Intermountain biogeography: A symposium,* 137–58. Great Basin Naturalist Memoirs no. 2.

Johnston, V. R. 1947. Breeding birds of the forest edge in Illinois. *Condor* 49: 45–53.

Kaplan, R. 1972. The dimensions of the visual environment: Methodological considerations. In *Environmental design: Research and practice,* ed. W. J. Mitchell, 6-7-1 to 6-7-5. Proceedings of the EDRA 3 conference, January, Univ. of California, Los Angeles.

Kaplan, S. 1973. Cognitive maps, human needs and the designed environment. Reprinted from *Environmental design research,* ed. W. F. E. Preiser. Stroudsburg, Pa.: Dowden, Hutchinson, and Ross. Supported by USDA Forest Service, Research Agreement 13–306. Univ. of Michigan, Dept. of Psychology, Ann Arbor.

Kaplan, S. 1976. Tranquility and challenge in the natural environment. In *Children, nature, and the urban environment.* Symposium proceedings, Northeastern Forest Experiment Station, USDA Forest Service, Upper Darby, Pa.

Kaplan, S., and J. S. Wendt. 1972. Preference and the visual environment: Complexity and some alternatives. In *Environmental design: Research and practice,* ed. W. J. Mitchell, 6-8-1 to 6-8-5. Proceedings of the EDRA 3 conference, January, Univ. of California, Los Angeles.

Karr, J. R., and K. E. Freemark. 1985. Disturbance and vertebrates: An integrative perspective. In *The ecology of natural disturbance and patch dynamics,* ed. S. T. A. Pickett and P. S. White, 153–68. Orlando, Fla.: Academic Press.

Klein, B. C. 1989. Effects of forest fragmentation on dung and carrion beetle communities in central Amazonia. *Ecology* 70:1715–25.

Knopf, F. L. 1986. Changing landscapes and the cosmopolitism of the eastern Colorado avifauna. *Wildlife Society Bulletin* 14:132–42.

Kushlan, J. A. 1979. Design and management of continental nature reserves: Lessons from the Everglades. *Biological Conservation* 15:281–90.

Lande, R., and G. F. Barrowclough. 1987. Effective population size, genetic variation, and their use in population management. In *Viable populations for conservation,* ed. M. E. Soule, 189. Cambridge: Cambridge Univ. Press.

Lay, D. W. 1938. How valuable are woodland clearings to wildlife? *Wilson Bull.* 50:254–56.

Leopold, A. 1933. *Game management.* New York: Charles Scribner's Sons.

Leopold, L. B. 1969. Landscape esthetics. *Natural History* 78:36–44.

Lomolino, M. V. 1982. Species-area and species-distance relationships of terrestrial mammals in the Thousand Island region. *Oecologia* 54:72–75.

Lomolino, M. V. 1984. Immigrant selection, predation, and the distributions of *Microtus pennsylvanicus* and *Blarina brevicauda* on islands. *American Naturalist* 123:468–83.

Lord, J. M., and D. A. Norton. 1990. Scale and the spatial concept of fragmentation. *Conservation Biology* 4:197–202.

Lynch, J. F., and D. F. Whigham. 1984. Effects of forest fragmentation on breeding bird communities in Maryland, USA. *Biological Conservation* 28: 287–324.

MacArthur, R. H., and E. O. Wilson. 1967. *The theory of island biogeography*. Princeton: Princeton Univ. Press.

MacClintock, L., R. F. Whitcomb, and B. L. Whitcomb. 1977. Island biogeography and "habitat islands" of eastern forest. Part 2, Evidence for the value of corridors and minimization of isolation in preservation of biotic diversity. *American Birds* 31:6–12.

Maehr, D. S. 1990. The Florida panther and private lands. *Conservation Biology* 4:167–70.

Marcot, B. G. In press. *Putting data, experience, and professional judgment to work in making land management decisions*. Proceedings of a workshop, Integrating Timber and Wildlife in Forest Landscapes: A Matter of Scale, October 1989, Eatonville, Wash. USDA Forest Service, Pacific Northwest Region and British Columbia Ministry of Environment.

Martin, T. E. 1981. Limitation in small habitat islands: Chance or competition? *Auk* 98:715–34.

Maser, C., J. M. Trappe, and C. Y. Li. 1984. *Large woody debris and long-term forest productivity*. Proceedings, Pacific Northwest Bioenergy Systems: Policies and Applications, May 10–11, Portland Oreg.

Maser, C., J. M. Trappe, and D. C. Ure. 1978. Implications of small mammal mycophagy to the management of western coniferous forests. *Transactions of the North American Wildlife and Natural Resources Conference* 43: 78–88.

Mealey, S. P., J. F. Lipscomb, and K. N. Johnson. 1982. Solving the habitat dispersion problem in forest planning. *Transactions of the North American Wildlife and Natural Resources Conference* 47:142–53.

Moore, N. W., and M. D. Hooper. 1975. On the number of bird species in British woods. *Biological Conservation* 8:239–391.

Newmark, W. 1986. Mammalian richness, colonization, and extinction in western North American national parks. Ph.D. thesis, Univ. of Michigan, Ann Arbor.

Norse, E. A., K. L. Rosenbaum, D. S. Wilcove, B. A. Wilcox, W. H. Romme, D. W. Johnston, and M. L. Stout. 1986. *Conserving biological diversity in our national forests*. Washington D.C.: Wilderness Society.

Noss, R. F., and L. D. Harris. 1986. Nodes, networks, and MUMs: Preserving diversity at all scales. *Environmental Management* 10:299–309.

Pickett, S. T. A., and J. N. Thompson. 1978. Patch dynamics and the design of nature reserves. *Biological Conservation* 13:27–37.

Pickett, S. T. A., and P. S. White, eds. 1985. *The ecology of natural disturbance and patch dynamics*. Orlando, Fla.: Academic Press.

Preston, F. W. 1962a. The canonical distribution of commonness and rarity. Part 1. *Ecology* 43:185–215.

Preston, F. W. 1962b. The canonical distribution of commonness and rarity. Part 2. *Ecology* 43:410–432.

Ranney, J. W., M. C. Bruner, and J. B. Levenson. 1981. The importance of edge in the structure and dynamics of forest islands. In *Forest island dynamics in man-dominated landscapes*, ed. R. L. Burgess and D. M. Sharpe, 67–96. New York: Springer-Verlag.

Reese, K. P., and J. T. Ratti. 1988. Edge effect: A concept under scrutiny. *Transactions of the North American Wildlife and Natural Resources Conference* 53:127–36.

Rothstein, S. I., J. Verner, and E. Stevens. 1980. Range expansion and diurnal changes in dispersion of the brown-headed cowbird in the Sierra Nevada. *Auk* 97:253–67.

Sammalisto, L. 1957. The effect of the woodland–open peatland edge on some peatland birds in south Finland. *Ornis Fennica* 34:81–89.

Samson, F. B. 1980. Island biogeography and the conservation of nongame birds. *Transactions of the North American Wildlife and Natural Resources Conference* 45:245–51.

Schoener, T. W., and A. Schoener. 1983. The time to extinction of a colonizing propagule of lizards increases with island area. *Nature* 302:332–34.

Schoener, T. W., and C. A. Toft. 1983. Spider populations: Extraordinarily high densities on islands without top predators. *Science* 219:1353–55.

Shugart, H. H., Jr. 1984. *A theory of forest dynamics*. New York: Springer-Verlag.

Shugart, H. H., Jr., and S. W. Seagle. 1985. Modeling forest landscapes and the role of disturbance in ecosystems and communities. In *The ecology of natural disturbance and patch dynamics*, ed. S. T. A. Pickett and P. S. White, 353–68. Orlando, Fla.: Academic Press.

Shugart, H. H., Jr., and D. C. West. 1981. Long-term dynamics of forest ecosystems. *American Scientist* 69:647–52.

Simonds, J. O. 1961. *Landscape architecture: The shaping of man's natural environment*. New York: McGraw-Hill.

Soule, M. E., ed. 1986. *Conservation biology*. Sunderland, Mass.: Sinauer Associates.

Soule, M. E., D. T. Bolger, A. C. Alberts, J. Wright, M. Sorice, and S. Hill. 1988. Reconstructed dynamics of rapid extinctions of chaparral-requiring birds in urban habitat islands. *Conservation Biology* 2:75–92.

Soule, M. E., B. A. Wilcox, and C. Holtby. 1979. Benign neglect: A model of faunal collapse in the game reserves of East Africa. *Biological Conservation* 15:259–72.

Stamps, J. A., M. Buechner, and V. V. Krishnan. 1987. The effects of edge permeability and habitat geometry on emigration from patches of habitat. *American Naturalist* 129:533–52.

Stiles, F. G., and D. A. Clark. 1989. Conservation of tropical rain forest birds: A case study from Costa Rica. *American Birds* 43(Fall):420–28.

Temple, S. A., and J. R. Cary. 1988. Modeling dynamics of habitat-interior bird populations in fragmented landscapes. *Conservation Biology* 2:340–47.

Thomas, J. W., C. Maser, and J. E. Rodiek. 1979. Edges. In *Wildlife habitats in managed forests*, ed. J. W. Thomas, 48–59. USDA Forest Service Agricultural Handbook no. 553.

Usher, M. B. 1985. Implications of species-area relationships for wildlife conservation. *Journal of Environmental Management* 21:181–91.

Walters, C. 1986. *Adaptive management of renewable resources*. New York: Macmillan.

Walters, C., and R. Hilborn. 1978. Ecological optimization and adaptive management. *Annual Review of Ecology and Systematics* 9:157–88.

Whitcomb, R. F. 1977. Island biogeography and "habitat islands" of eastern forest. Part 1, Introduction. *American Birds* 31:3–5.

Whitcomb, R. F., C. S. Robbins, J. F. Lynch, B. L. Whitcomb, M. K. Klimkiewicz, and D. Bystrak. 1981. Effects of forest fragmentation on avifauna of the eastern deciduous forest. In *Forest island dynamics in man-dominated landscapes*, ed. R. L. Burgess and D. M. Sharpe, 125–206. New York: Springer-Verlag.

Wilcove, D. S. 1987. From fragmentation to extinction. *Natural Areas Journal* 7: 23–29.

Wilcove, D. S., C. H. McLellan, and A. P. Dobson. 1986. Habitat fragmentation in the temperate zone. In *Conservation biology*, ed. M. E. Soule, 237–56. Sunderland, Mass.: Sinauer Associates.

Yahner, R. H. 1988. Changes in wildlife communities near edges. *Conservation Biology* 2:333–39.

Yahner, R. H., T. E. Morrell, and J. S. Rachael. 1989. Effects of edge contrast on depredation of artificial avian nests. *Journal of Wildlife Management* 53: 1135–38.

Yoakum, J., and W. Dasmann. 1971. Habitat manipulation practices. In *Wildlife management techniques*, ed. R. Giles, 173–231. 3d ed. Washington, D.C.: Wildlife Society.

Zonneveld, I. S. 1979. *Land evaluation and land(scape) science*. Enschede, The Netherlands: International Training Center.

Zube, E. H. 1986. The advance of ecology. *Landscape Architecture* 76:58–67.

4 Measuring Wildlife Habitat

Introduction

In this chapter we review five major aspects of measuring wildlife habitat: whom to measure, what to measure, how to measure it, when the measuring should be done, and where to do the measuring. We argue that a failure of any one of these five factors will likely result in a critical weakness in the data set and place severe limitations upon the applicability of results to various management situations. The use of guilds and indicator species in habitat analysis is also reviewed. Ingrained throughout our discussion is the ecological framework developed in the previous chapters; this framework could rightfully be called the "why" of habitat analysis.

In the opening section we develop the notion that it is usually better, first, to identify the problem in question (for example, one could choose to measure old-growth forests used by animals) and then to select the animal or animals that are the most appropriate test subjects: *whom* to measure. Our objective should be to focus measurements on the system, without expecting the organism of favor to respond in a predictable fashion; very little attention has been given to the subject of selecting an experimental organism.

Implicit in our argument is the correct identification of the *niche gestalt* of the animal. That is, as shown in Chapter 2, it is important to identify ways to determine what the animal perceives as its environment. Only then can we identify the factors of importance to the animal: the *what* question. Thus, we borrow from previous work on niche descriptions and resource partitioning and review theories of resource allocation to identify factors of importance in a study (for example, habitat and microhabitat factors, food by type and taxon).

Next we discuss various techniques used in measuring wildlife habitat: the *how* question. We identify major techniques, discuss their relationship to analysis of community structure, and outline various pros and cons regarding their usefulness. Techniques reviewed include the general area of experimental design and randomization (for example, simple random versus stratified random), the use of organism-centered plots versus the randomization procedures, and the like. We do not intend to provide an overview of all techniques; rather, we seek to discuss the proper use (including assumptions and limitations) of current techniques, and relate the methods chosen to our previous development of overall study objectives.

Consideration of *when* to measure involves two broad temporal scales: between-season and within-season designs. Several workers have clearly shown that resource requirements of animals in winter, for example, often bear little resemblance to requirements of the breeding season (see Conner 1981; Hutto 1981; Morrison et al. 1985). The use of differing winter and summer areas by big game and waterfowl is an obvious example of changes in areas of use and in food use. Further, species that reside permanently in an area commonly switch to different tree species and/or food types. Within a given time frame, resource requirements and activity patterns often vary with time of day, specific stage of the breeding cycle (weaning versus independent feeding stages of mammals, incubation versus nestling stages of birds), and other aspects of time-dependent resource requirements. Thus, one needs to have clearly identified the overall objective of the study within a framework of the temporal variation in activities exhibited by an animal.

The distribution and abundance of animals varies, of course, with variations in food supply, weather conditions, predator activity, and a host of other biotic and abiotic factors. As we show in this and previous chapters, these factors should influence the ways in which we measure habitat. Few studies, however, have recognized these variations in species distribution and abundances and instead have based analysis of theoretical considerations (like community structure) and management applications on site- and time-specific studies. We discuss the range of geographic areas sampled and their impact on study results: the *where* question.

Whom to Measure

When reading a publication on, say, the habitat used by a particular organism, have you ever wondered just why the author chose that particular subject for study? Why was Margaret Morse Nice drawn to the song sparrow (*Melospiza melodia;* Nice 1937, 1943) or Val Nolan, Jr., to the prairie warbler (*Dendroica discolor;* Nolan 1978)? Why did Olaus J. Murie detail the life history of the American elk (*Cervus canadensis;* Murie 1951) or Paul Errington the muskrat (*Ondatra zibithica;* Errington 1963)? We should be honest with ourselves and admit that certain animals—for whatever reason—raise certain personal feelings within us. We then *a posteriori* find justification for the study, as though a well-done, thorough study needs any.

There are reasons, however, why we must often set aside personal feelings and select the organism or organisms most suited for a particular study. That is, if we are asked to determine the effects of removal of old-growth forest on wildlife, we must first determine which species actually depend on this type of vegetation. But even if we discover, for example, that a certain bird is, to some degree, "old-growth dependent," it does not follow necessarily that this species is the most adversely affected by a change in the vegetation. Nor does it follow that we can even readily detect the effect of such a change on the species. Depending on the purpose of the study, the subject might better be one more sensitive to habitat changes or one that is less sensitive to environmental attributes not affected by management. The parameters chosen to measure "effect" may or may not be appropriate, depending on the study objectives. Take the pileated woodpecker (*Dryocopus pileatus*), for example. This is a large (42 cm long) species that occurs in highest densities in mature forests (Bull and Meslow 1977). Scientists initially thought that this species might classify as old-growth dependent. Subsequent research showed, however, that this highly mobile species used old trees primarily because of the presence of large (> 50 cm dbh) snags necessary for the species to excavate roosting and nesting cavities. Therefore, the presence of larg snags, not the unique vegetation microclimate, offered in old growth attracted the species. By providing large snags in younger forests, managers might be able to preserve the species outside the rapidly disappearing old-growth forest.

In contrast, studies on several species of salamanders have shown that these small, relatively nonmobile (compared with most birds and mammals) animals apparently depended upon the habitat and microclimate

Table 4.1. Percentage of species that differed in abundance between two substages within seral stages defined by the WHR matrix in the north coast–cascades zone of California

	Seral stage		
Species group[a]	Shrub—Sapling	Pole—Sawtimber	Mature—Old growth
Amphibians (8)	25	38	25
Reptiles (5)	20	0	0
Rodents (11)	27	0	0
Other mammals (11)	0	27	18
Passerine birds			
Spring (50)	28	14	12
Winter (31)	42	16	6
Other birds			
Spring (19)	32	5	5
Winter (11)	18	0	0
All species (146)	28	13	9

Source: Raphael and Marcot 1986, table 21.6.
[a]Values in parentheses are the numbers of species tested.

provided by mature forests (Raphael and Marcot 1986; see table 4.1). The thick duff layer and protecting cover of large trees provided the habitat necessary for these species. Thinning or removal of the overstory, site preparation by burning, and other disturbances would likely result in the extirpation of the species from a site.

In these two examples we see, then, the importance of selecting an appropriate species for studying an ecological community. (In fact, vertebrates may often be inappropriate subjects; see Morrison 1986.) Leaving only a few large snags may retain pileated woodpeckers in the stand, but it might eliminate other species in various vertebrate taxa. Numerous other examples in any type of community are possible: species usually select one or several of the specific features of the general habitat they occupy. This is often called the microhabitat or microenvironment of the organism.

How, then, does one select the proper organism for study? Although several authors and management agencies have tried to solve this problem by establishing guidelines for species selection, no single set of criteria are generally accepted for all conditions and objectives. Researchers cannot finally choose a modeling approach for assessing habitat data or representing species-habitat relationships until they select the species of interest. In a world free from budgetary worries, most biologists would attempt to develop fully validated models predicting the density and distribution of all species in an area (see Chapters 6 and 7). One of the reasons for developing these modeling approaches, however, is the realization that only a much-reduced subset of all possible species can be feasibly evaluated. Also, the models developed must be used with little or no testing, in a wide variety of situations. Therefore, it becomes critical that the correct species be selected.

The selection of species is multifaceted and influenced by biological, economic, and social considerations. The U.S. Forest Service has attempted to establish guidelines for the selection of such species for the purpose of assessing effects of management on wildlife communities (*Federal Register* 47(35):7690, Feb. 22, 1982). The regulation (36 CFR 219.19) implementing the National Forest Management Act of 1976 mandates that "fish and wildlife habitat shall be managed to maintain viable populations of existing native vertebrate species." The regulations then specify that "certain vertebrate species . . . shall be identified and selected as indicators of the effects of management." Designated Management Indicator Species (MIS), these species must be identified in all forest planning processes. As a whole, MISs are intended to represent the following categories: endangered and threatened plant and animal species identified on state and federal lists; species with special habitat needs that may be influenced significantly by planned management programs; species commonly hunted, fished, or trapped; and additional plant or animal species selected because their population changes are believed to indicate effects of management activities on other species.

In the above list, endangered, threatened, and game species are easily identified. Species with special habitat needs are identified as our knowledge of natural history increases. We have at least a fair understanding of habitat "generalist" versus "specialist" species of most terrestrial vertebrates; there is much overlap, however, between these two artificial categories, and this overlap varies throughout the range of a species. The selection of indicator species has received the most attention to date; we discuss this topic later in this chapter.

Fry, Risser, Stubbs, and Leighton (1986) proposed guides for selection of species to include in model-building processes that accurately represent a study area on a community level. That is, they advocated selecting species that would indicate the range of responses of all species of the community to some modification in the habitat caused by a management activity. Specifically, they evaluated the assumptions that selecting species from each taxonomic class (in the same proportions as they are known to occur) assures the best overall representation and that, within a taxonomic class, guilds offer an acceptable means of assessing impacts to many species by studying only a few. They began by listing seven steps involved in selecting species from a preexisting data base of wildlife-habitat relationships (they used a computerized data base, although the work can be done by hand).

Selecting Species

Step 1: Identify Habitat Types

Researchers should first identify all vegetation types within the study area. At first inspection, this may seem a rather straightforward step. But if vegetation types are not correctly identified, then the list of potential species may be incomplete. Further, certain elements of vegetation types not thought to be present in the area may actually occur, thus raising the potential of the presence of species that one would not normally expect. For example, some deciduous tree species usually associated with riparian areas may be distributed locally in wet depressions or other areas offering the proper amount of soil moisture.

Step 2: List All Species Suitable to Habitats

For all vegetation types identified in step 1, list all species that make optimal or suitable use of them for all life requisites at least seasonally. Here, of course, it is critical to properly determine the range of environments suitable for the species. Without in-depth studies of the natural history of a species, it is extremely difficult to correctly assign "quality" ratings to habitats. That is, the presence of a species does not necessarily indicate that the environment is suitable for successful reproduction (see Van Horne 1983).

Step 3: Eliminate Geographically Inappropriate Species

Delete species whose geographic range does not include the project area and species that require special habitat features not found within the project area. Again, one must take care to avoid accidentally deleting a species that actually occurs in the area. Especially important here are rare or seasonally rare species. Returning to our discussion in Chapter 2 of species that use an area for only brief periods (seasonal residents or transients on migration), the ramifications of missing a species could prove severe.

Step 4: Group Species Taxonomically

Arrange the species by taxonomic class (amphibians, reptiles, birds, mammals), and determine the percent composition of each class. Representation among taxonomic classes based solely on number of species (species richness) may not adequately describe the ecological functions of the animals in the community. Other approaches, requiring more detailed survey or census data, include weighting each species by its standing crop biomass, its population size, or its density in the area of interest. One can also array species according to their energy transfer rates among trophic levels or their occurrence by the ecological role they play in the community as prey, predator, or symbiont.

Step 5: Assign Guild Descriptors

If guild diversification is desired, give each species a guild descriptor. In brief, it is critical that the researchers place each species within the proper guild. We use the term *guild* here in the sense that Root (1967) intended: as a grouping of species that share a common trophic level. Species of a guild can cross taxonomic lines (for example, ground-foraging, seed-eating lizards and birds).

Step 6: Consider Time and Cost

The complete list obtained by the first five steps would provide the ideal array of species to represent issues, habitat conditions, and management effects of concern. Typically, however, these lists are much too long to adequately evaluate. Thus, researchers must make decisions regarding the number of species to use in evaluating each environment, given time and cost constraints. Such decisions entail compromising the range of representative species from the community in a study.

Here we must assume that species within each guild are sufficiently alike ecologically to be expected to respond in a roughly similar fashion to perturbations. On gross scales, at least, this response can probably be expected: an example is the response of all foliage-gleaning birds to removal of the overstory. The problem should be rather evident, however, given that such changes are obvious: we surely do not require sophisticated modeling techniques to make predictions based on such gross changes in the habitat.

Step 7: Select Individual Species

Finally, researchers should select individual species to fill the taxonomic categories, with or without guild specification. For specifying guilds, either choose species from a stratified species list that represents a broad array of guilds or randomly select species, rejecting any whose guild has been previously selected if unrepresented guilds still remain within the class. If guild representation is not desired, randomly select species within the taxonomic classes without considering guilds.

Given the time and funding constraints evident in most studies, following a stepwise procedure as outlined by Fry, Risser, Stubbs, and Leighton at least provides a systematic and at least grossly repeatable method of species selection. We say *grossly* repeatable because of the qualitative nature of most of the selection procedures. Comparing the two approaches of species selection described in step 7, Fry and his colleagues concluded that stratified random selection of evaluation species by taxonomic class was generally preferable to selecting species according to guild membership or simple random selection. They noted that bias in species selection by the investigator was minimized by requiring proportional allocation of all taxonomic classes of wildlife represented within the area of study. Of concern, however, was their finding that the taxonomic composition of the species used in evaluation greatly affected the outcome of modeling efforts. They also noted that the number of species needed for study always depended upon the precision and level of confidence one needed to achieve. The larger and more diverse the area of concern, the larger and more diverse is the number of species needed to reflect changes in vegetation composition and structure. Although the group did not explicitly state their recommendations for the number of species, examination of their data indicates that at least fifteen to twenty species are necessary to obtain stable results (stability measured by small changes in standard errors with subsequent increases in sample size). Although the U.S. Fish and Wildlife Service (1980) has recommended using ten species per vegetation type for model development, no justification for this number was given, as Fry and his colleagues noted.

Proper selection of the number and kinds of organisms does not stop with identification of suitable species. Most studies—both at theoretical and applied levels—have not examined sexual differences in habitat use. Most studies of habitat use by birds have been conducted during breeding periods, and males are simply more conspicuous because of territorial activities (like singing from exposed perches). Sex, breeding, or social status in mammals often results in different patterns of habitat use. In birds, males of migratory species often arrive on breeding grounds prior to the arrival of females (Myers 1981; Francis and Cooke 1986). Similar patterns are evident in some ungulate species that migrate seasonally along elevational gradients. Whereas male birds may select habitats alone, females may choose either males, habitats, or both (Cody 1985). Thus, different adaptive processes may be involved in habitat selection between the sexes and among species or taxa.

Studies of intersexual differences in habitat use are difficult, of course, in those species that are not markedly sexually dichromatic or dimorphic. In such species, extensive marking—which requires labor- and time-intensive trapping—is often prohibitive (this includes, of course, trapping of all classes of vertebrates).

Studies of sexual differences in habitat use have shown, however, that males and females often differ in use of habitat and/or other resources. In birds, such differences often involve subtle subdivision of habitat substrates or resources of the same territory. Although some have proposed intraspecific resource partitioning—possibly a result of competition for resources between the sexes—as the reason for such division of resources (see Rand 1952; Selander 1966; Robins 1971), others have shown that this division is related to constraints placed on choice of foraging location by sex-related reproductive activities, such as male birds foraging near song perches and females near nest sites (Morse 1968; Williamson 1971; Morrison 1982; Holmes 1986).

Use of different resources and/or foraging sites might also be related to intersexual differences in morphology. Morrison and With (1987) showed that male and female hairy woodpeckers (*Picoides villosus*) and white-headed woodpeckers (*P. albolarvatus*) partitioned habitat by subtle differences in use of tree species and substrates in the mixed-conifer zone of the western Sierra Nevada. These differences were related to differences in bill morphology and likely had ramifications for the overwinter survival of the species and sexes: female white-headed woodpeckers fed extensively on small (< 20 cm dbh) incense cedar (*Calocedrus decurrens*), a tree with low commercial value that is usually removed during silvicultural practices like thinning (Morrison et al. 1989). Had the researchers simply combined data for males and females of the species, this female-cedar relationship would have been obscured, with possibly severe consequences for the species.

Similarly, for white-tailed deer (*Odocoileus virginianus*), McCullough, Hirth, and Newhouse (1989) found significant differences between sexes in winter and breeding seasons. Habitat diversity influenced the availability of suitable food and cover across seasons and use of different habitats by the sexes. Thus, an increase in habitat diversity—a common recommendation for increasing numbers of deer—must be more carefully examined in light of such sex-specific differences in habitat and food needs.

In the next section we discuss the use of guilds and indicator species: these two concepts, although developed independently by different workers, are often used in combination in management contexts. We first review briefly the development of guilds and indicators and then discuss the methods and rationale for comparing animal communities and predicting the response of these communities (and species) to habitat alteration. Starting in the late 1970s and continuing into the 1980s, several authors attempted to modify the classic concept of guilds and indicators for use in management situations; we will compare and contrast these modifications and evaluate their usefulness for research and management.

The Guild Concept

Richard B. Root, in his now classic 1967 paper on resource use by the blue-gray gnatcatcher (*Polioptila caerulea*), defined *guild* as "a group of species that exploit the same class of environmental resources in a similar way. This term groups together species, without regard to taxonomic positions, that overlap significantly in their niche requirements." In an important paper published in 1981, Fabian Jaksic reviewed the literature related to ecological studies that used guild analysis, noting that most regularly violated Root's definition of the guild. Jaksic noted that most authors considered a guild to comprise taxonomically related species found in the same area. The researcher chose—within rather loose taxonomic boundaries—the group of organisms to be studied. But Root's definition of the guild explicitly states that a guild comprises species that exploit the same class of resources in a similar way *without regard* to their taxonomic similarities to each other. Thus, as Jaksic pointed out, studies that confine themselves to one taxonomic group are not studies of guilds at all. Rather, they are studies of syntopic co-occurring taxonomic assemblages. Jaksic extended this same criticism to studies of animal "communities." According to Jaksic, assemblages "consist of groups of organisms taxonomically related at the level chosen by the investigator (e.g., lizards, a suborder; small mammals, several orders)." Jaksic noted that not all strong interactions occur within taxonomic assemblages. For example, granivorous ants interact with heteromyid rodents more intensively than the latter do with other syntopic rodents. Some hawks and eagles interact more intensively with co-occurring foxes than with other syntopic falconiformes. In addition, predation by a variety of vertebrates affects the pattern of habitat use by rabbits (see Jaksic 1981 for a review of these studies).

Clearly, a great deal of confusion exists in the literature regarding guilds. Of particular concern is how "similar" the use of resources by species must be in order for the species to be considered members of a guild. Jaksic followed MacMahon and his colleagues (1981) in basing the criterion for "similar manner" on the effect of resource use on the resource itself. MacMahon and his group stated, "It does not matter whether an organism removes a tree leaf for nesting material, for food, or as a substrate to grow fungi which in turn are eaten; the leaf is gone and the leaf users belong to a common guild." Acceptance of this criterion means that guilds (the "community guild" of Jaksic) should be recognized on the bases of the use of investigator-defined resources, such as food.

Jaksic noted that an alternative to his community guild would be the recognition of guilds within taxonomic assemblages. Given that most communities are too complex to permit full guild analysis, the "assemblage guild" can also be a valuable field of study. Indeed, most workers have dealt at the level of assemblage guilds, even though most failed to recognize the difference between their approaches and the original definition given by Root (1967). Assemblage guilds likely do not include all of the significant interactions that affect members of the trophic guild. Jaksic stated that as-

Table 4.2. Classification of birds in the Yosemite Valley, California, according to the location and method of foraging

Species	Classification
American kestrel (*Falco sparvarius*)	Ground-insect
Spotted sandpiper (*Actitis macularia*)	Ground-insect
Band-tailed pigeon (*Columba fasciata*)	Foliage-seed (Apr.–Oct.); ground-seed (Nov.–Mar.)
White-throated swift (*Aeronautes saxatalis*)	Air
Northern flicker (*Colaptes auratus*)	Ground-insect (Apr.–Nov.); bark-drilling (Dec.–Mar.)
Hairy woodpecker (*Picoides villosus*)	Bark-drilling
Western wood-pewee (*Contopus sordidulus*)	Air
Violet-green swallow (*Tachycineta thalassina*)	Air
Red-breasted nuthatch (*Sitta canadensis*)	Bark-searching
Golden-crowned kinglet (*Regulus satrapa*)	Foliage-insect
Warbling vireo (*Vireo gilvus*)	Foliage-insect
Purple finch (*Carpodacus purpureus*)	Foliage-seed (Apr.–Oct.)
Chipping sparrow (*Spizella passerina*)	Ground-insect

Source: Modified from Salt 1953:271.

semblage guilds "contribute to the understanding of how related species partition (if at all) the available resources, but they are not likely to clarify completely the mechanisms by which this partitioning is attained. This is because they fail to consider organisms in other assemblages that might otherwise affect the outcome of purely intra-assemblage interactions. . . . It seems, then, that the study of guild structure within taxonomic assemblages is only a preliminary step for understanding the role of guilds in the organization of communities" (1981:398). Our concern is this: If assemblage guilds provide only a partial answer to community structure, then how useful are the predictions about the responses of guild members (or whole guilds) to a planned perturbation based on such guilds? Later in this section we show the clear ramifications this has for resource managers.

Workers prior to Root (1967) classified animals, usually in the same taxonomic assemblages (and usually working with birds), into groups based on similarities in the use of certain resources. For example, Salt (1953), in an ecological analysis of birds in California, classified species into groups based on location and method of foraging (table 4.2). He later applied this system to birds in a Wyoming study (1957). Bock and Lynch (1970) applied Salt's system to a study of birds in burned and unburned forests in the Sierra Nevada. Thomas and his coworkers (1979) grouped all known species of amphibians, reptiles, birds, and mammals of the Blue Mountains of Oregon and Washington into different "life-forms." These life-forms were based on habitats thought to be required for breeding and feeding. In reviewing the work of Thomas, Verner (1984) thought that the life-forms "could also have been called guilds." Although we do not wish to debate Verner's statement here, it should be evident that making *life-forms* synonymous with *guild* without attention to Root's definition and Jaksic's criticisms is a tenuous step at best. We think that workers have generally failed to recognize the distinction between categories of, say, foraging habits as used by Salt and others, and classification into guilds, as discussed by Jaksic. Categorizing by foraging behavior, for example, is an interesting and useful means of conveying the general activity patterns of birds to a reader; but such categories also fall under the heading of "assemblage guilds." In most cases the researcher is not really dealing with a true community guild at all, but rather only with a small group of taxonomically similar species that may be affected by a host of species not even under consideration. The concept is useful but must be clearly defined by each user.

Development of Guilds

Whether researchers develop community guilds or assemblage guilds, they must emphasize a basic and critical dichotomy between two methods of forming guilds: defining them *a priori* and defining them *a posteriori. A priori* guilds are those formed by a researcher without the benefit of empirical data on resource use by the species in the area of interest. Data on species present in the area may be available (such as species lists and abundance data), or nothing more than range maps and general information on habitat requirements may be consulted. Possibly a more descriptive term for such guilds would be *qualitative guilds. A priori* implies that data will be collected; this is often not the case.

Guilds formed *a posteriori* might also be called *quantitative guilds:* guilds formed after data on resource use are obtained. Clearly, such quantitative guilds remove much of the bias associated with guilds formed primarily by a worker's opinion (that is, qualitative guilds). Forming guilds based on actual resource use is not a completely unbiased procedure, of course. The variables selected for study, the season(s) of data collection, and the statistical methods used to define guilds will all influence the resulting guild classifications. The proper scientific procedure would be to collect data on resource use without assuming that the animals could later be divided into guilds. The null hypothesis should be that no guild structure exists. In addition—and of prime importance—are the species selected for collection of data. If one restricts the species under study to those thought to be similar, then the study will likely show that they are indeed "similar." Most multivariate classification methods, however, have been designed to create groups and classes artificially even if true biological differences are small (see Chapter 7). One must interpret such data and analyses in light of study objectives and methods.

We will now briefly review several methods used to develop *a posteriori,* or quantitative, guilds. We will not detail steps used to form *a priori,* or qualitative, guilds.

Quantitative Guilds

Verner (1984) surveyed how researchers defined guilds. His review of the avian literature in which guilds were given substantial attention—back to 1976—showed "a disquieting lack of consistency in how various workers assign bird species to guilds" (1984:1). Verner found that among more than fifty papers reviewed, nearly all described guilds based on some aspect of foraging behavior, be it by substrates used, strategies employed, food consumed, or some combination of these. Several workers also described both foraging and nesting guilds. Verner also found that different workers studying avian communities in similar habitats sometimes grouped the same species into different guilds. Therefore, one must be cautious when comparing guild analyses between studies. Verner's caution is not restricted, of course, to avian studies, although the use of guilds is less common in studies of other taxa.

The rationale for basing guilds on foraging behavior is centered on the belief that food type, food abundance, the methods used to obtain food, and the locations from which food is taken are of critical importance to the survival of an animal. Given the conspicuousness of most species of birds, especially compared with other vertebrates (like small mammals), it is not surprising that birds and their foraging behaviors have been the focus of most guild studies, especially for management purposes (see Chapter 5). As we shall see shortly, studies of small mammals, reptiles, and other nonavian taxa are usually based on location of animals in the habitat (such as trap sites) or on food items collected from fecal or stomach samples.

Holmes, Bonney, and Pacala used the foraging behavior of twenty-two species of insectivorous birds in New Hampshire to "provide an objective classification of the similarities and differences among syntopic bird species in terms of how they exploit food resources, and to identify those measured parameters associated with feeding that are most important in determining the structure of this forest bird community" (1979:512). They performed several multivariate statistical analyses on a data matrix consisting of the twenty-two bird species and the twenty-seven foraging characteristics that they measured. (The reader should refer to the original article for more details on the methods used.)

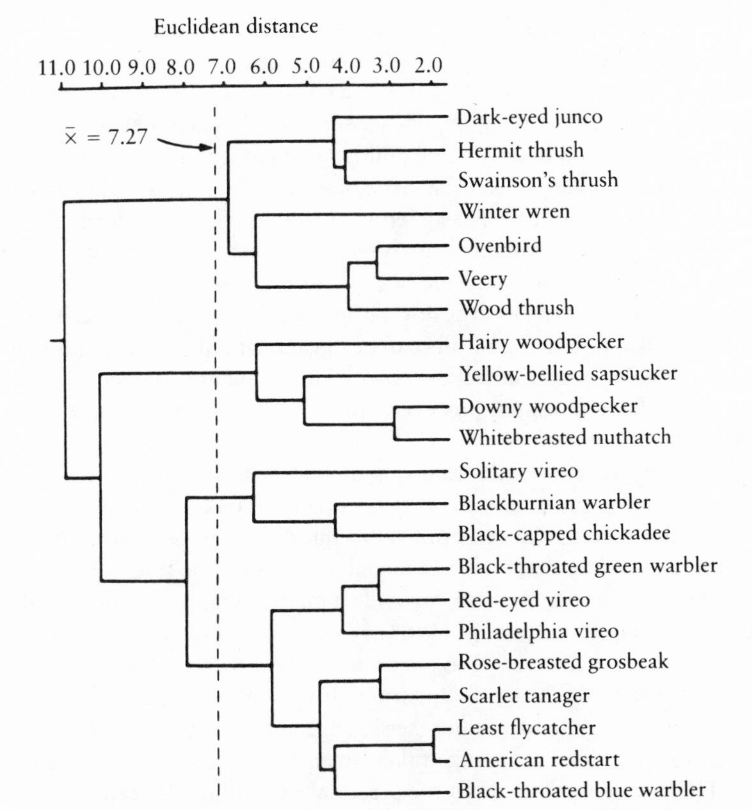

Figure 4.1. Community dendrogram of Euclidean distances between all species, indicating foraging relationships. Dashed line indicates the mean Euclidean distance between all combinations of species pairs in the Hubbard Brook bird community. (From Holmes, Bonney, and Pacala 1979, fig. 1.)

In one analysis, Holmes, Bonney, and Pacala used cluster analysis to create a two-dimensional representation of the twenty-two bird species in the twenty-seven-dimensional space. The diagram resulting from this analysis is reproduced in figure 4.1. According to Holmes, Bonney, and Pacala, "the species in the . . . bird community are separated into a number of distinct groupings which we consider as foraging guilds (cf. Root 1967) because they contain species that exploit food in similar ways" (1979:515). Note, however, that these are actually the assemblage guilds of Jaksic (1981) and not true community guilds, if we follow the strict definition.

Holmes and his coworkers defined their guilds as those species that were separated from one another by Euclidean distances greater than the mean distance between all species ($\bar{x} = 7.27$ in their example); this resulted in the formation of four guilds. (Euclidean distance is a measure of how similar each pair of species is; the greater the distance value, the less similar the species.) They noted, however, that no quantitative precedents existed for determining exactly where along the cluster diagram guilds should be identified. By referring to figure 4.1, we see that a Euclidean distance of 9.0 would have resulted in only three guilds, whereas a distance of 5.5 would have resulted in nine guilds. Also note that cluster analysis will eventually result in each species being assigned to its own guild (Euclidean distance < 2.0). Thus, while statistical analysis of quantitative data on foraging behavior is certainly a useful method—and one less biased than defining species to guilds *a priori*—it is not without subjectivity. Holmes, Bonney, and Pacala also noted that results can be biased not only by the number and type of variables measured but also by variables that are not included. They strengthened their results, however, through the use of several different statistical techniques.

Wagner (1981) analyzed foraging behavior in her attempt to describe the structure and organization of birds occurring in oak woodlands of California. She collected data in both fall/winter and spring in an attempt to determine if guild membership changed seasonally. She prepared dendrograms using cluster analysis to "provide a pictorial representation of niche overlaps among bird species" (1981:975). She used the unweighted pair-group method on arithmetic averages to analyze the guild of foliage- and bark-gleaning birds she studied. Although we will not detail her analyses, examination of her dendrograms (see fig. 4.2) indicates that temporal variation exists in species groupings taken from different seasons in the same area. This is apparently due to the seasonal occurrence of certain species. For example, the yellow-rumped warbler (*Dendroica coronata*) occurred during fall/winter but not during spring. In contrast, the orange-crowned warbler (*Vermivora celata*) occurred during spring but not during fall/winter. Wagner's results present a simple but important example of a point we raised earlier: guild membership (structure) depends on the species included in the analysis, which is largely determined by the season of study. This will become a very important consideration when we discuss the use of guilds in management applications.

Analysis of guild structure has not been restricted to the study of avian communities. Pianka (1980) presented a useful summary of potential methods of guild analysis using lizards. He rested his analyses on the assumption that niche relationships provided the most appropriate arena in which to search for guild structure, a rationale used by most students of quantitative guilds. Pianka asked several interesting questions that bear repeating here. Although he noted that he could not answer all of these questions, consideration of them reveals the extent to which guild structure is thought to be an important force in the structure and organization of animal communities. He asked:

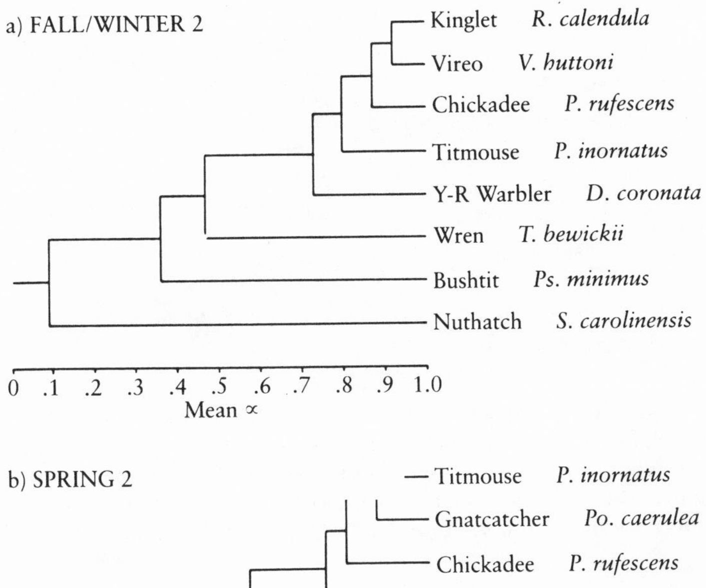

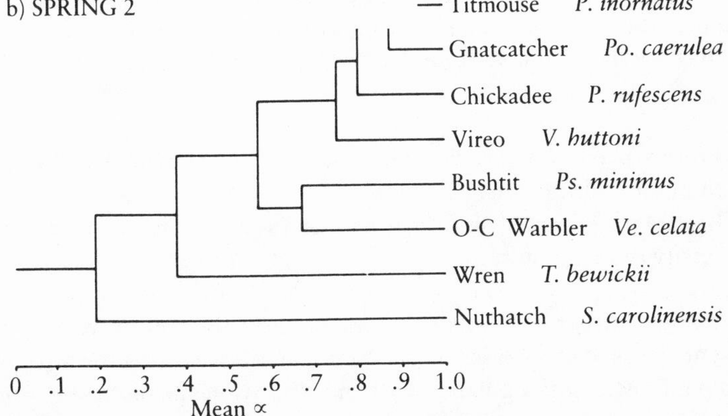

Figure 4.2. Dendrograms based on niche overlaps of foliage- and bark-gleaning guilds. Fall/Winter 2 represents October 1975–January 1976. Spring 2 represents April–early June 1976. (From Wagner 1981, fig. 2.)

1. How many and which niche dimensions separate species?
2. To what extent are species spread out evenly in niche-resource space? Or do clusters of functionally similar species (guilds) exist?
3. How can such guild structure be detected and measured? What are its components?
4. Are such guilds merely a result of built-in design constraints on consumer species, and/or do guilds simply reflect natural gaps in resource space? Or, can guild structure evolve when resources are continuously distributed as a means of reducing diffuse competition? (A community without guild structure would presumably have greater diffuse competition than one with guild structure.)

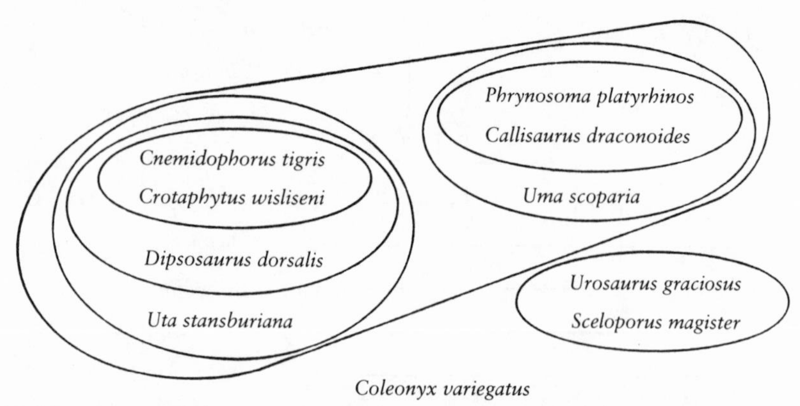

Figure 4.3. Microhabitat guild structure in North American desert lizards as revealed by single-linkage cluster analysis. The two-species arboreal guild sets off a seven-species terrestrial guild, which includes two subguilds, one composed of lizards that frequent open spaces and one composed of those species that stay close to shrubs. Only one species is not a member of any guild. (From Pianka 1980, fig. 5.)

5. Do more diverse communities have more guild structure than simple communities do?
6. What are the effects of guild structure on the assembly, structure, and diversity of communities?

As we noted earlier and as Pianka summarized, the basic raw data set for most analyses of niche overlap and community structure is the resource matrix, which is a rectangular m by n matrix (using Pianka's notation) indicating the amount (or rate of use) of each of m discrete resource states such as food types or microhabitats used by n different consumer species. Pianka discussed several possible approaches to summarize guild and community structure and provided an example based on one of the methods. Briefly, Pianka and his associates recorded data on microhabitat use and time of activity, collecting animals to determine stomach contents and reproductive conditions. He subjected the data on microhabitat use to the single-linkage method of cluster analysis; we reproduce the results of this analysis in figure 4.3. We see that a two-species guild of arboreal species is set off from a seven-species guild of terrestrial species, within which are nested two smaller "subguilds" (Pianka's term), one comprising three species that frequent open spaces between plants (*Phrynosoma platyrhinos, Callisaurus draconoides,* and *Uma scoparia*), and the other, those species that are usually found closer to cover adjacent to shrubs (*Cnemidophorus tigris, Crotaphytus wislizeni, Dipsosaurus dorsalis,* and *Uta stans-*

buriana). The remaining species (*Coleonyx variegatus*) is nocturnal and not considered a member of either guild.

Pianka's use of the term *subguild* indicates the difficulty in determining exactly where to draw the line between species groups. The scale upon which species locations are plotted in some dimensional space that is being used to represent resource use will in large part determine how "near" or "far apart" the species appear to our eye, while they are still the same distance apart in measurable units. Pianka's inclusion of a nocturnal species also proved enlightening. Remember that the definition of *community guild* implies simply that a resource is being used and does not address when the use takes place; if the resource is gone, then it is not available to another species. Another interesting analysis of guilds—here including both amphibians and reptiles—was conducted by Inger and Colwell (1977).

Intraguild Analysis

Once guilds are defined, the researcher is sometimes interested in examining interactions among species within a guild. By dealing with species that have been isolated based on similarities in resource use, the researcher can then concentrate on those groups via more detailed methods of data collection and analysis. Such analyses are pertinent if the objective is to manage species based on guilds: without such analyses, a manager can have little confidence in decision-making.

MacNally (1983) recognized two types of empirical information that have been used to describe within-guild structure: data on the functional organization of guilds and descriptive data on guild structure. Data on the functional organization of guilds are derived from the perturbation or manipulation of natural guilds. In contrast, most empirical data are descriptive; researchers collect data without referring to the predictions of various theories of guild structure and so usually analyze the data in a post hoc fashion that permits only discussions about consistencies with theories. Such tests are, of course, weak. Recall the questions raised by Pianka: Does guild structure result from design constraints and/or gaps in resource space? Does guild structure influence coexistence and allow diversity? Clearly, more than descriptive data are needed to answer such important, but hard to analyze, questions. Descriptive data, however, can be used to identify inconsistencies in various competing hypotheses concerning guilds and community structure.

Management Applications and Indicator Species

The guild concept has been touted as a means of streamlining the tasks of environmental assessment and resource monitoring (R. A. Johnson 1981;

Severinghaus 1981; Verner 1984). Resource managers have used guilds to predict effects of impacts on a species without specifically studying the species. Economically, the potential savings in costs are tremendous, since one could study only a few species per guild to determine the impact on all members of the guild. This is the guild-indicator concept.

Verner (1984) reviewed several approaches used to construct guilds (or guildlike categories) primarily for management purposes. These approaches included the guilds of Severinghaus (1981), the life-forms of Thomas (1979), and the "guild blocks" of Short and Burnham (1982). Verner detailed five basic reasons why the guild-indicator approach for management would be unlikely to work (see also Szaro 1986; Block, Brennan, and Gutierrez 1987; Patton 1987). First, guild members are not necessarily alike in all the ways in which they use zones of the habitat for various purposes. Consequently, members of the same guild may be affected differently by certain habitat changes. Second, thorough studies of niche differentiation have shown that species subdivide the habitat by specializing in diets, foraging substrates, foraging times, and the like. Thus, all "foliage insectivores" do not necessarily obtain the same food at the same time or in the same manner, and individuals may be unequally affected by a change in their environment (probably not counting, however, such gross changes as removal of the entire overstory). Third, animals in the same guild may change their behaviors within or between seasons in a dissimilar fashion. Thus, even if a proper indicator could be found for one place and time, it is unlikely that the same indicator would be reliable for more than a short period. Fourth, geographic variations in species' behaviors would require placing the same species in different guilds in different parts of its range. Of course, this problem affects guild delineations for both the guild-indicator and whole-guild approaches. It does show, however, the necessity of making guilds location-specific. Finally, ways in which species use their environment can vary even over short distances when habitat attributes differ.

Thus, the use of an indicator species to predict response of other species is a tenuous procedure (see also Mannan, Morrison, and Meslow 1984; Landres, Verner, and Thomas 1988). Simply too many variables must be considered. Further, given that species are differentiated as species because of differences in morphology, behavior, resources used, competitive interactions, and the like, we should not expect that one species could very precisely reflect the behavior and ecology of another, let alone a whole group of species. Morrison (1986) observed that the problem with using indicators is separating the myriad of factors that can cause changes in animal populations. Further, animals are usually affected by changes that are two or more steps removed from the initial perturbation.

Thus, if guilds or indicators are to be useful management tools, they must be selected on bases that are specific to time, location, and habitat. Most likely, different guilds will comprise different species based on the season, geographic area, and environment of interest. Analyzing how animals respond to their environment—either singly or in groups—is a worthwhile endeavor that will lead toward greater understanding of species and their environments. That is the basis of community ecology. It is unwise to suppose, however, that these processes are simple or that they can be accurately assessed by grouping species into user-defined units. The researcher must clearly address the goal of the endeavor, including the numerous potential problems outlined in this section, before attempting to use guilds or indicator species for management purposes.

What to Measure

Spatial Scale

We must first view habitat analysis from the broadest scale, working our way down to the finest level of scale necessary to answer the question of interest. This approach is the cornerstone of model development. That is, if our question simply concerns the potential or general distribution of waterfowl by broad environmental classes (for example, marshlands versus deciduous woods), then measurement of water depth, plant species composition, and the like are simply unnecessary. Further, correct species identification and density estimates are not even required.

Green (1979) asserted that it is not the degree to which a model meets perfection, but rather its adequacy for fulfilling a prescribed purpose, that renders it valid. There may be simply no reason for high precision if the biological change to be detected is large or if the study's purpose is to identify potential or candidate habitats. Green asked, "What criteria should be applied to choice of a system and variables in it for use in applied environmental studies?" (1979:10). He answered his question with four criteria: spatial and temporal variability in biotic and environmental variables that would be used to describe or predict impact effects, feasibility of sampling with precision and a reasonable cost, relevance to the impact effects and a sensitivity of response to them, and some economic or aesthetic value, if possible. As we will see, the temporal and spatial variability evident in all environmental systems makes adherence to Green's first point extremely difficult; his remaining three points are well worth considering in the development of any model of habitat relationships.

The scales at which wildlife-habitat relationships can be examined along a continuum are not unlike the model of how an animal selects "proper" habitat (see fig. 2.8). The relation of the way an animal perceives its environment and the way we relate these perceptions to some organized method of study have the utmost importance in model development. We can examine habitat use at its broadest, or biogeographic, scale, passing through successively finer-scale evaluations until we reach the level of the individual. Here we are more interested in finer-scale models that concern species-specific habitat requirements within or between a few study areas. We must not forget, however, that these "finer-scale" models will likely vary considerably between locations and certainly between populations. The magnitude of these variations determines the generality of a model (the ability of a model developed in one place and time to be reliable at another place and time). The literature on ecological scale is expanding (see Wiens 1989).

Controversy exists, however, even over the scale at which species-specific models should be developed. Irwin and Cook (1985) developed habitat suitability models for pronghorn (*Antilocapra americana*) on winter ranges in several western states (see also Cook and Irwin 1985). They concluded that field testing of habitat models would be most reliable if used with regional rather than local information, asserting that the wide range of environmental conditions necessary for evaluating a large number of variables generally cannot be obtained locally. An example is variation in winter precipitation: such a measure will vary widely among various locations but will take on an overall stable average within a restricted geographic area. Irwin and Cook related their work to an earlier study by Lancia, Miller, Adams, and Hazel (1982), who had concluded that local, intensive studies of habitat selection were highly useful for field-testing habitat models. At odds with this conclusion, Irwin and Cook stated that such a test precludes reliable use of the model except in areas similar to those in which it was developed and that users must adapt the model to satisfy local, unique conditions (see Chapters 6 and 7 for further discussion of model validation).

We do not think, however, that a dichotomy actually exists between the conclusions of Irwin and Cook (1985) and Lancia, Miller, Adams, and Hazel (1982). Rather, we think that proper model development begins with analysis of the wide variety of factors that can affect a population. These should include variations in abiotic and biotic factors across the range of the species, as well as the response of the animal to such variations within the sites where these factors are directly felt by the individual. Irwin and Cook acknowledged the need for intensive, localized studies when they stated that "users must adapt the model to satisfy local, unique conditions." We argue that one must include such local studies in order even to identify these "unique conditions." But Irwin and Cook correctly noted that a study confined to one site and/or time will likely miss those factors that can be identified only on a broader scale. Thus, the general process involving what to measure is not as controversial as it may seem. Rather, many researchers misunderstand the overall view of the different phases of model development and the scale at which they are formulated and applied.

This discussion does raise the question, however, of the emphasis on intensive versus extensive models. Both theoretical constructs and empirical studies indicate that, for the development of useful predictive relationships of a species and its environment, one must decipher site-specific aspects of an animal's use of habitat. Although extensive studies may be able to differentiate unused from used environments for a species on a broad geographic scale (for example, the use of young growth versus mature forest), they fail when one needs to track density on a more site- or population-specific scale.

We do not think that Irwin and Cook's and Lancia, Miller, Adams, and Hazel's studies are proper examples of extensive and intensive studies, respectively. Irwin and Cook analyzed rather detailed aspects of the habitat use of the pronghorn (cover of forbs, grazing intensity, cover of graminoids). Their argument concerned, we think, the range over which such variables were recorded.

By the term *extensive* we refer to models that are useful in identifying features of the environment predicting distribution by plant association (like mixed-conifer) and other equally broad categories, similar to the life zone concept of Merriam (1898). As we argued in Chapter 2, however, this level of refinement is critical to the proper development of a habitat model. It tells us how the species (or its ancestor) got to the area in the first place, and it likely lends insight into why an animal occupies the broad environmental classification that it does. Questions of an animal's "preadaptation" to habitat fall under this category and are often an important aspect of habitat analysis.

The extensive approach cannot tell us such things as how an animal reacts to changes in slope, litter depth, stocking density of trees by species, or density and composition of predators. Thus, the question of extensive versus intensive models really involves two very different modeling objectives and techniques. But these techniques are complementary in that both can be blended into a single model—or two related models—that seek to describe, explain, and ultimately predict the distribution and habitat use patterns of a species.

To develop either an extensive or an intensive model requires that the means by which an animal perceives its environment be known. Returning to our conceptual model of the process of habitat "selection" (fig. 2.8), we see that the factors responsible for an animal's choosing an area (the settling response) follow a pattern of increasing specificity. Once the level

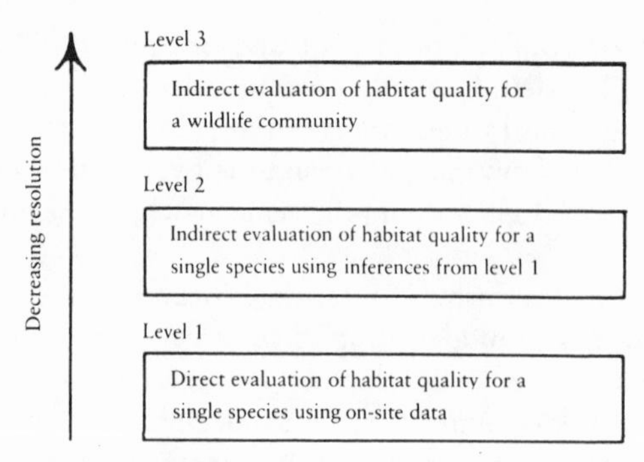

Figure 4.4. Hierarchical description of habitat quality assessments. (From Van Horne 1983, fig. 1 © 1983 The Wildlife Society).

of specificity is determined, then the researcher can determine the type of data collection required (fig. 4.4).

We cannot determine here the level of detail required to develop the proper model for every occasion. What we are trying to develop is the thought process that researchers must use on a study-specific basis. We will present and discuss, however, several examples of the types of variables collected by researchers attempting to develop predictive models. The list of potential studies is long, and our choices should not be taken as indicating only the best or even the most common techniques. Rather, we think it is important that readers see specifically—through the use of examples—the range of variables being collected by workers in the field.

Measurements of the Animal

We have repeatedly alluded to the question of how refined our estimates of the numbers of an animal population need be for model development: Must we estimate at the level of presence-absence, index of abundance (such as numbers seen per unit time), or absolute density (number per unit area)? A whole class of habitat models are based on correlating animal numbers to some feature or features of the animal's biotic and abiotic environment. The purpose of these models is to develop adequate predictions of the presence, abundance, or density of the species based on environmental features (see Chapters 6 and 7). The methods vary based on the available data: techniques for handling presence-absence data are not necessarily the same as those required for density data (for example, logistic regression versus multiple regression).

This discussion implies that our estimates of the numbers of animals present must be reliable. That is, are our absolute or even relative indices of abundance a fair reflection of the numbers of animals actually present? If we use multiple observers, are their estimates comparable? Can estimates derived by one observer or group of observers in one study area be used to validate a model developed by other people at a different location? Throughout this book we discuss varying effects of observer error, errors encountered because of known or hidden biases, and other factors that work against development of accurate models. Here we are particularly interested in errors associated with estimating numbers of animals. This is, quite obviously, a large and complicated problem: although the problems generally apply to all taxa, the methods used to estimate population cause particular problems for each group of animals. We simply want to remind the reader of the multitude of problems associated with counting animals and briefly to review the detailed publications available on estimating numbers of animals. (For literature on censusing animal populations, see Schemnitz 1980; Miller and Gunn 1981; Ralph and Scott 1981; Davis 1982; and Bury and Raphael 1983.)

In contrast with the correlative approach of model development are various "focal-animal" methods: models based on focal-animal techniques use the presence (or absence) of an animal as an indication of the habitat used (or not used) by the species. No correlation between density and habitat is involved. Rather, a species (or several species) is sampled across a range of conditions, and the "preferred" habitat is determined through various statistical correlation techniques. It is important to remember that the way in which data are collected constrains the analytical methods that can be applied. It is often truly stated that all researchers should consult a biostatistician prior to beginning data collection. Beware, however, that numerous methods and numerous statisticians exist; statisticians do not always agree on the appropriate approach. Careful consideration of their views, in the context of your knowledge of the biology of the species (and your own statistical knowledge), should result in a strong and justifiable study design.

Researchers must first determine exactly what aspects of the animal's behavior to measure, both its outright behaviors and the location(s) at which these behaviors take place. Focal-animal surveys most often use the location of the animal as the center of a circular plot—or set of circular plots—used to measure habitat parameters. Measurements taken within these plots are thought to indicate the habitat preferences of the animal. Animals do not, however, spend equal amounts of time at each activity (and location) during the day. Feeding, drinking, resting, defending territory, mating, grooming, and the like each consumes different amounts of time (and energy). The amount of time spent at any one activity may not indicate the importance of that activity to the animal: drinking may take only a few minutes of each day, but without water the animal is unlikely to survive (this relates, of course, to the concept of limiting factors; see Odum 1971). Thus, a proper research design must stratify these behaviors by the locations in which they occur.

For example, it is well known that male birds usually sing from exposed perches during the breeding period. Many studies of bird-habitat use have used the location of such singing males as the centers of plots defining the habitat of the species (see James 1971; Holmes 1981; Morrison 1984). But how well do such plots indicate the species' habitat, or even the individual territory? Collins (1981) examined habitat data collected at perch sites and at nest sites for several species of warblers. He found that 29 percent of the nest sites had vegetation structures that significantly differed from the corresponding perch sites within a territory. Not surprisingly, he found that basing habitat analysis only on perch sites overestimated the tree component of the habitat. Collins examined only structural, not floristic, aspects of the habitat. His study showed, however, that a proper study design would include at least recognition of the differences among behaviors reflected in the results of habitat analysis.

Numerous studies, of both birds and mammals, have sought to describe foraging behaviors (see Chapter 5). Analysis of these behaviors gives insight into the specific methods used by an animal to obtain food. For example, saying that a bird forages in pine tells us little about the methods used to obtain the food or the type of food being taken. A bird could glean insects from the tops of needles, hover-glean and take prey from the undersides of limbs or needles, probe bark, flake bark, flycatch in or near the tree, or perform these or different maneuvers in the act of feeding. Many birds use numerous strategies depending upon the availability of prey (see Morrison et al. 1985). In addition, modes of obtaining food will vary based on the size and age structure of the habitat: as trees grow, for example, bark thickens, thus changing the insect fauna and the ways in which birds obtain it. In the western United States, a species of scale insect (*Xylococculus macrocarpae*) is found under the thin bark of small (< 25 cm dbh) incense cedar. The scales are able to penetrate the thin bark with their mouthparts, thus obtaining nutrients from the tree. During winter, birds feed heavily on these insects (see Morrison et al. 1985; Morrison et al. 1989). As the trees begin to exceed 25 cm dbh, the bark becomes too thick for penetration by the scales. Thus, the mere presence of incense cedar does not guarantee an adequate food supply for birds. This relationship between incense cedar, scale insects, and birds was determined by observing numerous species of birds probing, pecking, and flaking the bark of small cedar. Thus, the behaviors used to obtain food tell us a great deal about the habitat requirements of the animal.

Although behaviors of animals may give a good indication of patterns of habitat use, they do not in themselves give information on finer-scale aspects regarding the food consumed. That is, a bird gleaning among leaves or a deer feeding in a meadow may be selecting certain types and sizes of food. Indeed, the size and type of food consumed by an animal plays an important role in theoretical constructs that seek to explain methods of species co-occurrence through resource partitioning. In a classic work, Schoener (1974) outlined the five major factors affecting the ways animals can avoid competing for limited resources: differences between food type and habitat; food type and time; habitat and time; habitat and habitat (vertical versus horizontal segregation); and food and food (type and taxon).

Habitat structure, foraging location, and prey consumption all determine the utilized area or habitat of a species. What they do not tell us, however, is the "quality" of the habitat for survival and, ultimately, reproductive success (and survival of the offspring to breed). *Quality* is a relative term that has no agreed-upon definition. Quality varies along a continuum: for example, can we say that a habitat in which an individual produces two offspring is twice as "good" as one in which only one young is produced? To answer this question we must know more about the health of the offspring, their survivorship and reproduction in the future, and the variability of reproduction over time. Van Horne (1983) popularized the notion that simply quantifying the area used by an animal may have little to do with survival and reproductive output. An animal, she observed, may occupy "marginal" habitats because more "optimal" ones are already fully saturated by superior individuals. Thus, the behaviors exhibited by individuals in these marginal areas may not indicate the status of the population as a whole or the fitness of the individuals.

Measurements of the Vegetation

To select the proper set of habitat variables to study and model, we must strive to determine what aspects of the environment the animal recognizes as relevant. This is, of course, the central point of the previous section. As Krebs noted, "We must be careful here to define the perceptual world of the animal in question before we begin to postulate the mechanisms of habitat selection" (1978:40). Krebs identified two basic kinds of factors that must be kept separate in discussing habitat selection: evolutionary factors, conferring reproductive fitness and survival value on habitat selection; and behavioral factors, giving the mechanisms by which animals select habitat

(1978:42). The proper behavioral factors which have direct ties to the survival of the species must be identified. Thus, we will be interested primarily in behaviors that result from stimuli from landscape and terrain; breeding, display, and feeding sites; food; and other animals. This list is a modification of that given by Hilden (1965) and revised by Krebs (1978:42).

Of course, the specific measurements taken on vegetation in a study depend upon the objectives and intended use of the investigation. It is possible, however, to develop rather detailed guides regarding the refinement necessary to adequately describe the habitat of an animal.

Dueser and Shugart (1978), in a well-known study of microhabitat utilization by small mammals, outlined four criteria that guided their selection of habitat variables for measurement: each variable should provide a measure of the structure of the environment which is either known or reasonably suspected to influence the distribution and local abundance of the species; each variable should be quickly and precisely measurable with nondestructive sampling procedures; each variable should have intraseasonal variation that is small relative to interseasonal variation; and each variable should describe the environment in the immediate vicinity of the animal. Dueser and Shugart noted that the final criterion reflected their concern for describing the environment in sufficient detail to detect subtle differences among microhabitats which appeared to be grossly similar. If we look closely at the list of Dueser and Shugart, we see that their first criterion tells us that previous natural history information, plus a good dose of biological common sense, will help narrow the choice of variables. We must be careful here, however, not to follow past mistakes or misinformation regarding the natural history of the animal and, further, not to let our preconceived notions and biases eliminate potentially important variables. It is often worthwhile to let a biologist that is, say, familiar with the area but not the species, review your list of variables: mammalogists may be able to offer valuable advice to herpetologists or ornithologists planning a study. Dueser and Shugart were apparently concerned with "nondestructive" sampling procedures because of their desire to make repeated observations at a site (here, the trap site). Further, it would be hard to repeat a study in the same site if the habitat had been substantially altered. Some sampling, however, is of necessity destructive, even if researchers simply observe the animals. A case in point is that of intensive searches for amphibians and reptiles in which down wood, litter, and duff substrates are thoroughly scoured in order to census all animals.

Whitmore (1981) also outlined general criteria for selecting variables. He listed three general categories that involved "practical aspects" of variable selection: variables should be measurable to the desired level of precision; variables should be biologically meaningful; and variables should be relevant to the species in question. As an example of his second category, Whitmore asked what possible direct meaning measurements on tree roots would have to a canopy foraging bird such as the cerulean warbler (*Dendroica cerulea*). Regarding his third category, he asked how directly important the percentage of grass cover or the ratio of grasses to forbs could be to a bark forager such as the brown creeper (*Certhia familiaris*). It can be argued that these variables may have some indirect relationship to a variable of obvious importance to an animal; for example, root condition could relate to canopy volume or condition, which in turn might relate to insect density and thus bird behavior. Indirect measurements are only preferred, however, if the variable directly responsible for the activity or response by the animal cannot be measured or can only be measured at great expense. The further one moves away from the primary influence, the more the opportunity for error arises.

As we showed earlier in this book, the distribution of plants is tied closely to regimes of temperature, soil, and moisture. Animals, in turn, are linked in some fashion to these plants, for shelter, for the food they offer, or both. But to what specific aspect of vegetation are animals responding? What are the stimuli causing the behaviors that we call resource use?

With the rise in interest of studying animal "diversity," researchers developed various relationships that sought to relate the number and kinds of animals to some measure of the gross structure of the vegetation. Most famous is the foliage height diversity–bird species diversity (FHD–BSD) constructs of MacArthur and MacArthur (1961). In figure 4.5 we see that the diversity of birds rises as vegetation becomes increasingly stratified. A plethora of studies followed the early work of the MacArthurs, with most researchers finding significant statistical relationships between FHD and BSD; these findings were not universal, however (see Roth 1976). Karr and Roth (1971) suggested that the scatter in a FHD-BSD relationship likely resulted from important, but unmeasured, variables influencing the avian community.

In vertically simple vegetation, such as brushlands and grasslands, FHD would not be expected to provide a good indicator of animal diversity. Recognizing this problem, Roth (1976) developed a method by which the dispersion of clumps of vegetation—shrubs—forms the basis for a measure of habitat "heterogeneity." Roth was able to relate BSD to this measure of habitat heterogeneity, or patchiness. Other workers also built relationships between measures of horizontal vegetation development and animal communities (see Wiens 1969; Morrison and Meslow 1983).

Unfortunately, measures such as diversity indices sacrifice complexity for simplicity. That is, FHD, BSD, and other such indices collapse detailed information on plants and animals into a single number. Remember that

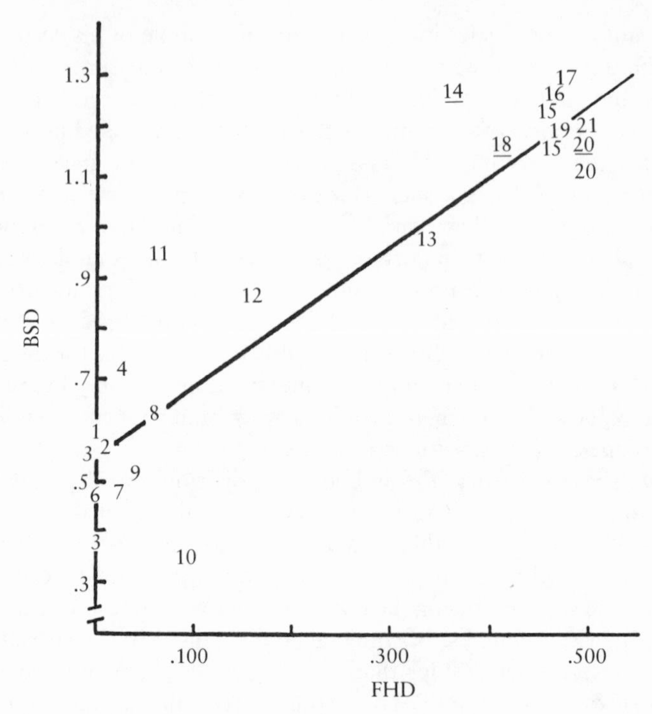

Figure 4.5. Foliage height diversity (FHD) versus bird species diversity (BSD). (From Willson 1974, fig. 1.)

for the indices to have any meaning at all, the data that is used to construct them must be accurate. Further, such relationships are mostly useful at scales above the individual, single species, and local levels. They are adequate for comparing locations in a general sense but tell nothing about the species compositions of the areas. Students examining diversity indices soon discover that two samples with entirely different species compositions can have identical index values.

The FHD–BSD relationships do indicate, however, that species respond to complexity in their environment. But to what specifically do they respond? The observer can increase FHD by simply changing the number of categories included in the calculations. Dividing tree heights into 1 m, 2 m, or 5 m intervals, for example, would likely change the resulting picture of foliage structure for an area. Thus, FHD, while giving some indication of foliage strata and the response of animal communities to such vegetation development, is much too arbitrary and general to result in more than a gross and relative examination of communities.

A more useful measure of foliage structure may be foliage volume, the actual surface area of foliage available for consumption or as substrates

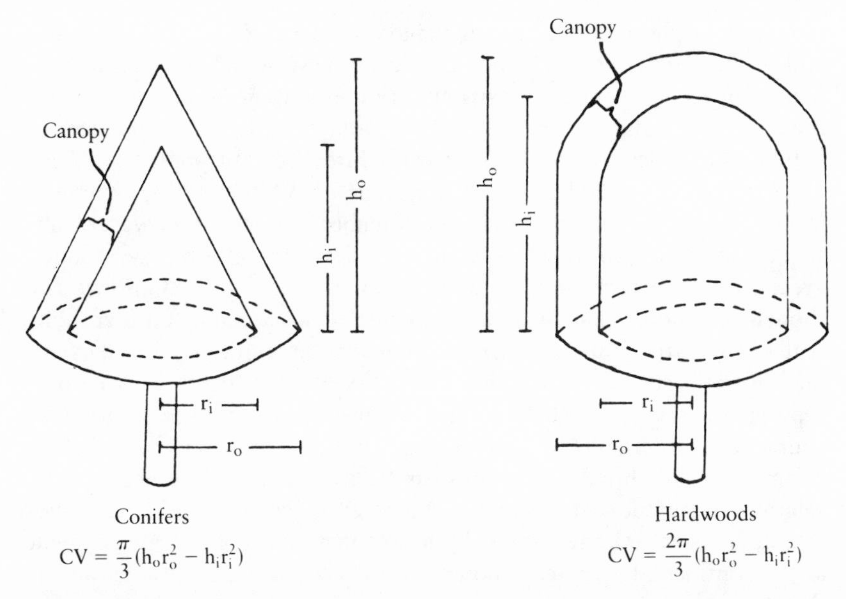

Figure 4.6. Tree shapes assumed and formulas used to calculate canopy volume (CV). (From Sturman 1968, fig. 2.)

for insects and other prey. It is extremely tedious, however, to measure actual foliage volume. Plant ecologists have done so by cutting down plants and then measuring and counting their leaves, needles, and other parts. Such data are necessary for accurate predictions of photosynthetic rates and measurements of tree growth (see Carbon, Bartle, and Murray 1979). After ecologists have completed such tedious work, they can develop statistical models that relate some easily measurable aspect of the plant to foliage volume (Van Deusen and Biging 1984). Plant ecologists also use the dry weight of plant material—leaves, grass stems, and other parts—as a measure of plant volume.

Researchers have developed even more indirect measures of foliage volume. In a comparison of habitats of the chestnut-backed chickadee (*Parus rufescens*) and the black-capped chickadee (*P. atricapillus*), Sturman (1968) used equations for shapes that approximated the structure of conifers and hardwoods (fig. 4.6). Methods such as Sturman's (see Mawson, Thomas, and DeGraaf 1976) can be viewed as compromises between the labor-intensive techniques that involve sampling whole trees and the qualitative estimates of FHD by height categories. We know of no study that has adequately quantified and contrasted these various methods, although such an investigation is clearly needed.

Two basic and obvious aspects of vegetation can be distinguished: the structure, or physiognomy, and the taxon of the plant, or floristics. Many authors have initially concluded that vegetation structure and habitat configuration (size, shape, and distribution of vegetation in an area), rather than particular plant taxonomic composition, most determine patterns of habitat occupancy by animals, especially birds (see Hilden 1965; Wiens 1969; James 1971; Anderson and Shugart 1974; Willson 1974; James and Wamer 1982; Rotenberry 1985). As Rotenberry observed, however, subsequent studies have shown that plant species composition plays a much greater role in determining patterns of habitat occupancy than previously thought. He noted that such physiognomic versus floristic considerations really involve the scale at which the animal communities are being examined. Rotenberry postulated that birds species–plant taxon associations, especially within similar habitat types, are mediated by the specific food resources that different plant taxa provide. These differences in the apparent nature of bird-habitat associations can be attributed to differences in the relative geographic scale over which they were measured. The same species that appears to respond to the physical configuration of the environment at the continental level may show little correlation with physiognomy at the regional or local level. Thus, animals may be differentiating between gross habitat types on the basis of physiognomy (that is, they may occupy a general area that is "proper" in its structural configuration), with further refinement of the distribution (and thus, abundance) within this habitat type based on plant taxonomic considerations. Note that this scenario of Rotenberry's relates closely to Hutto's hypothesized mode for the process of habitat "selection" in animals (Hutto 1985; see fig. 2–8). Rotenberry quantified his ideas using data from the shrubsteppe studies of Wiens and his coworkers. Rotenberry's analysis of these data indicate that, when the correlation between physiognomy and flora is (statistically) separated, a significant relationship remains between bird abundance and flora, but not between birds and physiognomy (fig. 4.7).

Thus, we again return to an emerging theme in model development: the variables measured, and the intensity of those measurements, are based on the scale involved and the level of model refinement required. Simple presence-absence studies of species at regional or broader scales likely do not require analysis of vegetation on a taxonomic level. Broad categorization by physiognomy—probably including differentiation no lower than "deciduous," "evergreen," and other such general classes—is probably adequate. Plant taxonomy becomes increasingly important, however, as our studies become increasingly site-specific. As a general rule, then, one would always be safe in collecting data on as fine a scale as time and budget allow; lumping is always possible. A better approach is to begin a study with a preliminary evaluation of the variables and sampling methods

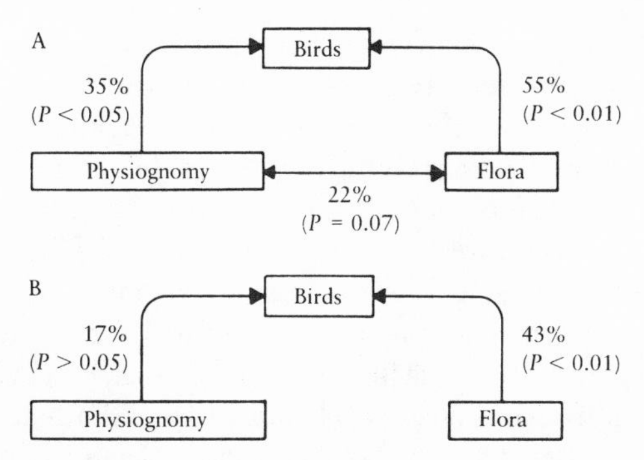

Figure 4.7. *A:* Coefficients of determination ($r^2 \times 100$) between similarity and distance matrices based on avian, floristic, and physiognomic composition of eight grassland study sites. Significance levels of association given in parentheses. *B:* Partial coefficients of determination, as above. Correlation between physiognomy and flora has been partialled out. (From Rotenberry 1985, fig. 1.)

necessary to achieve the desired level of model refinement. That is, if one seeks to determine the habitat use patterns of a particular species across its range, then preliminary sampling from a number of sites will likely give a good indication of the type and number of variables required; necessary sample sizes can also be predicted from this preliminary sample.

We turn now to specific examples of variables collected by researchers seeking to describe habitat use patterns of animals. The examples we use come from widely read and cited papers. We offer these few examples as good starting points for students planning similar evaluations of habitat use by a particular species. We strongly recommend that researchers concentrating on a particular taxon not restrict themselves to literature relating to that group only. For example, a researcher designing a study to examine habitats of ground-foraging birds would likely gain valuable information on types of variables and sampling design by reviewing papers on small mammals and reptiles. We will also give brief summaries of the objectives of each study: remember that the variables selected should relate closely to the objectives.

Dueser and Shugart (1978) were among the first workers to quantify microhabitat use patterns of small mammals in a multivariate sense. The variables they selected, however, would apply regardless of the analytical techniques used. Dueser and Shugart had as their goal the description of microhabitat differences among the small mammal species of an upland forest in eastern Tennessee. Their specific objectives were to characterize and compare microhabitats of species within the forest and to examine the relationship of species distributions and abundances to the relative availability of preferred microhabitats. Habitat information was collected for six strata at each capture site: overstory, understory, shrub level, herbaceous level, forest floor, and litter-soil level. Table 4.3 lists the variables Dueser and Shugart examined: note that they did not record species-specific information on plants beyond designations of "woodyness," "evergreenness," and the like. They did record, however, the number of woody and herbaceous species. They paid special attention to features of the forest floor, such as litter-soil compactability, fallen log density, and short herbaceous stem density. They found that certain of these soil variables played a significant role in describing the microhabitats of the species studied (see also Rosenzweig and Winakur 1969). Except for the lack of detailed information on plant taxa, we consider Dueser and Shugart's study an excellent example of a very detailed set of variables used to differentiate among species of co-occurring animals. We know of no study in the ecology of ground-foraging birds that has included such detail.

Van Horne (1982) described the niche occupancy patterns of deer mice (*Peromyscus maniculatus*) in order to determine whether hypothesized intraspecific competitive interactions result in utilization patterns that differ between age classes. The concept of the niche used by Van Horne included both the description of habitat occupancy envisioned by Grinnell (1928) and a more functional description with emphasis on food habits (Elton 1927). Van Horne's study was a synthesis of food and habitat utilization patterns as they relate to population densities and processes. In contrast with Dueser and Shugart (1978), Van Horne initially collected floristic information as well as brief data on physiognomy (table 4.4). Because plant species of similar morphotype showed high correlation coefficients, Van Horne combined the percentage cover values of species within morphotypes to form a reduced data set for analysis. Combining species-specific data after collection is the preferred method: this gave Van Horne the option of running analyses either separately or after lumping data. We do not know if the results of Dueser and Shugart would have been different had they included plant taxa data.

Reinert (1984) designed a study to determine if differential habitat utilization provided significant separation of populations of timber rattlesnakes (*Crotalus horridus*) and northern copperheads (*Agkistrodon contortrix mokeson*), which occur sympatrically in temperate deciduous forests of eastern North America. His approach was a multivariate habitat description consistent with the Hutchinsonian definition of *niche* (Green 1971) and based on the concept of the *niche gestalt* (James 1971). Reinert was

Table 4.3. Designation, descriptions, and sampling methods for variables measuring forest habitat structure, used by Dueser and Shugart

Variable	Methods
1. Percent canopy closure	Percentage of points with overstory vegetation, from 21 vertical ocular tube sightings along the center lines of 2 perpendicular 20 m^2 transects centered on trap
2. Thickness of woody vegetation	Average number of shoulder-height contacts (trees and shrubs), from 2 perpendicular 20 m^2 transects centered on trap
3. Shrub cover	Same as (1), for presence of shrub-level vegetation
4. Overstory tree size	Average diameter (in cm) of nearest overstory tree, in quarters around trap
5. Overstory tree dispersion	Average distance (m) from trap to nearest overstory tree, in quarters
6. Understory tree size	Average diameter (cm) of nearest understory tree, in quarters around trap
7. Understory tree dispersion	Average distance (m) from trap to nearest understory tree, in quarters
8. Woody stem density	Live woody stem count at ground level within a 1.00 m^2 ring centered on trap
9. Short woody stem density	Live woody stem count within a 1.00 m^2 ring centered on trap (stems $\leq$ 0.40 meters in height)
10. Woody foliage profile density	Average numbers of live woody stem contacts with a 0.80 cm diameter metal rod rotated 360°, describing a 1.00 m^2 ring centered on the trap and parallel to the ground at heights of 0.05, 0.10, 0.20, 0.40, 0.60 . . . 2.00 meters above ground level
11. Number of woody species	Woody species count within a 1.00 m^2 ring centered on trap
12. Herbaceous stem density	Live herbaceous stem count at ground level within a 1.00 m^2 ring centered on trap
13. Short herbaceous stem density	Live herbaceous stem count within a 1.00 m^2 ring centered on trap (stems $<$ 0.40 meters in height)

(*table continued on following page*)

Table 4.3. (*Continued*)

Variable	Methods
14. Herbaceous foliage profile density	Same as (10), for live herbaceous stem contacts
15. Number of herbaceous species	Herbaceous species count within a 1.00 m^2 ring centered on trap
16. Evergreenness of overstory	Same as (1), for presence of evergreen canopy vegetation
17. Evergreenness of shrubs	Same as (1), for presence of evergreen shrub-level vegetation
18. Evergreenness of herb stratum	Percentage of points with evergreen herbaceous vegetation, from 21 step-point samples along the center lines of 2 perpendicular 20 m^2 transects centered on trap
19. Tree stump density	Average number of tree stumps $\geq$ 7.50 cm in diameter, per quarter
20. Tree stump size	Average diameter (cm) of nearest tree stump $\geq$ 7.50 cm in diameter, in quarters around trap
21. Tree stump dispersion	Average distance (m) to nearest tree stump $\geq$ 7.50 cm in diameter, in quarters around trap
22. Fallen log density	Average number of fallen logs $\geq$ 7.50 cm in diameter, per quarter
23. Fallen log size	Average diameter (cm) of nearest fallen log $\geq$ 7.50 cm in diameter, in quarters around trap
24. Fallen log dispersion	Average distance (m) from trap to nearest fallen log $\geq$ 7.50 centimeters in diameter, in quarters around trap
25. Fallen log abundance	Average total length (+0.50 m) of fallen logs $\geq$ 7.50 centimeters in diameter, per quarter
26. Litter-soil depth	Depth of penetration ($<$ 10.00 cm) into litter-soil material of a hand-held core sampler with 2.00 cm diameter barrel
27. Litter-soil compactability	Percent compaction of litter-soil core sample (26)
28. Litter-soil density	Dry weight density (g/cm^2) of litter-soil core sample (26), after oven-drying at 45°C for 48 h

Table 4.3. (*Continued*)

Variable	Methods
29. Soil surface exposure	Same as (18), for percentage of points with bare soil or rock

Source: Dueser and Shugart 1978, appendix.

Table 4.4. Description of habitat variables used by Van Horne

Variable	Source
Herb cover	*Athyrium felix-femina* + *Dryopteris dilatata* + *Thelypteris phegopteris* + *Gymnocarpium dryopteris* + *Pteridium aquilinium* + *Blechnum spicant* + *Cornus canadensis* + *Rubus pedatus* + *Coptis asplenifolia* + *Carex* sp. + *Moneses uniflora* + *Tiarella trifoliata* + Gramineae + *Lysichiton americanum* + *Lypocodium* sp. + *Epilobium angustifolium* + *Listera cordata* + *Linnaea borealis* + *Epilobium hornemanii* + *Caltha palustris;* all measured below 25 cm
Total canopy	Cover above 160 cm
Low canopy	(Cover above 40 cm) − (Cover above 160 cm)
Low log	Log below 25 cm
High log	Litter and log above 25 cm and below 150 cm
Shrub	*Vaccinium* sp. + *Menziesia ferruginea* + *Rubus spectabilis* + *Ribes* sp. + *Echinopanax horridum* + *Sambucus racemosa;* all measured below 25 cm
High herb	*Athyrium felix-femina* + *Dryopteris dilatata* + *Pteridium aquilinum* + *Blechnum spicant* + *Carex* sp. + *Lysichiton americanum* + *Epilobium angustifolium;* all measured between 25 and 150 cm
Seedlings	*Picea stichensis* + *Chamaecyparis nootkatensis* + *Thuja plicata* + *Tsuga mertensiana* + *T. heterophylla;* seedling cover below 150 cm
Trees	*Picea sitchensis* + *Chamaecyparis nootkatensis* + *Thuja plicata* + *Tsuga mertensiana* + *T. heterophylla;* tree cover below 150 cm
Water	Bog or creek

Source: Modified from Van Horne 1982, table 1.

interested not only in features of the ground but also in the climatic conditions immediately surrounding the snakes. He measured temperature and humidity at several locations near a snake, as well as the structure of the surrounding vegetation (table 4.5). Here again, no information on plant taxa was included.

Birds have received by far the most attention with regard to the analysis of habitat use patterns. This is likely a reflection of the conspicuousness of most birds: most are active during the day, most give at least some vocalizations during all parts of the year, and they are inexpensive to observe (one only needs a pair of binoculars and a notebook). James (1971) conducted one of the first and most famous studies quantifying bird-habitat relationships. In Chapter 2 we discussed her development of the niche gestalt and the concept's importance to the way in which we view—and thus analyze—habitat; note that Reinert made special mention of the niche gestalt in his study on snakes, thus showing its wide applicability. James used fifteen measures of vegetation structure—here again, ignoring floristics—to describe the multidimensional "habitat space" of a bird community in Arkansas. She followed closely the methods she developed with Herman Shugart, Jr. (James and Shugart 1970). The general analytical techniques (multivariate analysis of focal-bird observations) and conceptual framework (niche gestalt) she used led to a plethora of studies that expanded upon her basic ideas.

Anderson and Shugart designed a study to analyze "the relationships of the spatially heterogeneous distributions of 28 habitat variables to the distributions of 28 breeding bird species" in Tennessee (1974:828). They rationalized their techniques by noting that the "simultaneous study of 28 species precludes detailed study of the behavioral mechanisms of any single species, and we focus on identifying habitat variables that have measurable species responses (in terms of nonrandom species distributions on the habitat variables)" (1974:823). They considered the surface of their study area a mosaic in which each element was described by twenty-eight quantitative variables. Their goal was to determine in what way the mosaic elements associated with the activities of a given bird species could be discriminated from other mosaic elements. Data on plant taxa were not presented; rather, the researchers concentrated on biomass and density data. They divided each type of variable into rather fine classes based on dbh (table 4.6). They later used correlation analysis to eliminate the confounding effects of intercorrelations among habitat variables. Thus, they followed the procedure of recording data on a fine scale and relied on later analyses to determine the level of lumping possible.

As these examples show, workers have placed a strong emphasis on measuring the structure, or physiognomy, of the habitat; little attention has

Table 4.5. Structural and climatic variables used by Reinert

Mnemonic	Variable	Sampling method
ROCK	Rock cover	Coverage (%) within 1 m^2 quadrant centered on snake location
LEAF	Leaf litter cover	Same as ROCK
VEG	Vegetation cover	Same as ROCK
LOG	Fallen log cover	Same as ROCK
WSD	Woody stem density	Total number of woody stems within 1 m^2 quadrant
WSH	Woody stem height	Height (cm) of tallest woody stem within 1 m^2 quadrant
MDR	Distance to rocks	Mean distance (m) to nearest rocks ($>$ 10 cm max. length) in each quarter
MLR	Length of rocks	Mean max. length (cm) of rocks used to calculate MDR
DNL	Distance to log	Distance (m) to nearest log ($\geq$ 7.5 cm max diameter)
DINL	Diameter of log	Max. diameter (cm) of nearest log
DNOV	Distance to overstory tree	Distance (m) to nearest tree ($\geq$ 7.5 cm dbh [diameter at breast height])
DBHOV	dbh of overstory tree	Mean dbh (cm) of nearest overstory tree within each quarter
DNUN	Distance to understory tree	Same as DNOV (trees $<$ 7.5 cm dbh $>$ 2.0 m height)
CAN	Canopy closure	Canopy closure (%) within 45° cone with ocular tube
SOILT	Soil temperature	Temp (°C) at 5 cm depth within 10 cm of snake
SURFT	Surface temperature	Temp (°C) of substrate within 10 cm of snake
IMT	Ambient temperature	Temp (°C) of air at 1 m above snake
SURFRH	Surface relative humidity	Relative humidity (%) at substrate within 10 cm of snake
IMRH	Ambient relative humidity	Relative humidity (%) 1 m above snake

Source: Reinert 1984, table 1.

Table 4.6. Habitat variables used by Anderson and Shugart

Habitat variable
Foliage biomass of trees 1.2–8.4 cm dbh
Foliage biomass of trees 8.5–22.8 cm dbh
Foliage biomass of trees > 22.8 cm dbh
Branch biomass of trees 1.2–8.4 cm dbh
Branch biomass of trees 8.5–22.8 cm dbh
Branch biomass of trees > 22.8 cm dbh
Bole biomass of trees 1.2–8.4 cm dbh
Bole biomass of trees 8.5–22.8 cm dbh
Bole biomass of trees > 22.8 cm dbh
Number of trees 1.2–8.4 cm dbh
Number of trees 8.5–22.8 cm dbh
Number of trees > 22.8 cm dbh
Foliage biomass of average tree 1.2–8.4 cm dbh
Foliage biomass of average tree 8.5–22.8 cm dbh
Foliage biomass of average tree > 22.8 cm dbh
Branch biomass of average tree 1.2–8.4 cm dbh
Branch biomass of average tree 8.4–22.8 cm dbh
Branch biomass of average tree > 22.8 cm dbh
Bole biomass of average tree 1.2–8.4 cm dbh
Bole biomass of average tree 8.4–22.8 cm dbh
Bole biomass of average tree > 22.8 cm dbh
Stump biomass of average tree 1.2–8.4 cm dbh
Stump biomass of average tree 8.4–22.8 cm dbh
Stump biomass of average tree > 22.8 cm dbh
Percentage of slope of plot
Solar radiation reaching plot (kcal)
Solar radiation reaching plot / expected solar radiation at this latitude
Number of saplings

Source: Modified from Anderson and Shugart 1974, table 1.
Note: All biomass data is in dry weight.

been given to floristic components. Earlier in this chapter we developed the rationale for this emphasis on structure, the FHD–BSD relationships, and, further, the rationale involving how an animal perceives its environment, the niche gestalt. Although James (1971) did not exclude considerations of floristics from her development of the niche gestalt, it appears that most workers have, nevertheless, concentrated on issues of volume, layering, density, and other such measures of physical structure in developing relationships between animals and their environments.

As Rotenberry (1985) pointed out, floristics is indeed an important measure of habitat, the magnitude of its importance often depending upon the scale at which one is working. Few authors have included detailed analysis of floristics in their development of habitat models, however.

Eddleman, Evans, and Elder (1980) provided a fine example of a detailed analysis of both structural and floristic habitat. Using the Swainson's warbler (*Limnothlypis swainsonii*) in Illinois, they calculated size-class distributions of vegetation overstory (dbh, basal area, average stems per ha, and percentage of overstory stems, using eight size classes) and calculated the relative density, dominance, and frequency by individual species of six overstory trees, eight understory plants, and five ground cover plants. They found that Swainson's warblers required habitat with a high canopy closure and an understory composed mainly of giant cane (*Arundinaria gigantea*). They thus present us with an example of a species whose habitat description requires both structural (canopy closure) and floristic (cane density) characteristics. Failure to include both of these characteristics in a model describing the habitat of this warbler would likely result in very poor predictive power.

How to Measure

Wildlife Habitat

Now we wish to review some of the standard methods used to measure wildlife habitat. Our goal is not a complete survey of the literature for all taxa, since Cooperrider, Boyd, and Stuart (1986) provided an overview of basic sampling techniques for all major groups of wildlife. Rather, we want to encourage further appreciation of the methods and thought processes involved in the selection and subsequent measurement of habitat variables. For continuity, we will concentrate on the studies we examined when discussing variable selection.

Wildlife ecologists must turn to plant ecologists for advice on measuring habitat characteristics. Indeed, even a cursory review of the methods sections in various wildlife-related publications will reveal a reliance on standard, classical methods of quantifying the floristics and structure of vegetation: point quarter, circular plots and nested circular plots, sampling squares, line intercept, and the like. These methods are used for good reason: they have been developed and extensively tested by plant ecologists. This is not to say we should not be innovative. Standard methods do, however, provide an excellent starting point from which the wildlife ecologist can adapt a specific method to meet specific needs. Students are advised to review one or more of the comprehensive books available on vegetation ecology (such as Daubenmire 1968; Mueller-Dombois and Ellenberg 1974; Greig-Smith 1983; Cook and Stubbendieck 1986; Bonham 1989).

Sampling Principles

Noon (1981) suggested that, in designing a sampling protocol, the more apparently homogeneous the habitat, the finer it will need to be sampled in order to detect its inherent heterogeneity. For example, grasslands are thought to require sampling vegetation at a much finer level than would forests. We must add a caveat to this assumption: it seems to us that determination of the level of "fineness" at which vegetation is measured depends primarily upon the level of model refinement desired and the scale at which the model will be applied. For example, animal species inhabiting the forest understory may very well be differentiating habitat at the same level of resolution that species in a grassland do. Thus, both forest and grassland would require a very fine level of sampling in the understory. Alternatively, determining the species composition and density of coniferous trees does not require as detailed a sampling effort as does identifying the species composition of grasses and shrubs. Thus, this general assumption must be tempered by the researcher's objectives in analyzing the community of plants and animals.

We must remember that the variables we measure—and the means by which we measure them—will themselves play a substantial role in determining the results of our analyses: if we do not determine plant species composition, then floristics can have no direct predictive power in the resulting models. We can begin, of course, by thoroughly reviewing the literature in hope of gaining a general perception of the niche gestalt of the species in question—our perception of how the animal perceives its environment. We must be careful, of course, not to simply accept as fact all statements concerning habitat use patterns given in the literature. First—and this is more likely to occur in books than in scientific journals—authors tend to repeat what previous authors have written; these citations are known as secondary, tertiary, and so on, based on how far removed they are from the original citation. Second, we must realize that most studies directly apply only to the time and place in which the study was conducted. For example, a study conducted during the fall in the Northwest may provide little useful guidance for the study conducted in California in the summer. And finally, although repeatability and corroboration are cornerstones of scientific understanding, there is simply no overriding reason to basically repeat what another worker has done.

No matter how many studies they review or people they consult, researchers will still be uncertain about how finely species are discriminating habitat (see Noon 1981). The solution, it appears, is to sample features of the general geographic area of interest—often called macrohabitat variables—as well as at the microhabitat level. It is true that one can never go wrong by "oversampling" an area, as long as an adequate sample size is obtained. If researchers counted and measured every plant, then they would be assured of capturing all the predictive power that vegetation density and floristics could provide. Lumping of variables after data collection is usually performed for simplicity of understanding and presentation when it is accompanied by little loss of, or even an increase in, predictive power. For example, one may find that describing tree density by intervals of 5 cm dbh gains little as compared with describing the relationship by 10 cm intervals. Our problem is that neither time nor money usually allows such a fine level of sampling on a large scale. Many workers concerned with this problem have solved it by subsampling an area or two using a series of transects, plots, and subplots that give a rather refined picture of the vegetation—and the response of the animal to this vegetation—over the area(s). Then, preliminary data analyses can be conducted to determine which variables appear to be useful in describing the habitat of the animal. These methods can then be applied to the study as a whole.

Before discussing specific procedures used by researchers to define wildlife habitat, we should first address the question of what determines the utilized (or used) area of an animal in comparison with the nonutilized (or unused) area. We previously discussed the problems associated with the use of different areas by an animal to perform different tasks. Thus, where samples are taken depends upon the activity pattern of the animal within the area, although random sampling within a utilized area is often used to measure the vegetation available to the animal. Note that we do not refer to "habitat availability," because the term *habitat* is being restricted for description of the characteristics of the environment that the animal actually uses; these terms are often used interchangeably, and thus improperly.

Several problems are readily apparent in determining availability. First, we must note a contrast between what vegetation is actually available for the animal to use in some manner and the actual density of that vegetation in the area. For example, the terminal buds of sapling conifers are certainly present in the forest; they can be quantified. But are they available as a foraging substrate by a 200 g bird? It is doubtful that such a small tree could support such a bird at the tip of a branch. The animal might exploit this potential resource if it possesses the necessary behaviors. Can the bird hover-glean? Vegetation beyond the reach of a deer is likewise unavailable for it to feed upon, even though such material may be very similar to the foliage that it can reach. It should be apparent, then, that what is considered available to one animal may not be likewise available for another. This obviously complicates studies of entire animal communities.

A general commonality can be found in most studies of wildlife habitat. Sampling is usually centered on the location of a trap or track detection (such as for small mammals), a foraging bird or large mammal (or even a snake), or on the point used to count birds or vocal mammals. In all sampling schemes one must consider the *independence* of each sample from all other samples. Independence of samples is important because it is necessary for most statistical analyses and predictive models. That is, small mammal traps or bird census points must be sufficiently spaced so that the vegetation sampling does not overlap between points. Even if spacing of plots is not a problem, it is nevertheless critical that the plot (or transect, or set of plots) measure only the area utilized by the animal. For example, if the researcher is using plots of a 30 m radius (fixed plots) in which birds are counted (see Morrison, Timossi, and With 1987), then vegetation measurements should be contained within that radius. Alternatively, vegetation measured in, say, a 10 m radius around such a census point might not adequately describe the vegetation actually present in the census area (and thus, not yield good predictions of bird abundance). The varying modes of habitat use among species means that different intensities of sampling should be used. In practice, researchers try to strike a balance between intensity of sampling and the number of plots sampled. Unfortunately, this raises the real possibility that no species will be sampled adequately. Adequacy is based, of course, on the objectives of the study. Although we cannot provide an answer for this apparent dilemma, researchers should at least be fully aware of the problems involved in developing a sampling protocol.

Habitat Sampling Methods

Since the late 1960s and early 1970s studies have approached wildlife-habitat relationships from a multivariate standpoint based on the concept of the multidimensional niche, as defined by Hutchinson (1957, 1978). Naturally, each variable measured can be analyzed univariately and, indeed, will lend insight into the habitat relationships of an animal or group of animals. In either case, it is important to realize the rationale behind the variables selected and the techniques used to measure them.

No two studies—even those conducted in similar habitats—use the same techniques (unless by design, which is rare). This applies also, of course, to the methods used to estimate the number of animals in the area. We will discuss specific problems of observer variability and bias later in this section.

James and Shugart (1970) compared four of the methods recommended by plant ecologists for making quantitative estimates of vegetation. These were *plotless* methods, such as the quarter method (Cottam and Curtis 1956; Phillips 1959) and the wandering quarter method (Catana 1963), and *areal* methods, such as arm-length transects (Rice and Penfound 1955; Penfound and Rice 1957) and circular plots (Lindsey, Barton, and Miles 1958). James and Shugart compared the average work accomplished using

Table 4.7. Average work accomplished in thirty minutes of field effort, recording the species and diameters of trees in an upland Ozark forest in Arkansas

Sampling method	Number of units	Number of trees identified and measured
Quarter method	12 quarters	48
Wandering quarter method	40 trees	40
Tenth-acre circles	2 circles	57
Hundredth-acre rectangles	6 rectangles	19

Source: James and Shugart 1970, table 1.

these four sampling methods in thirty minutes of field effort by one observer, assuming that he or she was familiar with the method and species of trees present in the study area (see table 4.7). Naturally, the amount of work accomplished will vary with the density of the vegetation and the number of observers involved. James and Shugart found that results from the two plotless methods (quarter method and wandering quarter method) tended to overestimate the total tree density and to underestimate the tree density by species. Results from the two areal methods they used—hundredth-acre rectangles and tenth-acre circles—gave fairly accurate estimates of total density and density by single species. They presented detailed descriptions of their preferred method, sampling many small circular plots for forest vegetation.

We have discussed the importance of James's 1971 paper to our conceptualization of how animals perceive their environment—the niche gestalt. The methods James adopted to gather field data related to her ideas also have had a pronounced influence on analysis of wildlife habitat. Simply put, most studies subsequent to hers followed her lead.

James (1971) sought to devise field techniques that would give quantitative measurements of the vegetation within the breeding territories of individual birds. To do this she obtained estimates of the characteristics of vegetation by sampling one 0.1 acre (0.04 ha) circular plot within the territory of each singing male bird. The methods she adopted are the result of the analyses she conducted with Shugart in their Arkansas study areas. She felt that the size of plot would give an adequate description of the vegetation within the individual's territory. She acknowledged a potential bias in concentrating on song perches but felt this would be a minor problem in a forest habitat. She apparently did not test this assumption, however. As we have shown, this assumption is likely faulty in most situations (see also Collins 1981). But regardless of the problems inherent in concentrating

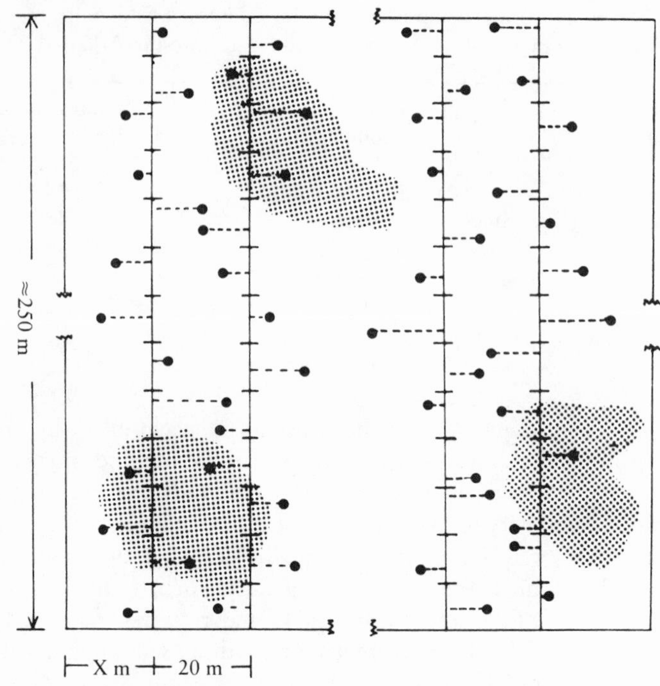

Figure 4.8. Line transect system with randomly located sampling points indicated. Territories are superimposed over sample locations. (From Noon 1981, fig. 3.)

only on one behavior of an animal, the methods used by James have played a substantial role in the development of wildlife-habitat relationships.

Circular plots are easy to establish, mark, measure, and relocate, and estimates of animal numbers within such plots can be statistically related to vegetation data in a straightforward manner. Plots provide for the sampling of vegetation and animals at specific locations in time and space. If plots can be considered independent data points (a function of the sampling design), then one's sample size is simply equal to the number of plots sampled. Or if plots are used to sample from a single small study area, then plot samples can be averaged and associated measures of variance calculated. Noon (1981) presented a useful description and example of both transect and areal plot sampling systems for analysis of bird habitat (see figs. 4.8 and 4.9).

In summary, fixed-area plots and transects can be used to provide site-specific, detailed analysis of animal-habitat relationships. Small transects are often more useful to quantify grass, forb, and shrub cover within each plot. Most sampling methods used in the 1970s and 1980s to measure wild-

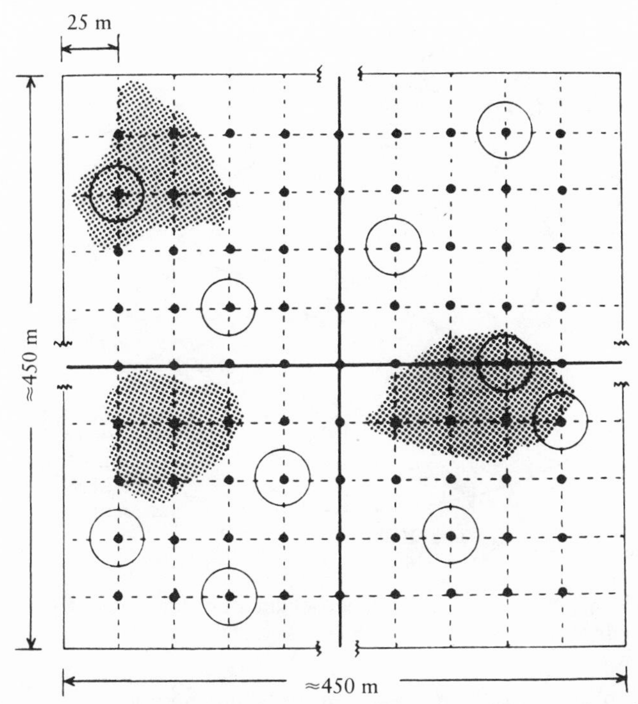

Figure 4.9. Areal plot system with randomly located 0.04 ha circular plots indicated. Territories are superimposed over sample locations. The 25 m grid system illustrated here would only be necessary in very closed habitats. (From Noon 1981, fig. 4.)

life habitat—for subsequent multivariate analysis—use fixed-area circular (or sometimes square or rectangular) plots as the basis for development of a sampling scheme that may then include subplots, sampling squares, and transects.

Dueser and Shugart (1978) devised a detailed sampling scheme that combined plots of various sizes and shapes, as well as small transects (see fig. 4.10). Although designed for analysis of small mammal habitat, the techniques easily can be adapted for most terrestrial vertebrates. Dueser and Shugart established three independent sampling units, centered on each trap: a 1.0 m^2 ring, two perpendicular 20 m^2 arm-length transects, and a 10 m radius circular plot. The meter-square circular plot provided a measure of vertical foliage profile from ground level to a 2 m height, for both herbaceous and woody vegetation. Also, four replicate core-sample estimates of litter-soil depth, compactability, and dry weight density were made on the perimeter of this central ring. The two arm-length transects

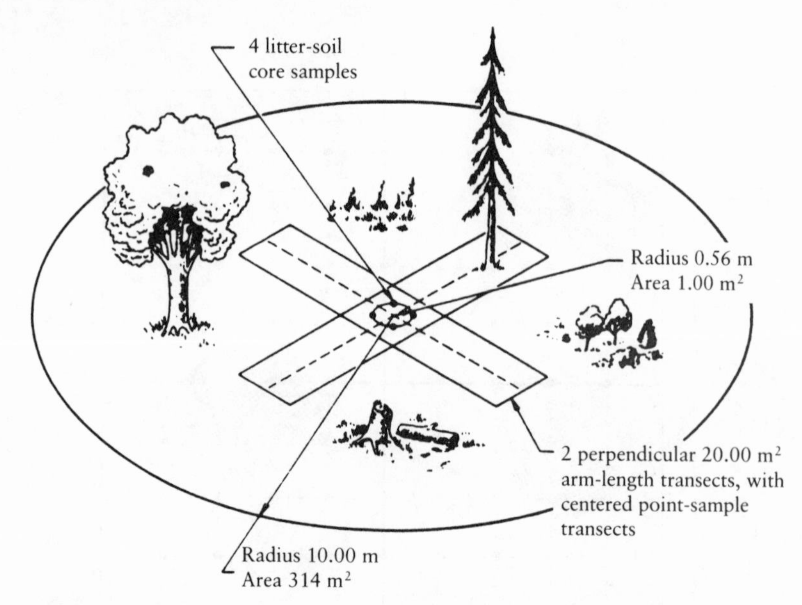

Figure 4.10. Habitat variable sampling configuration used by Dueser and Shugart in their study of small mammal habitat use. (From Dueser and Shugart 1978, fig. 1.)

provided measures of cover type, surface characteristics, and density and evergreenness of the four strata of vegetation. Data recorded for each quarter of the 10 m radius plot included the species, dbh, and distance from the trap to the nearest understory and overstory trees, numbers of stumps and fallen logs, basal diameter and distance of nearest stump and fallen log, and total length of fallen logs.

Van Horne (1982), in her study of small mammal (*Peromyscus*) populations, did not use the circular plot scheme of Dueser and Shugart. Rather, sampling was conducted in 10 m squares centered on each livetrap station. Groundcover as well as percentage cover of plant species below 25 cm in height were estimated using a 20 cm × 30 cm sampling frame placed at two random locations in each 5 m × 5 m quarter of the sampling square. Dead wood smaller than 2.6 cm in diameter was classified as litter, whereas larger pieces were termed logs. Conifers smaller than 2.6 cm in diameter were classified as seedlings, those larger than this but smaller than 10.1 cm were termed saplings, and the rest were considered trees. (We should note here than Van Horne's categorization of size classes by fractions of centimeters was apparently due to an *a posteriori* conversion from English to metric: 2.6 cm = 1 in.; 10.1 cm = 4 in. She gave no indication that her diameter measurements in the field were precise to tenths of a centimeter, and she probably should have rounded the converted values to the closest unit.) Percentage cover of each plant species from 25 cm to 150 cm above the ground was estimated visually for the entire sampling square. Measures of canopy cover were taken at heights of 40 cm and 160 cm using a section of polyvinyl chloride (PVC) pipe 3 cm in diameter and 14 cm in length with a 90 degree bend at its midpoint. Van Horne placed a small mirror at the pipe's elbow so that, by looking horizontally into one end of the pipe, she could look out the other end at the round image representing a vertical cone-shaped field of view, in which the percentage canopy cover could be quantified. (Later in this section we discuss problems in visually estimating percentage cover and other characteristics of vegetation.) Finally, Van Horne estimated overall stand characteristics using the point-centered quarter technique (Cottam and Curtis 1956) with samples taken at each location.

Unless one actually compares the efficiency and accuracy of these various techniques for specific studies and study objectives, it is impossible to comment fairly on the desirability of one method over another. We think these studies were well conceived, conducted, and analyzed. One could not, of course, make direct comparisons between studies because of variations in the methods. The methods are applied properly, with adequate sample sizes. A useful review of many of these and other methods of measuring habitat is given by Cook and Stubbendieck (1986).

In his analysis of snake populations, Reinert (1984) adopted techniques similar to those used by James (1971) and Dueser and Shugart (1978), with the exception that he used photographs to evaluate percentage of surface cover (leaf litter, rock, log, and vegetation in the 1 m^2 plots). Using a 35 mm camera equipped with a 28 mm wide-angle lens, Reinert took color slide photographs encompassing the 1 m^2 plot from directly above the location where the snake was observed. Reinert then determined the various surface cover percentages by superimposing each slide onto a 10 × 10 square grid. Reinert, then, more rigorously quantified his measure of cover values than most workers, who use ocular estimates. The specific sampling scheme he used, as well as the categories and characteristics measured, are summarized in figure 4.11. Such rigor may not be necessary if the researcher can be trained to recognize cover classes, such as those of 10 or 20 percent.

Habitat Diversity and Heterogeneity

Researchers have used indices of foliage height diversity and habitat heterogeneity to describe the diversity and dispersion of vegetation within an area. Such indices might provide additional insight beyond measures of plant density and size into how animals perceive their surroundings. To determine foliage height diversity (FHD), workers most often stratify the vegetation in an area into height classes and use some method to deter-

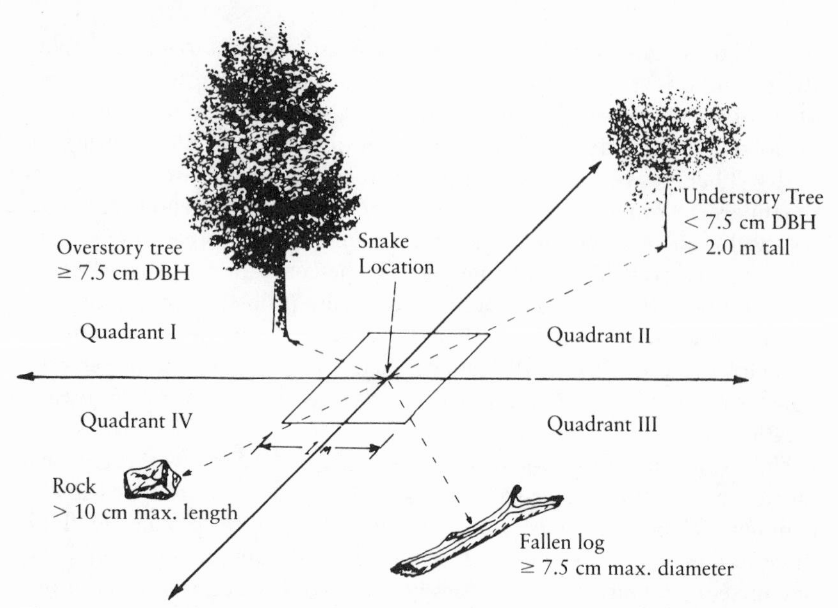

Figure 4.11. Sampling arrangement for snake locations. (From Reinert 1984, fig. 1.)

mine where live foliage intercepts a line extended vertically through these strata. In vegetation with a low canopy, researchers can physically place a long pole with height demarcations at various locations—usually systematically along a transect—and record where the live foliage touches the pole (see Wiens 1969; Willson 1974). In taller vegetation, workers often ocularly estimate such interceptions (Kitchings and Levy 1981; Collins, James, and Risser 1982; Morrison 1982; Engstrom, Crawford, and Baker 1984). We described how Van Horne (1982) used a curved section of PVC pipe and a mirror to record the foliage height structure of her study areas. Although seldom done, such recordings can be made for each species of plant intercepting the line (or pole). Using this method, a researcher tries to obtain yet another method of quantifying how the animal perceives its environment: Is it the density of conifers, or is it the distribution of conifer foliage by height strata and species, that determines the animal species and community present?

Determination of FHD was originally applied to between-site comparisons (see MacArthur and MacArthur 1961; Karr and Roth 1971). The technique also applies—as shown by Van Horne (1982)—to within-site analysis of vegetation use. Transects can be established in conjunction with circular plots centered on traps, nest or perch sites, or any other area of animal activity (see Morrison 1982). The resulting index, however, is largely a function of the height intervals chosen by the researcher. FHD is not strictly a measure of foliage volume; it indexes the distribution as well as the volume of foliage available. Therefore, accurate quantification of total foliage volume would not replace this index (and vice versa). Foliage volume can, of course, be determined by height strata; Morrison, Timossi, and With 1987 did so in a study of bird communities in the western Sierra Nevada. Again, all such measures or indices simply offer additional means of trying to decipher the variables primarily responsible for patterns of vegetation use.

The strata chosen are, of course, a function of the inherent height profile of the area and involve the ever-present question of scale, or how "finely" the habitat is sampled. To humans, a grassland may have one layer of vegetation relative to a multilayered forest. But the animal inhabiting the grassland may perceive its environment on a much finer horizontal and vertical scale. Here again, we can only recommend that as fine a subdivision of the habitat as seems reasonable should be measured; you are free to try various combinations of data later.

Workers have also sought to quantify the horizontal dispersion or "patchiness" of vegetation. The impetus for such measures came from workers in grasslands and shrublands, who felt that FHD did not likely play a substantial role in describing animal communities in those situations; Roth (1976) reviewed studies failing to find meaningful correlations between BSD and FHD. In fact, MacArthur, MacArthur, and Preer (1962) concluded that patterns of vegetation in the horizontal plane were the principal factors affecting bird species diversity and that the effects of the vegetation in this plane were more significant than those of additional vegetation layers. It appears, in fact, that habitat patchiness plays a more important role in describing bird communities at the within-site level than does FHD, with FHD performing fairly well at the broader scale of between-site comparisons (Roth 1976).

Researchers have developed several measures of habitat heterogeneity or patchiness. Roth (1976) based his measure of heterogeneity on the point-centered quarter technique, using the coefficient of variation of the point to plant distance as his measure of heterogeneity. In early growth clear-cuts of western Oregon, Morrison and Meslow (1983) applied Roth's method, defining a *patch* as a clump of trees or shrubs greater than 2 m tall; such vegetation protruded above the extensive low shrub cover on the study sites. All of these measures can be calculated on a per trap, census point, or individual animal basis within a study area. It is important to note, however, that a patch will vary depending upon the definition applied to it by the user.

Used alone, indices of diversity or heterogeneity sacrifice much detail for the sake of simplicity. When used in combination with other measures of the environment, however, they are likely to add insight into how an animal perceives its environment.

Analysis of Community Structure

Ecologists often wish to compare the communities in two or more areas, or the same community at different times. Because identifying all organisms in a community is rarely possible, only a subset of the community, usually the organisms in a given taxa (such as birds, mammals, beetles, fish) are chosen for examination. The purposes of comparing communities vary. Sometimes researchers make comparisons to aid management decisions. For example, biologists might compare selected assemblages of animals in various potential sites for a construction project as part of the process of deciding where the construction should occur. In this hypothetical case, the area with the least diverse fauna and the fewest uncommon species might be the best site for development, from a biological standpoint. An analysis of community structure at different times also might be useful in assessing the effects of an environmental disturbance such as pollution in a riverine environment (see Patrick, Hohm, and Wallace 1954).

The structure of a community also has theoretical significance. An examination of the patterns of community structure across an environmental gradient can provide insights into how different communities function and perhaps why they developed as they did. Comparisons of this kind can be made between areas at all levels of environmental classification, from broad to relatively narrow. For example, one could assess and compare the assemblages of species in a given taxa between tropical forests and tundra, between temperate marshlands and temperate coniferous forests, between different coniferous forest types within a region, or between different seral stages within a forest type.

The initial levels of analysis of community structure usually involve an assessment of the number and identity of species and sometimes their relative abundances. These analyses fall under the most general level of habitat evaluation. Here we will review some of the techniques for analyzing the structure of communities, specifically measures of diversity and similarity, and evaluate the usefulness of these kinds of measures in habitat analysis.

Species Diversity

The work and ideas of MacArthur (1957, 1958), MacArthur and MacArthur (1961), and Hutchinson (1959) stimulated interest in the patterns of species diversity in ecological systems. Most biologists probably have an intuitive understanding of diversity, but Peet (1974) noted that the concept was defined and used in a number of different ways (see also Hurlbert 1971) and that measuring it was not as simple as might be expected.

The concept of species diversity generally encompasses two components: the number of species in a community, or species richness, and the evenness of distribution of individuals among species, sometimes called equitability (Peet 1974). Incorporating measures of equitability when describing the structure of a community is considered important because communities with the same number of species and individuals can be quite different. For example, there are obvious differences between a community with ten species, each with ten individuals, and another community, also with ten species and one hundred individuals, but with the distribution arranged so that nine species have one individual each and the tenth has ninety-one.

Species richness and equitability can be measured separately or combined into a single index. Measures of diversity that reflect both number of species and evenness of distribution of individuals among species are called heterogeneity indices (Peet 1974). Our discussion focuses on measures of species richness and heterogeneity. Readers should consult the article by Peet (1974) for a review of measures of evenness or equitability (see also Ludwig and Reynolds 1988).

Researchers must make several assumptions and decisions before generating measures of diversity (Peet 1974). First, the subject matter to be used in the measure must be placed into clearly defined classes. When measuring species diversity, species taxa are used as classes. But height classes of vegetation, for example, could be used to calculate diversity of vegetation height. Second, all individuals assigned to a specific class are assumed to be equal. This assumption probably does not hold for most animal communities because, for example, breeding individuals probably have a greater influence on community structure than nonbreeders. Third, all classes (that is, species) are assumed to be equally different. Fourth, some measure of importance of each class or species must be available or generated. When measuring species diversity, the abundance of each species is frequently used as the measure of importance, although other measures, such as biomass, are possible.

Enumerating all the species and individuals in a community, even those restricted to a given taxa, is impractical (impossible?) because the numbers change with ingress and egress. Thus, most measures of diversity are calculated from samples. The samples must clearly represent the structure of the community being examined before the diversity measures can be meaningful. Furthermore, the number of species encountered will increase as the area or the number of individuals sampled increases. Most measures of diversity, therefore, depend on the size of the sample. Comparisons of diversity between communities are appropriate only when the samples from which the measures were generated are equivalent.

Species counts. The simplest, and perhaps the most effective, measure of species richness is the number of species found in a sample of individuals, or a complete census of an area. This measure, as noted above, heavily depends on sample size, and sampling must be done with care to ensure that comparisons are valid.

Indices. Other indices exist for estimating species richness that are independent of sample size. All of these indices assume a functional relationship between the expected number of species and the size of the sample. To use these kinds of indices, one must know the relationship between the expected number of species and sample size, and it must be constant for the communities being compared (Peet 1974).

One example of this kind of index was developed by Preston (1948). Preston postulated that for naturally occurring biotic communities, numbers of individuals among species were log-normally distributed. His index,

$$N = n_0 a(2\pi)^{1/2}$$

where N is the total number of species theoretically available for observation, n_0 is the number of species in the modal octave (that is, the "crest" of the graph of the number of species plotted against the number of individuals per species on a log scale), and a is the logarithmic standard deviation, can be used to reveal how many species were missed when sampling occurred. Patrick and Strawbridge (1963) used Preston's approach to study the effects of pollution on diatom communities because the distribution of individuals among species in these communities closely fit the log-normal curve. Unfortunately, the patterns of distribution of individuals among species is rarely known for most groups of animals and cannot necessarily be expected to remain constant among communities. Therefore, use of this kind of index is restricted.

Heterogeneity Indices

Use of heterogeneity indices requires no assumptions about the distribution of individuals among species. These indices merely reflect community structure by changing with species richness, equitability, and sometimes density. Heterogeneity indices are calculated directly from relative or absolute abundances of species in a sample.

Simpson (1949) proposed a heterogeneity index that measures the probability that two individuals selected at random from a sample will belong to the same species. The index is calculated by the following formula, which is adjusted for a finite sample:

$$D = 1 - \sum_{i=1}^{s} \{[n_i(n_i - 1)]/[N(N - 1)]\}$$

where D is the diversity index, N is the total number of individuals of all species in the sample, n is the number of individuals of species i, and s is the number of species in the sample. This index varies from 0, for a community in which all individuals belong to the same species, to 1, for a community in which each individual belongs to a different species. Peet (1974) divided heterogeneity indices into two groups: those that are most sensitive to changes in rare species (Type I), and those that are most sensitive to changes in common species (Type II). Simpson's index was classified as a Type II index.

Another index of diversity that has been used frequently in ecology is one that is based on information theory (Shannon and Weaver 1949). This index, called the Shannon-Weaver function, measures the degree of uncertainty of predicting the species of an individual picked at random from a community. Uncertainty clearly increases as the number of species and equitability increases. Uncertainty (H') in a community can be estimated by the following formula:

$$H' = -\sum_{i=1}^{s} (n_i/N)\log(n_i/N)$$

where n_i, N, and s are the same as described above. This index varies from 0, for a community in which all individuals belong to the same species, to a relatively high number for a community with many species and an even distribution of individuals among species. The Shannon-Weaver function is most sensitive to changes in rare species (Tramer 1969) and is, therefore, a Type I index according to Peet's classification.

Hurlbert (1971) proposed another heterogeneity index based on the probability of interspecific encounters (PIE). Hurlbert argued that the concept of species diversity was burdened with semantic, conceptual, and technical problems. His criticisms centered on the lack of a precise and unambiguous definition of species diversity, the inability to assign appropriate relative weights to the two components of species diversity (species richness and equitability), the nonconcordance of various indices, and the lack of a biological basis for the indices based on information theory. Most of Hurlbert's criticisms have merit. The index he proposed,

$$PIE = \sum_{i=1}^{s} (n_i/N)[(N - n_i)/(N - 1)]$$

where n_i, N, and s are the same as described above, however, offers little improvement over other indices. In fact, PIE is simply a variation of the Simpson index and is equal to the measure of evenness that can be derived from the Simpson formula scaled to theoretical maximum diversity. Moreover, the assumptions upon which the index is based—random distribution of individuals in space and random encounters among individuals—rarely hold in nature.

Application of Diversity Indices

The problems associated with heterogeneity indices make their use in habitat analysis somewhat suspect. Peet (1974) concluded that selection and use of a diversity index should be based on the elements of the community that are of interest (for example, changes in common versus rare species), the ease with which the index can be applied, and the ease with which the index can be interpreted. We urge caution in the use of heterogeneity indices because of the difficulty of interpreting what a change or difference in an index means to community function. Furthermore, the user generally must return to the original data to decipher what component (richness or equitability) caused the difference.

Conversely, diversity indices (both richness and heterogeneity measures) have a potential value as signals of environmental problems. Human-induced changes in natural environments often cause a decrease in species richness and an increase in dominance of a few species, changes that heterogeneity indices reflect. Assessment of changes in diversity can be evaluated at a variety of levels, from an entire landscape to a patch of one plant community at a particular seral stage. All diversity measures are only a first step toward understanding why a change has taken place in a community or why a community is structured the way it is. Only by studying the requirements of individual species and the relationships of those species to their environments can we begin to decipher how community structure relates to functional attributes.

Similarity Indices

One of the major attributes of diversity indices is that they are not affected by changes in species composition. For example, if an exotic species replaced a native species after a disturbance (like a housing development), most measures of diversity would not reflect this change if the abundance of the native and exotic species were nearly equal. Differences in species composition are probably important in any comparison of community structure, but they are critical when one is assessing the effect of a disturbance on a biotic community. Similarity indices can be used to quantify the similarity or dissimilarity of the species composition, and sometimes the relative abundances of species, in two communities.

Researchers have proposed numerous similarity indices, which can be divided into two groups. The first group includes indices that reflect only the presence or absence of species. The second group includes indices that incorporate actual or relative abundances of individual species. The relative degree of similarity of two communities (that is, samples) is assessed by comparing the index with its maximum theoretical value (the value of the index when the two communities are identical) or with its maximum expected value (the value of the index for two random samples from the same fauna) (Wolda 1981).

Huhta (1979) compared the responses of a variety of indices (table 4.8) with changes in assemblages of species of spiders and beetles with forest succession. Wolda (1981) examined the effects of sample size and community diversity on the expected maximum values of a similar group of indices. The following information is a summary of their findings.

Binary indices. Mountford's (1962) index (table 4.8, no. 2) was sensitive only when the communities being compared had a high percentage of species in common (Huhta 1979; Wolda 1981), whereas Yule's (1912) index of association (no. 4) responded in the opposite manner. The other four binary indices (nos. 1, 3, 5, 6) showed the expected successional trends, but Huhta (1979) noted that the index of Baroni-Urbani and Buser (1976; no. 6), although considered mathematically faultless, required knowledge of the number of potential species present in an area. This information is often not known, and thus the index is likely to be impractical for many applications.

Wolda (1981) found that all binary indices produced higher values of similarity as the size of the sample (from either community) increased. The relationship between sample size and the index value grew stronger as the number of species increased in the communities being compared. These relationships again call for careful consideration of sample size when generating and interpreting indices associated with community structure.

Indices including measures of relative abundance. Morista's (1959) index (no. 11), the measure of Odum (no. 8), the squared Euclidian distance index (no. 10), and, to some extent, the percent similarity index (no. 7) all responded relatively erratically to the data sets analyzed by Huhta. Transforming the data with a logarithmic transformation before calculation improved the performance of most of these indices by reducing the influence of changes in the dominant species (Huhta 1979). The performance of the Canberra metric (no. 9) was unaffected by transformation, but it was not as sensitive as other measures (such as nos. 7 and 8 with transformed data) at low index values. Other measures were, in general, acceptable, with Kendall's *tau* coefficient (no. 16) and the coefficient of correlation with transformed data (no. 14) producing good results.

Huhta (1979) concluded that six of the indices he examined were appropriate for use. His tentative order of preference was Kendall's *tau*, the Odum measure with transformed data, percent similarity with trans-

Table 4.8. Some proposed indices for quantifying the similarity of biotic communities

Number	Name	Formula[a]
1	Sorenson's quotient of similarity (1948)	$QS = \frac{2c}{a+b}$
2	Mountford's index (1962)	$I = \frac{2c}{2ab - (a+b)c}$
3	Dice's association index (1945)	$M = \frac{c}{\min(a, b)}$
4	Yule's coefficient of association (1912)	$Y = \frac{ad - bc}{ad + bc}$
5	Correlation coefficient χ^2 method	$\Phi = \frac{\lvert ad - bc \rvert - N/2}{[(a+b)(c+d)(a+c)(b+d)]^{1/2}}$
6	Baroni-Urbani and Buser's index (1976)	$S = \frac{(ad)^{1/2} + a}{(ad)^{1/2} + (a+b+c)}$
7	Percent similarity	$PS = \Sigma \min(p_{1i} p_{2i})$
8	Odum's similarity measure (1950)	$O = \frac{2 \Sigma \min(n_{1i} n_{2i})}{N_1 + N_2}$
9	"Canberra metric" of Clifford and Stephenson (1975)	$CM = 1/k \Sigma \frac{\lvert n_{1i} - n_{2i} \rvert}{(n_{1i} + n_{2i})}$
10	Squared Euclidean distance	$D^2 = \Sigma(p_{1i} - p_{2i})^2$
11	Morista's index (1959)	$C_\lambda = \frac{2 \Sigma n_{1i} n_{2i}}{(\lambda_1 + \lambda_2) N_1 N_2}$ where $\lambda_1 = \frac{\Sigma n_{1i}(n_{1i} - 1)}{N(N_1 - 1)}$ $\lambda_2 = \frac{\Sigma n_{2i}(n_{2i} - 1)}{N_2(N_2 - 1)}$
12	Grassle and Smith measure (1976)	(See Grassle and Smith 1976)

Table 4.8. (*Continued*)

Number	Name	Formula[a]
13	Horn's measure of overlap (1966)	$R_\circ = \frac{H'_{max} - H'_{obs}}{H'_{max} - H'_{min}}$ where $H'_{max} = [H'(N_1) + H'(N_2)]/2 + \ln_2$ $H'_{min} = \frac{[N_1 H'(N_1) + N_2 H'(N_2)]}{(N_1 + N_2)}$, and $H'_{obs} = H'(N_1 + N_2)$
14	Coefficient of correlation	(See a statistics text)
15	Spearman's rank correlation coefficient	(See a statistics text)
16	Kendall's *tau* coefficient (adopted by Ghent 1963)	(See a statistics text)

Source: Huhta (1979).

[a]Description of symbols:

1. a = number of species in sample A.
   b = number of species in sample B.
   c = number of species common to samples A and B.
2. Symbols as in 1.
3. Symbols as in 1.
4. a = number of species common to samples in A and B.
   b = number of species present in sample A only.
   c = number of species present in sample B only.
   d = number of species absent from both samples (requires that the potential number of species be known).
5. Symbols as in 4.
6. Symbols as in 4.
7. p_{1i} = the proportion of the ith species in sample A.
   p_{2i} = the proportion of the ith species in sample B.
8. N_1 and N_2 are the total number of individuals, and n_{1i} and n_{2i}, the number of individuals in the ith species in two samples, respectively.
9. k = the number of species excluding those absent from both samples. Other symbols as in 8.
10. Symbols as in 7.
11. Symbols as in 8.
13. H' = the Shannon-Weaver function.

formed data, the coefficient of correlation with transformed data, the Canberra metric, and Horn's (1966) measure of overlap (no. 13). Interestingly, Morista's index, the only similarity measure that Wolda (1981) found to be independent of sample size and level of diversity, was not on Huhta's preferred list. Results such as these suggest that there is probably not one "best" index that will meet the needs of every data set. The index of choice will depend on the question being asked and the kinds of data available.

Related Measures

Understanding the structure of communities requires an understanding of the habitat requirements of and interspecific relationships among the species that make up the community. It should not be surprising, then, that investigations of niches often go hand in hand with, or follow, studies of species diversity (Pianka 1974). Diverse communities may be supported by a large quantity or a great variety of resources (many niches), include many species that are highly specialized (small niches), or both. Efforts to quantify the dimensionality of niches, and thus partially explain the diversity of communities, have focused on two measures: niche breadth and niche overlap.

Niche breadth is a measure of the diversity of resources used by an animal. Niche breadth can be used, for example, to assess the relative degree of specialization of species in a community or to assess the diversity of resources used by species under different environmental conditions. (Niche breadth should, in theory, be wider in the absence of competitors and predators—fundamental niche—than in their presence—realized niche; Hutchinson 1959.) Indices used to measure niche breadth, often the same as those used to measure species diversity, include the Shannon-Weaver information statistic (H') and the reciprocal of Simpson's diversity index (Levins 1968; Petraitis 1979). When using these indices as measures of niche breadth, different resources used by the animal (such as different food items) are considered classes, and some measure of the use of these resources (such as percentage of a food item in the diet) indicates the relative importance of each class.

Niche overlap is the degree to which two species use the same resources. Measures of overlap are important in ecology because of the basic idea that "there is an upper limit on how similar the ecologies of two species can be and still allow coexistence" (Pianka 1981:171). Researchers can use overlap indices to evaluate concepts such as the principle of competitive exclusion, limiting similarity, and species packing. Measures of niche overlap also have been used as indices of competition. Considerable overlap between two species has been interpreted as indicating intense competition. It can be easily argued, however, that the reverse is true. Considerable overlap in the use of a given resource may indicate an abundance of that resource and little competition, whereas little overlap may suggest that past competition for the resource has driven the two species to partition the resource between them. Clearly, the use of overlap measures as indices of competition requires caution. Methods of calculating multivariate measures of the niche are also available (see Chapter 7).

The indices of community similarity are among the functions that have been used as measures of niche overlap. A justifiable criticism of some of these indices is that they do not measure the abundance of the resources used by the species being compared or the degree of preference that each species has for the resource in question. Hurlbert (1978), Petraitis (1979), and Lawlor (1980) discussed other indices of overlap that accounted for resource abundance and species preferences. Not surprisingly, data used to generate measures of niche breadth and overlap are often the same as those required to identify the habitat of an animal. For example, identification of preferred foods and the sites used when foraging, raising young, and avoiding predators are critical considerations for designers of habitat management and research activities. Knowledge of how the presence of competitors and predators alters habitat use also is important, but it is not common. Ludwig and Reynolds (1988) provide a thorough treatment of indices used in ecological research.

Sample Size and Observer Bias

Regardless of the species and/or sexes selected for study, the types of variables chosen and the methods used to sample them, and any other aspect of habitat analysis one cares to mention, all is for naught if an insufficient number of observations are made. To most, this must seem an obvious statement. Unfortunately, very little attention has been paid in the scientific literature to this fundamental question of statistical analyses. We are not sure why and can only advance the suggestion that researchers have generally tried to collect the largest samples their budgets would allow. All of us have heard this, and several of us must admit to following this "strategy." But if we are to construct meaningful models of wildlife-habitat relationships—models that can withstand the test of scientific, and possibly even legal, review—we must address this basic question of experimental design. If researchers must limit sampling because of budgetary constraints, then they would be better advised to limit the scope of the study (for example, the number of species sampled) and produce one good model, rather than many weak ones.

Most general statistics texts discuss the determination of sample size necessary to perform certain tasks (to detect differences, to fit a certain confidence interval). These methods vary based on the type of data being collected (such as continuous or binomial). Steel and Torrie (1960:154) outlined calculations of the number of replicates required in an experimental design. The requisite number depends on an estimation of the population variance, the size of the difference to be detected, the assurance with which one desires to detect the difference (Type II error), the level of significance to be used in the actual experiment (Type I error), and whether a one- or two-tailed test is required. Steel and Torrie (1960:154, equation 8.17) and Sokal and Rohlf (1981:263, box 9.13) gave formulas and examples of how to conduct such an analysis; we will not repeat their material here. It is important to note, however, that these procedures require an initial estimate of the population variance. In practice, one can use an estimate of the population variance obtained from a previous study. This estimate is best obtained from a preliminary analysis of the population of interest. Steel and Torrie also gave a formula for and an example of how to calculate the necessary sample size to estimate a mean by a confidence interval guaranteed to be no longer than a certain prescribed length. The procedure—called Stein's two-stage sample—involves taking a sample, estimating the variance, and then computing the total number of observations necessary (stage 1). The additional observations are then obtained, and a new mean based on all observations is computed (stage 2).

Another method of obtaining an estimate of necessary sample size involves continually evaluating the data as increasing sample sizes are obtained. This technique is especially useful when one has little or no idea of the true population variance, is gathering data on numerous variables—variables that will be analyzed multivariately—for numerous species, and is concerned that the associated variance may vary widely in time and space. Further, continual evaluation of data gives the added benefit of providing in-depth familiarity with the data set, familiarity that may identify unexpected or interesting facts about the species in time to modify the sampling protocol. Mueller-Dombois and Ellenberg (1974) suggested plotting standard error as a function of the number of samples and denoting an adequate sample size when the curve decreased to less than a 5–10 percent gain in precision for a 10 percent increase in sample size.

Douglas H. Johnson, a statistician for the U.S. Fish and Wildlife Service, gave interesting and insightful commentary on the handling of questions of sample size. He wrote:

> As a consulting statistician, the first question I am usually asked is how many observations are necessary. I generally respond with one of three numbers: 1, 50, or "great gobs." The former answer, 1, is given occasionally when I believe the hypothesis is

> stated incorrectly, or in too much generality; a single observation is likely to refute it. The latter answer, great gobs, is given when no hypothesis is at hand, or, if there is one, it is so slippery as to evade capture and possible rejection. Neither answer satisfies the biologist, of course, but either serves as a necessary prelude to sitting down and thinking about an appropriate hypothesis. The middle answer, 50, or possibly 100 or 10 or 30, is given when the hypothesis is well stated and the experimental procedures thoughtfully described. (D. H. Johnson 1981:55)

Block, With, and Morrison (1987) conducted a study that included analysis of both necessary sample size and interobserver variability in making estimates of mean values of vegetation attributes used by birds in the mixed-conifer zone of the western Sierra Nevada. The locations of foraging birds served as the center of 0.04 ha circular plots in which structural and floristic habitat characteristics were quantified. An interesting twist to the study was the comparison of two common methods for measuring vegetation: visual (ocular) estimates and more objective techniques involving the use of standard measuring devices (diameter tape for tree diameters, clinometer for tree heights, measuring tape for distances). For each variable the researchers randomly selected, with replacement, ten subsamples of five, ten, fifteen, twenty, thirty, forty, fifty, and sixty plots. They then calculated the mean for each sample and then the mean of the means for each subset. Stability of the estimates was defined as the point where the estimates of the mean remained within one standard deviation of subsequent estimates and there was little variation in the magnitude of the confidence interval (CI). The stability of CIs, however, does not necessarily indicate that population means are stable. The technique helps to identify adequate sample sizes for stable CIs.

Block, With, and Morrison found that the type of variable collected, whether estimated or measured, affected the sample size required. Minimum sample size requirements for measured variables ranged from 20 (average tree dbh) to 50 (average tree height), and those for estimated variables ranged from 20 (average tree dbh) to more than 75 (average height to first live branch). In addition, the final point estimates of variables frequently differed between estimates and measurements (fig. 4.12).

Gotfryd and Hansell (1985) noted that, in both laboratory and field research, sampling often involved multiple workers. Researchers routinely lump data collected by several observers into one set for analysis, ignoring potential hazards of variability between observers. As Gotfryd and Hansell stated, "Ignoring observer-based variability may lead to conclusions being precariously balanced on artifacts, spurious relations, or irreproducible

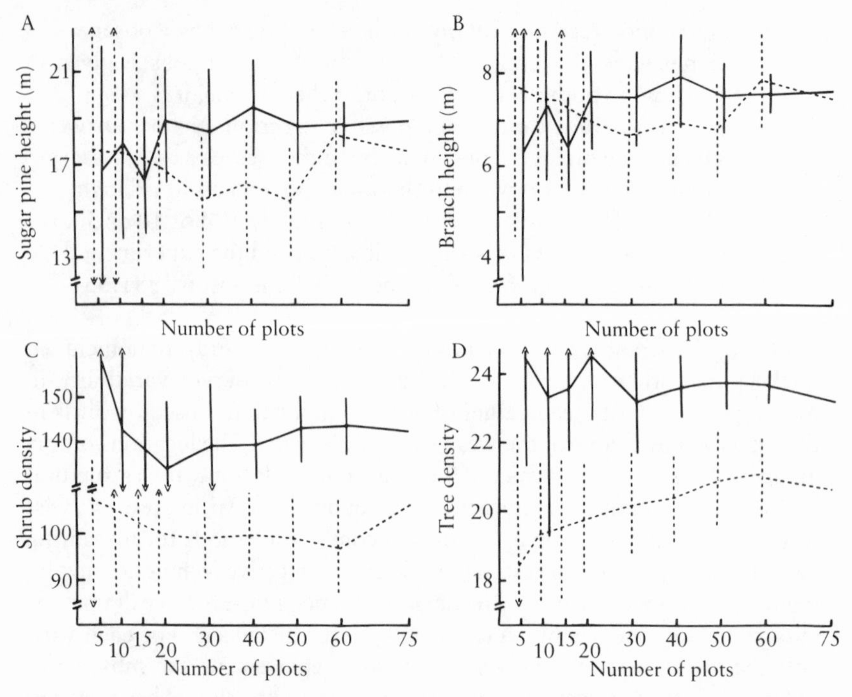

Figure 4.12. Influence of sample size on the stability of estimates (dashed horizontal lines) and measurements (solid horizontal lines) of bird-habitat characteristics. Dashed or solid vertical lines represent 1 SD from point estimates for estimates and measurements, respectively. Variables shown are *A*, average height of sugar pine; *B*, average height to the first live branch of sugar pine; *C*, average number of shrubs within sample plots; and *D*, average number of trees within sample plots. (From Block, With, and Morrison 1987, fig. 2.)

trends" (1985:224). They thus designed a study to assess the sensitivity of a standard ecological study method to observer bias. They followed the methods detailed by James and Shugart (1970).

Gotfryd and Hansell used four observers to independently sample eight plots located within a patch of oak-maple forest near Toronto, Canada. The variable set they used is given in table 4.9. They found that observers differed significantly in their measurements on eighteen of the twenty vegetation variables (table 4.10). Transformations of variables and nonparametric techniques did not improve results. As these authors recognized, their analysis addressed precision of estimates between observers; no measure of the accuracy of their results was available. As we saw in the previous example from Block, With, and Morrison (1987), however, measured and

Table 4.9. Vegetational habitat variables and their mnemonics, used by Gotfryd and Hansell

Mnemonic	Variable
TRSP	Number of tree species
SHRSP	Number of shrub species
*SDEN	Density of woody stems < 7.6 cm diameter at breast height (DBH)
*CC	Canopy cover
*GC	Ground cover
BAA	Basal area (BA) of trees 7.6 to 15.2 cm DBH
BAB	BA of trees 15.2 to 23 cm DBH
*BA1	BAA + BAB
*BA2	BA of trees 23 to 53 cm DBH
*BA3	BA of trees > 53 cm DBH
CH1	Maximum canopy height
CH4	Maximum canopy height in quadrant having lowest canopy
*CHAV	Average of canopy height maxima by quadrant
*CHRNG	Ch1 − CH4
*CHCV	CHRNG/CHAV
DTR1	Distance to nearest tree > 15.2 cm DBH
DTR4	Maximum of by-quadrant nearest tree distances
*DAV	Average of by-quadrant nearest tree distances
*DRNG	DTR1 − DTR4
*DCV	DRNG/DAV

Source: Gotfryd and Hansell 1985, table 1.
*Variables used in multivariate analyses.

ocular estimates may yield quite different results, even after stability in the value of the estimates has been attained.

Estimation of distances to objects is especially critical for some bird census techniques, such as the variable circular plot (Reynolds, Scott, and Naussbaum 1980) and the line-transect (Emlen 1971) methods. In a field experiment, Scott, Ramsey, and Kepler (1981) found that observers, after training, could estimate the distance of singing birds with a 10–15 percent accuracy. They noted, however, that the errors involved in nonexperimental conditions were unknown; the possibility exists that observers, knowing that they were not being "tested," would not make measurements as carefully as they would under experimental conditions. Errors in, and variance of, distance estimates increase with increasing distance (B. G. Marcot, unpublished data).

Researchers can reduce interobserver variability by following closely a

Table 4.10. Summary of univariate statistical analysis on vegetation variables used by Gotfryd and Hansell

	ANOVA effect[b]			
Variable[a]	Obs	Site	Rep	Obs × Site
TRSP	**	**	ns	**
SHRSP	**	**	ns	ns
SDEN	**	**	ns	**
CC	**	**	ns	**
GC	**	**	*	ns
BAA	**	**	**	*
BAB	**	**	ns	**
BA1	**	**	*	**
BA2	**	**	ns	**
BA3	**	**	ns	**
CH1	**	**	*	**
CH4	**	**	ns	**
CHAV	**	**	ns	*
CHRNG	**	**	ns	*
CHCV	**	**	ns	**
DTR1	**	**	ns	**
DTR4	ns	**	ns	ns
DAV	**	**	ns	*
DRNG	ns	**	ns	*
DCV	*	**	ns	*
Totals				
**	17	20	1	11
*	1	0	3	6
ns	2	0	16	3

Source: Gotfryd and Hansell 1985, table 2.
[a]Mnemonics given in table 4.9.
[b]These four columns report on the significance of observers, sites, replications, and the observer by site interaction.
*$P < 0.05$.
**$P < 0.01$.

set of well-defined criteria for selecting and training observers. Although designed for bird censusing, the steps outlined by Kepler and Scott (1981) could be applied generally to most types of sampling: Carefully screen applicants initially to eliminate the more obvious visual, aural, and psychological factors that increase observer variability. Organize a rigorous observer training program, which will further reduce inherent variation but will not eliminate it. Such training must place heavy emphasis on distance estimation. Also, periodic training sessions with observers should be conducted so they can "recalibrate" their estimates with some known values (see Block, With, and Morrison 1987).

Block, With, and Morrison (1987) used several univariate and multivariate procedures to test for differences among three observers in estimating plant physiognomy and floristics. They found that the ocular estimates by the three observers differed for thirty-one of the forty-nine variables they measured. Perhaps the most confounding aspect of using multiple observers they found was the unpredictable nature of the variation among observers. Multiple comparisons of estimates for the thirty-one significant variables resulted in all possible combinations of observers. Thus, when samples from different observers are pooled, sampling bias can increase. In Chapter 7 we discuss the influence of sample size and interobserver variability on results of multivariate analyses of habitat and behavioral data.

Plant ecologists have long recognized differences among data collection techniques (see Cooper 1957; Lindsey, Barton, and Miles 1958; Schultz, Gibbens, and Debano 1961; Cook and Stubbendieck 1986; Hatton, West, and Johnson 1986; Ludwig and Reynolds 1988). Block, With, and Morrison (1987) asserted that the objective of using measurement techniques is to reduce the magnitude of human error, and although these methods contain their own biases, it is expected that they are more accurate (more precise or less biased) than ocular estimates (see Schultz, Gibbens, and Debano 1961). Although the cost of measuring plant physiognomy and floristics in an adequate number of plots is great in both time and money, the ramifications of not following a rigorous sampling design are severe.

When to Measure

Throughout this book we refer to the importance, in studies of animal behavior, of assessing the influence of temporal variations in resource use. Too many researchers ignore temporal variation in habitat use, in at least two main ways: While often recognizing that temporal variations do occur, they sample from such a narrow time period (for example, by collecting all samples during the breeding period) that their resulting habitat relationships are so specific to time and location that they apply only minimally to other situations. Second, they may sample from across some broad time period (like the "summer season") and then "average-out" the relationships over the period. Given adequate samples, we are not faulting the first strategy with regard to the detail of the data collected; applicability is the question. Our preferred study design would be repetition of the first strategy for every appropriate biological period (prebreeding, breeding,

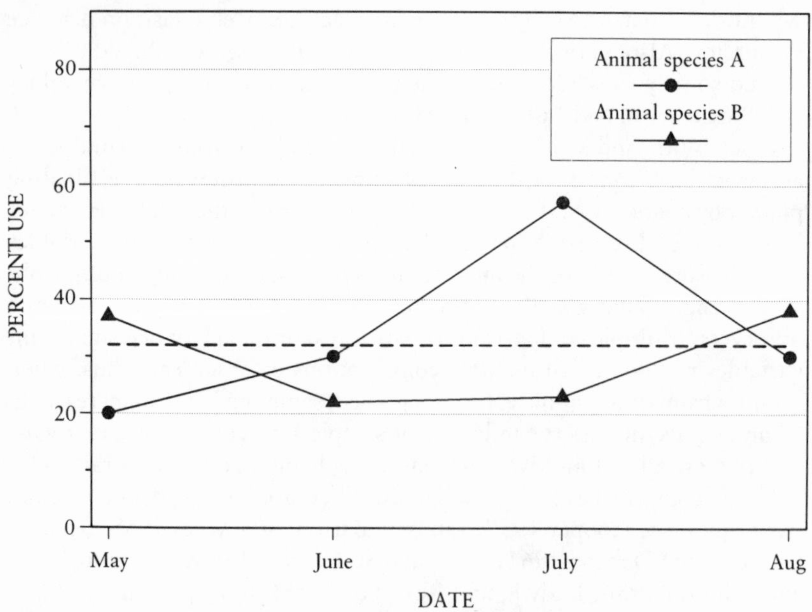

Figure 4.13. Use of a species of tree by two hypothetical animal species during "summer." Dashed horizontal line represents approximate average of use values for both species (calculated separately) across time.

early postbreeding). For the second strategy, it is easy to conceptualize what "averaging" may mean to resulting descriptions of animal-habitat relationships: in figure 4.13 we see that the average use of a species of tree over time by two hypothetical species is not, in fact, a close approximation of the actual use. The average indicates, however, that the species use this species of tree basically identically. Referring back to our sections on whom to measure, it is easy to see how combining data for males and females would further complicate the matter. Here again, we see that the finer we stratify our sampling, the greater the number of levels of resolution we have available for subsequent analysis. The question concerning how finely to measure is a study-specific problem, and further, within a particular study, the level of resolution will probably vary with time (for example, between seasons). It is essential that a program of continuous evaluation of the relationship between sample size and data output be conducted during studies. This allows collection of adequate samples to answer questions. Further, it also guards against oversampling.

Sampling within Seasons

Most studies of wildlife-habitat relationships address the concerns we have raised here about assessing temporal variation in activity patterns by animals. Often, however, the "solution" is to concentrate sampling within a single "season." But as we saw in figure 4.14, the definition of a season is often a question of user-defined scale. For example, analyses of bird populations are usually conducted in the "breeding season" or the "winter season." The breeding season, however, contains subperiods that differ markedly in activity patterns and, in many cases, types of habitat used.

Thus, we see that we should apply a good deal of caution when reviewing studies that average across "seasons." We must recognize that the statistical average of what an animal does may not give a valid indication of the animal's specific requirements for resources. A good deal of the variations (error) evident in our models of habitat relationships likely involve the sampling problems noted here.

Sampling between Seasons

No distinct separation actually exists between sampling wildlife populations within or between seasons (intra- or interseasonal, respectively). True, the breeding season can be defined to end when parental care of young ends or when pair bonds break, or at some other time based on species-specific behavior. (Note that when a season begins and ends is species-specific even within closely related taxa; care should be taken in designing community-level studies that broadly define seasons. The amount of care given young by parents varies widely among species. For example, in some species the male may have no role in the care of the young, often moving into areas and habitat different from the female's.) We separate intra- and interseasonal concerns here to emphasize the problem of sampling scale. Because of limited funding, availability of personnel, and other problems related to management of studies, most researchers are forced into either "breeding season" or "winter season" work; most study has been concentrated during the summer (or breeding) periods. But an increasing body of literature indicates that the survival of the animal may depend on the nonbreeding period. Intuitively, we would expect that the fall and winter periods, when populations are at their greatest numbers (because of offspring), resources are declining (trees and insects are going dormant), and the weather is becoming more harsh, would be the most difficult times for an animal. Further, because of these changes in environmental conditions, the habitat itself changes. Large-scale migration by numerous species of animals provides a good indication of the lengths to which animals must go to find favorable living conditions.

The importance of nonbreeding *periods* (our use of the plural of *period* is not trivial) to animals was popularized in a 1972 monograph by Stephen D. Fretwell. In *Populations in a Seasonal Environment*, Fretwell examined models of population size as they related to variations in the environment across seasons. Further, he analyzed the roles of both winter and breeding habitats in the regulation of animal numbers: Does winter habitat regulate breeding densities, or the converse? He concluded that most species in a seasonal environment were limited primarily by winter seasons. Fretwell also developed various models that related the carrying capacity of the habitat to population regulation within and between seasons.

This is not to say that Fretwell was the first worker to recognize the importance of the nonbreeding periods to animals. For some reason that has not been fully rationalized or debated, however, researchers have conducted most studies of wildlife-habitat relationships during "breeding" or "summer" periods. We can best evaluate this propensity for the breeding or summer periods as follows: Funding often limits our choice of the length of the study, and because of often-limited access to study areas during winter, workers choose to concentrate sampling efforts during summer periods. Some have also discussed the possibility that if one supplies the habitat needs of a species evident during summer, then the needs during the nonbreeding periods will be accounted for by default (see Verner 1983, 1984). If an animal does not change its geographic range and pattern of habitat use and no externally generated changes are applied to the area, then such a rationale is probably appropriate, assuming that the habitat is actually supplying the necessary requirements for survival. But if an animal changes its pattern of habitat use or a change occurs in the habitat itself, then a summer-based study cannot be used with any confidence to predict the subsequent response of the animal.

Long-Term Temporal Changes

We have discussed the need to evaluate habitat relationships between seasons. We have seen several examples of "summer versus winter" evaluations of habitat use by animals. The magnitude of the temporal variation, however, is a question of scale, and the scale can run across as well as within years. Thus, relationships developed over a period of a few years may not hold for a different period of years. In Chapter 2 we discussed the ideas of Wiens (1977) regarding bottlenecks of extremes in resource availability or other environmental conditions that drastically influence a population. An animal's reaction to such conditions will provide the most important data with which one can evaluate habitat relationships. We realize, of course, that funding for such long-term studies is extremely limited. Several studies have been and continue to be conducted, and their results are important to consider in evaluating the results of shorter-term studies.

Although we have mentioned the important study of "bottlenecks" or "ecological crunches" by Wiens (1977), conducted in the arid shrub-steppe environment of the southwestern United States, we should note that Wiens's views are not universally accepted. In a longer-term study in the northeastern United States, Holmes, Sherry, and Sturges (1986) proposed an opposite model. They asserted that food resources may be chronically low for several to many successive years and may become superabundant only during relatively brief outbreaks of prey. In periods between outbreaks, food may frequently and perhaps regularly limit reproductive output of populations in northern temperate forests, and effects of predators, weather, and other mortality factors contribute to highly variable reproductive success.

The results of Holmes, Sherry, and Sturges led them to propose a general model of bird community structure in northern temperate forests. Their model likely applies, in general, to other taxa and other geographic areas. Its implications for evaluation of habitat relationships are clearly stated:

> Each bird species seems to respond to its environment in a unique way, as determined by its evolutionary history (*cf.* Sabo and Holmes 1983) and by a combination of different processes and factors that act on its populations. Some factors take place on a local scale (e.g., vegetation structure, food abundance), while others operate over larger areas (e.g., some weather effects and conditions in areas where species winter). Additionally, some factors represent selective forces that act over evolutionary time on the morphology, behavior, and life history patterns of species, whereas others produce responses in ecological time. (Holmes, Sherry, and Sturges 1986:215)

This model, general as it is, recognizes the importance of temporal and spatial scale as well as evolutionary history and ecological events.

Thus, we want to emphasize the importance of examining the results of any study of habitat relationships from the perspective of known environmental extremes that could have caused the results obtained. For example, it is well known that weather varies (droughts, El Niño years), that insects go through periodic irruptions, and that certain species exhibit marked cycles in their numbers (see Keith 1963). We can also search the historical literature—natural history notes of early scientists, for example—and examine the current distribution and abundance of animals in light of past information. True, most such comparisons, based on mostly qualitative information, must be speculative, but they can offer useful insights into changes in distribution and the biology of the species. Nest and egg records, stored and cataloged by museums such as the Western Foundation of Vertebrate Zoology and the U.S. Museum of Natural History, are a rich source of mostly unpublished data on nesting phenology and geographic range. Forced to conduct mostly short-term studies, we must review our results by considering the overall, long-term environment confronting the animal.

Can short-term studies, then, tell much about animal communities? Wiens (1984:202–203) thought not, answering ". . . a short-term approach is likely to produce incomplete or incorrect perceptions of a complex reality. . . . perhaps obtaining results that are superficial and quite possibly incorrect." Holmes, Sherry, and Sturges (1986) concluded that, to increase our understanding of population regulation and community structure, we must conduct long-term monitoring, along with more critical examination of the relative importance of factors influencing spatial and temporal variability in distribution and abundance. To meet these goals, they recommended intensive studies on single plots over a series of years, on replicated plots within the same habitat, and on replicate plots across habitats. In addition, they hoped for an increase in experimental manipulations of habitat, food resources, competitors, predators, and other parameters, coupled with demographic studies that considered year-round mortality. These and other recommendations made in this chapter are indeed "tall orders," but available evidence clearly indicates the necessity of their implementation if the development of wildlife-habitat relationships is to advance.

Where to Measure

In discussing when to measure habitat, we referred to seasonal changes in habitat occupancy by a wide range of vertebrates. These included changes in habitat use by animals permanently residing in an area, as well as by animals that undertook migratory movements. Obviously, animals that migrate change habitats, even though habitats occupied seasonally may be superficially similar to each other. Indeed, the reason for an animal's leaving an area in the first place usually involves the impending inappropriateness of the area for continued survival; insect prey disappears, snow covers remaining forage, weather conditions become severe. Thus, where we measure wildlife habitat largely depends upon when the measuring is done. In this section we will mostly restrict our discussion to core versus edge areas of habitat and differential use of habitat by different segments of a species' overall population.

The decision on where one measures habitat is also closely tied to the objectives of the study. In an environmental impact assessment, where the area of interest may be quite small, then deciding where to sample is simpler than in a study in which a model describing the overall habitat use of an animal is desired (see Stewart-Oaten, Murdoch, and Parker 1986). Of course, the selection of a proper nontreated, or control, site in an impact assessment is critical. Although we are primarily interested in the development of widely applicable models, the reader should bear in mind that our comments also apply to selection of control sites in impact assessment.

As Collins (1983) observed, researchers have confined most analyses of avian habitat to single-site comparisons. Although community-level comparisons between sites are common (see Karr 1971; Rabenold 1978; Rotenberry and Wiens 1980), workers based such studies upon a defined habitat structure so that they could compare bird species diversity, niche parameters, and behavior within a similar type of habitat.

Using the wide-ranging black-throated green warbler (*Dendroica virens*), Collins (1983) gathered data in Maine, New Hampshire, New York, and Minnesota to determine whether the structure of the habitat used by this warbler varied at several points in its range and, if differences did exist, which structural variables differed between sites. Collins noted that knowledge of the range of habitat structure may facilitate development of regional models predicting the future extent of a species' habitat, clearly an important consideration in habitat management. Collins found that six habitat variables (canopy height, percentage of conifers, and four diameter classes of trees) differed significantly between the five study sites (see fig. 7.3, p. 291). Collins concluded that the habitat of the black-throated green warbler differed in three-dimensional structure as well as plant species composition. Further, sites with similar plant community types did not necessarily have the same habitat structure. Thus, habitat descriptions based only upon dominant plant species are not always useful, because habitat structure may be dissimilar at different sites within any widely distributed vegetation type. These findings have obvious implications for the development of habitat models based on broad and generalized categories of habitat types.

We have mentioned the difficulty involved in selecting the proper location to measure habitat. Habitat varies within and between seasons, between sexes, and even within an individual's territory depending upon the focal point of the measurement (such as perch or nest site). Social interactions within a population further complicate the issue.

Van Horne (1983) discussed these problems in an important paper on relationships between habitat quality and animal density. She noted that social interactions could prevent subdominant animals from entering what is actually high-quality habitat, thus suppressing reproduction in the high-quality habitat. The surplus individuals may then collect in habitat "sinks" where densities may fluctuate widely (see also Lidicker 1975). Animals in the low-quality sinks survive and/or reproduce poorly. Thus, in a good year, the source population may produce an excess of juveniles that will emigrate and increase to high densities in the sinks. Because juveniles are subdominant, no social interaction prevents high densities in the sink habitats, in contrast with the adult-dominated high-quality or source habitats. Densities in the lower-quality habitats may thus actually be greater than those in the high-quality habitats. Thus, modelers who seek to develop relationships between density (or some index thereof) and habitat or who use a focal-animal sampling scheme may base their conclusions on animals that occupy inferior habitat (that is, habitat incapable of supporting the population). Van Horne discussed numerous examples in which the location and density of animals gave misleading indications of habitat quality (such as States 1976).

Van Horne concluded her review by noting that "intensive multi-annual demographic study of a single species over the range of habitats being measured is needed to interpret the broader surveys. Without attention to demography, even multi-annual surveys or censuses will not necessarily be sufficient to distinguish 'source' from 'sink' habitats" (1983:900). Thus, we must qualify our previous remarks regarding the usefulness of long-term studies in habitat modeling: even such long-term efforts are complicated by the location within a population's range from which samples are drawn.

Literature Cited

Anderson, S. H., and H. H. Shugart, Jr. 1974. Habitat selection of breeding birds in an east Tennessee deciduous forest. *Ecology* 55:828–37.

Baroni-Urbani, C., and M. W. Buser. 1976. Similarity of binary data. *Systematic Zoology* 25:251–59.

Block, W. M., L. A. Brennan, and R. J. Gutierrez. 1987. Evaluation of guild-indicator species for use in resource management. *Environmental Management* 11:265–69.

Block, W. M., K. A. With, and M. L. Morrison. 1987. On measuring bird habitat: Influence of observer variability and sample size. *Condor* 89:241–51.

Bock, C. E., and J. F. Lynch. 1970. Breeding bird populations of burned and unburned conifer forest in the Sierra Nevada. *Condor* 72:182–89.

Bonham, C. D. 1989. *Measurements for terrestrial vegetation*. New York: John Wiley and Sons.

Bull, E. L., and E. C. Meslow. 1977. Habitat requirements of the pileated woodpecker in northeastern Oregon. *Journal of Forestry* 75:335–37.

Bury, R. B., and M. G. Raphael. 1983. Inventory methods for amphibians and reptiles. In *Renewable resource inventories for monitoring changes and trends*, ed. J. F. Bell, and T. Atterbury, 416–19. Corvallis: Oregon State Univ., College of Forestry.

Carbon, B. A., G. A. Bartle, and A. M. Murray. 1979. A method for visual estimation of leaf area. *Forest Science* 25:53–58.
Catana, A. J., Jr. 1963. The wandering quarter method of estimating population density. *Ecology* 44:349–60.
Clifford, H. T., and W. Stephenson. 1975. *An introduction to numerical classification*. New York: Academic Press.
Cody, M. L. 1985. An introduction to habitat selection in birds. In *Habitat selection in birds*, ed. M. L. Cody, 3–56. New York: Academic Press.
Collins, S. L. 1981. A comparison of nest-site and perch-site vegetation structure for seven species of warblers. *Wilson Bulletin* 93:542–47.
Collins, S. L. 1983. Geographic variation in habitat structure of the black-throated green warbler (*Dendroica virens*). *Auk* 100:382–89.
Collins, S. L., F. C. James, and P. G. Risser. 1982. Habitat relationships of wood warblers (Parulidae) in north central Minnesota. *Oikos* 39:50–58.
Conner, R. N. 1981. Seasonal changes in woodpecker foraging patterns. *Auk* 98: 562–70.
Cook, C. W., and J. Stubbendieck, eds. 1986. *Range research: Basic problems and techniques*. Denver: Society for Range Management.
Cook, J. G., and L. L. Irwin. 1985. Validation and modification of a habitat suitability model for pronghorns. *Wildlife Society Bulletin* 13:440–48.
Cooper, C. F. 1957. The variable plot method for estimating shrub density. *Journal of Range Management* 10:111–15.
Cooperrider, A. Y., R. J. Boyd, and H. R. Stuart, eds. 1986. *Inventory and monitoring of wildlife habitat*. USDI Bureau of Land Management Service Center, Denver, Colo.
Cottam, G., and J. T. Curtis. 1956. The use of distance measures in phytosociological sampling. *Ecology* 37:451–60.
Daubenmire, R. 1968. *Plant communities: A textbook of plant synecology*. New York: Harper and Row.
Davis, D. E., 1982. *Handbook of census methods for terrestrial vertebrates*. Boca Raton, Fla.: CRC Press.
Dice, L. R. 1945. Measures of the amount of ecologic association between spaces. *Ecology* 26:297–302.
Dueser, R. D., and H. H. Shugart, Jr. 1978. Microhabitats in a forest-floor small mammal fauna. *Ecology* 59:89-98.
Eddleman, W. R., K. E. Evans, and W. H. Elder. 1980. Habitat characteristics and management of Swainson's warbler in southern Illinois. *Wildlife Society Bulletin* 8:228–33.
Elton, C. 1927. *Animal ecology*. London: Sidgewick and Jackson.
Emlen, J. T. 1971. Population densities of birds derived from transect counts. *Auk* 88:323–42.
Engstrom, R. T., R. L. Crawford, and W. W. Baker. 1984. Breeding bird populations in relation to changing forest structure following fire exclusion: A 15-year study. *Wilson Bulletin* 96:437–50.
Errington, P. 1963. *Muskrat populations*. Ames: Iowa State Univ. Press.
Francis, C. M., and F. Cooke. 1986. Differential timing of spring migration in wood warblers (Parulinae). *Auk* 103:548–56.
Fretwell, S. D. 1972. *Populations in a seasonal environment*. Princeton: Princeton Univ. Press.
Fry, M. E., R. J. Risser, H. A. Stubbs, and J. P. Leighton. 1986. Species selection

for habitat-evaluation procedures. In *Wildlife 2000: Modeling habitat relationships of terrestrial vertebrates*, ed. J. Verner, M. L. Morrison, and C. J. Ralph, 105–8. Madison: Univ. of Wisconsin Press.

Ghent, A. W. 1963. Kendall's "tau" coefficient as an index of similarity in comparisons of plant or animal communities. *Canadian Entomologist* 95: 568–75.

Gotfryd, A., and R. I. C. Hansell. 1985. The impact of observer bias on multivariate analyses of vegetation structure. *Oikos* 45:223–34.

Grassle, J. F., and W. Smith. 1976. A similarity measure sensitive to the contribution of rare species and its use in investigation of variation in marine benthic communities. *Oecologia* 25:13–22.

Green, R. H. 1971. A multivariate statistical approach to the Hutchinsonian niche: Bivalve molluscs of central Canada. *Ecology* 52:543–56.

Green, R. H. 1979. *Sampling design and statistical methods for environmental biologists*. New York: John Wiley and Sons.

Greig-Smith, P. 1983. *Quantitative plant ecology*. 3d ed. Berkeley: Univ. of California Press.

Grinnell, J. 1928. Presence and absence of animals. *University of California Chronicles* 30:429–50.

Hatton, T. J., N. E. West, and P. S. Johnson. 1986. Relationships of error associated with ocular estimation and actual total cover. *Journal of Range Management* 39:91–92.

Hilden, O. 1965. Habitat selection in birds. *Annales Zoologici Fennici* 2:53–75.

Holmes, R. T. 1981. Theoretical aspects of habitat use by birds. In *The use of multivariate statistics in studies of wildlife habitat*, ed. D. E. Capen, 33–37. USDA Forest Service General Technical Report RM–87.

Holmes, R. T. 1986. Foraging patterns of forest birds: Male-female differences. *Wilson Bulletin* 98: 196–213.

Holmes, R. T., R. E. Bonney, Jr., and S. W. Pacala. 1979. Guild structure of the Hubbard Brook bird community: A multivariate approach. *Ecology* 60: 512–20.

Holmes, R. T., T. W. Sherry, and F. W. Sturges. 1986. Bird community dynamics in a temperate deciduous forest: Long-term trends at Hubbard Brook. *Ecological Monographs* 56:201–220.

Horn, H. S. 1966. Measurement of overlap in comparative ecological studies. *American Naturalist* 100:419–24.

Huhta, V. 1979. Evaluation of different similarity indices as measures of succession in arthropod communities of the forest floor after clear-cutting. *Oecologia* 41:11–23.

Hurlbert, S. H. 1971. The nonconcept of species diversity: A critique and alternative parameters. *Ecology* 52:577–86.

Hurlbert, S. H. 1978. The measurement of niche overlap and some relatives. *Ecology* 59:67–77.

Hutchinson, G. E. 1957. Concluding remarks. *Cold Spring Harbor Symposium on Quantitative Biology* 22:415–27.

Hutchinson, G. E. 1959. Homage to Santa Rosalia; or, Why are there so many kinds of animals? *American Naturalist* 93:145–59.

Hutchinson, G. E. 1978. *An introduction to population ecology*. New Haven: Yale Univ. Press.

Hutto, R. L. 1981. Temporal patterns of foraging activity in some wood warblers in relation to the availability of insect prey. *Behavioral Ecology and Sociobiology* 9:195–98.

Hutto, R. L. 1985. Habitat selection by nonbreeding, migratory land birds. In *Habitat selection in birds*, ed. M. L. Cody, 455–76. New York: Academic Press.

Inger, R. F., and R. K. Colwell. 1977. Organization of contiguous communities of amphibians and reptiles in Thailand. *Ecological Monographs* 47:229–53.

Irwin, L. L., and J. G. Cook. 1985. Determining appropriate variables for a habitat suitability model for pronghorns. *Wildlife Society Bulletin* 13:434–40.

Jaksic, F. M. 1981. Abuse and misuse of the term "guild" in ecological studies. *Oikos* 37:397–400.

James, F. C. 1971. Ordinations of habitat relationships among breeding birds. *Wilson Bulletin* 83:215–36.

James, F. C., and H. H. Shugart, Jr. 1970. A quantitative method of habitat description. *Audubon Field Notes* 24:727–36.

James, F. C., and N. D. Wamer. 1982. Relationships between temperate forest bird communities and vegetation structure. *Ecology* 63:159–71.

Johnson, D. H. 1981. How to measure habitat: A statistical perspective. In *The use of multivariate statistics in studies of wildlife habitat*, ed. D. E. Capen, 53–57. USDA Forest Service General Technical Report RM–87.

Johnson, R. A. 1981. Application of the guild concept to environmental impact analysis of terrestrial vegetation. *Journal of Environmental Management* 13:205–222.

Karr, J. A. 1971. Structure of avian communities in selected Panama and Illinois habitats. *Ecological Monographs* 41:207–33.

Karr, J. R., and R. R. Roth. 1971. Vegetation structure and avian diversity in several New World areas. *American Naturalist* 105:423–35.

Keith, L. B. 1963. *Wildlife's ten-year cycle*. Madison: Univ. of Wisconsin Press.

Kepler, C. B., and J. M. Scott. 1981. Reducing count variability by training observers. *Studies in Avian Biology* 6:366–71.

Kitchings, J. T., and D. J. Levy. 1981. Habitat patterns in a small mammal community. *Journal of Mammalogy* 62:814–20.

Krebs, C. J. 1978. Ecology: The experimental analysis of distribution and abundance. New York: Harper and Row.

Lancia, R. A., S. D. Miller, D. A. Adams, and D. W. Hazel. 1982. Validating habitat quality assessment: An example. *Transactions of the North American Wildlife and Natural Resources Conference* 47:96–110.

Landres, P. B., J. Verner, and J. W. Thomas. 1988. Ecological uses of vertebrate indicator species: A critique. *Conservation Biology* 2:316–28.

Lawlor, L. 1980. Overlap, similarity, and competition coefficients. *Ecology* 61: 245–51.

Levins, R. 1968. *Evolution in changing environments*. Princeton: Princeton Univ. Press.

Lidicker, W. Z., Jr. 1975. The role of dispersal in the demography of small mammals. In *Small mammals: Their productivity and population dynamics*, ed. F. B. Golley, K. Petusewicz, and L. Ryszkowski, 103–28. New York: Cambridge Univ. Press.

Lindsey, A. A., J. D. Barton, and S. R. Miles. 1958. Field efficiencies of forest sampling methods. *Ecology* 39:428–44.

Ludwig, J. A., and J. F. Reynolds. 1988. *Statistical ecology: A primer on methods and computing*. New York: John Wiley and Sons.

MacArthur, R. H. 1957. On the relative abundance of bird species. *Proceedings of the National Academy of Sciences* 43:293–95.

MacArthur, R. H. 1958. Population ecology of some warblers in northeastern coniferous forests. *Ecology* 39:599–619.

MacArthur, R. H., and J. W. MacArthur. 1961. On bird species diversity. *Ecology* 42:594–98.

MacArthur, R. H., J. W. MacArthur, and J. Preer. 1962. On bird species diversity. Part 2, Prediction of bird census from habitat measurements. *American Naturalist* 96:167–74.

McCullough, D. R., D. H. Hirth, and S. J. Newhouse. 1989. Resource partitioning between sexes in white-tailed deer. *Journal of Wildlife Management* 53: 277–83.

MacMahon, J. A., D. J. Schimpf, D. C. Andersen, K. G. Smith, and R. L. Bayn, Jr. 1981. An organism-centered approach to some community and ecosystem concepts. *Journal of Theoretical Biology* 88:287–307.

MacNally, R. C. 1983. On assessing the significance of interspecific competition to guild structure. *Ecology* 64:1646–52.

Mannan, R. W., M. L. Morrison, and E. C. Meslow. 1984. Comment: The use of guilds in forest bird management. *Wildlife Society Bulletin* 12:426–30.

Mawson, J. C., J. W. Thomas, and R. M. DeGraaf. 1976. *Program HTVOL: The determination of tree crown volume by layers*. USDA Forest Service Research Paper NE–354.

Merriam, C. H. 1898. *Life zones and crop zones*. USDA Biological Services Bulletin no. 10.

Miller, F. L., and A. Gunn, eds. 1981. *Symposium on census and inventory methods for populations and habitats*. Northwest Section, The Wildlife Society. Forest, Wildlife and Range Experiment Station Contribution no. 217, Univ. of Idaho, Moscow.

Morista, M. 1959. Measuring of interspecific association and similarity between communities. *Memoirs of the Faculty of Science*, Kyushu Univ., ser. E, 3: 65–80.

Morrison, M. L. 1982. The structure of western warbler assemblages: Ecomorphological analysis of the black-throated gray and hermit warblers. *Auk* 99: 503–13.

Morrison, M. L. 1984. Influence of sample size and sampling design on analysis of avian foraging behavior. *Condor* 86:146–50.

Morrison, M. L. 1986. Bird populations as indicators of environmental change. In *Current ornithology*. Vol. 3, ed. R. F. Johnston, 429–51. New York: Plenum Press.

Morrison, M. L., D. L. Dahlsten, S. M. Tait, R. C. Heald, K. A. Milne, and D. L. Rowney. 1989. *Bird foraging on incense-cedar and incense-cedar scale during winter in California*. USDA Forest Service Research Paper PSW–195.

Morrison, M. L., and E. C. Meslow. 1983. Bird community structure on early-growth clearcuts in western Oregon. *American Midland Naturalist* 110: 129–37.

Morrison, M. L., I. C. Timossi, and K. A. With. 1987. Development and testing of linear regression models predicting bird-habitat relationships. *Journal of Wildlife Management* 51:247–53.

Morrison, M. L., I. C. Timossi, K. A. With, and P. N. Manley. 1985. Use of tree species by forest birds during winter and summer. *Journal of Wildlife Management* 49:1098–102.

Morrison, M. L., and K. A. With. 1987. Interseasonal and intersexual resource partitioning in hairy and white-headed woodpeckers. *Auk* 104:225–33.

Morse, D. H. 1968. A quantitative study of foraging of male and female spruce-woods warblers. *Ecology* 49:779–84.

Mountford, M. D. 1962. An index of similarity and its application to classificatory problems. In *Progress in soil zoology*, ed. P. W. Murphy, 43–50. London: Butterworths.

Mueller-Dombois, D., and H. Ellenberg. 1974. *Aims and methods of vegetation ecology*. New York: John Wiley and Sons.

Murie, O. J. 1951. The elk of North America. Harrisburg, Pa.: Stackpole Co.; Washington, D.C.: Wildlife Management Institute.

Myers, J. P. 1981. A test of three hypotheses for latitudinal segregation of the sexes in wintering birds. *Canadian Journal of Zoology* 59:1527–34.

Nice, M. M. 1937. Studies in the life history of the song sparrow, part 1. *Transactions of the Linnaean Society of New York* no. 4.

Nice, M. M. 1943. Studies in the life history of the song sparrow, part 2. *Transactions of the Linnaean Society of New York* no. 6.

Nolan, Val, Jr. 1978. The ecology and behavior of the prairie warbler, *Dendroica discolor*. *Ornithological Monographs* no. 26.

Noon, B. R. 1981. Techniques for sampling avian habitats. In *The use of multivariate statistics in studies of wildlife habitat*, ed. D. E. Capen, 42–52. USDA Forest Service General Technical Report RM–87.

Odum, E. P. 1950. Bird population of the Highlands (North Carolina) Plateau in relation to plant succession and avian invasion. *Ecology* 31:587–605.

Odum, E. P., ed. 1971. *Fundamentals of ecology*. 3d ed. Philadelphia: Saunders.

Patrick, R., M. H. Hohm, and J. H. Wallace. 1954. A new method for determining the pattern of the diatom flora. *Notulae Naturae* 259:1–12.

Patrick, R., and D. Strawbridge. 1963. Variation in the structure of the natural diatom communities. *American Naturalist* 98:51–57.

Patton, D. R. 1987. Is the use of "management indicator species" feasible? *Western Journal of Applied Forestry* 2:33–34.

Peet, R. K. 1974. The measurement of species diversity. *Annual Review of Ecology and Systematics* 5:285–307.

Penfound, W. T., and E. L. Rice. 1957. An evaluation of the arms-length rectangle method in forest sampling. *Ecology* 38:660–61.

Petraitis, P. S. 1979. Likelihood measures of niche breadth and overlap. *Ecology* 60:703–10.

Phillips, E. A. 1959. *Methods of vegetation study*. New York: Holt, Rinehart and Winston.

Pianka, E. R. 1974. *Evolutionary ecology*. New York: Harper and Row.

Pianka, E. R. 1980. Guild structure in desert lizards. *Oikos* 35:194–201.

Pianka, E. R. 1981. Competition and niche theory. In *Theoretical ecology*, ed. R. M. May, 167–96. Sunderland, Mass.: Sinauer Associates.

Preston, F. W. 1948. The commonness, and rarity, of species. *Ecology* 29:254–83.

Rabenold, K. N. 1978. Foraging strategies, diversity, and seasonality in bird communities of Appalachian spruce-fir forests. *Ecological Monographs* 48: 397–424.

Ralph, C. J., and J. M. Scott, eds. 1981. Estimating numbers of terrestrial birds. *Studies in Avian Biology* no. 6.

Rand, A. L. 1952. Secondary sexual characters and ecological competition. *Fieldiana-Zoology* 34:65–70.

Raphael, M. G., and B. G. Marcot. 1986. Validation of a wildlife-habitat-relationships model: Vertebrates in a Douglas-fir sere. In *Wildlife 2000: Modeling habitat relationships of terrestrial vertebrates*, ed. J. Verner, M. L. Morrison, and C. J. Ralph, 129–38. Madison: Univ. of Wisconsin Press.

Reinert, H. K. 1984. Habitat separation between sympatric snake populations. *Ecology* 65:478–86.

Reynolds, R. T., J. M. Scott, and R. A. Naussbaum. 1980. A variable circular-plot method for estimating bird numbers. *Condor* 82:309–13.

Rice, E. L., and W. T. Penfound. 1955. An evaluation of the variable-radius and paired-tree methods in the blackjack-post oak forest. *Ecology* 36:315–20.

Robins, J. D. 1971. Differential niche utilization in a grassland sparrow. *Ecology* 52:1065–70.

Root, R. B. 1967. The niche exploitation patterns of the blue-gray gnatcatcher. *Ecological Monographs* 37:317–50.

Rosenzweig, M. L., and J. Winakur. 1969. Population ecology of desert rodent communities: Habitats and environmental complexity. *Ecology* 50:558–72.

Rotenberry, J. T. 1985. The role of habitat in avian community composition: Physiognomy or floristics? *Oecologia* 67:213–17.

Rotenberry, J. T., and J. A. Wiens. 1980. Habitat structure, patchiness, and avian communities in North American steppe vegetation: A multivariate analysis. *Ecology* 61:1228–50.

Roth, R. R. 1976. Spatial heterogeneity and bird species diversity. *Ecology* 57: 773–82.

Sabo, S. R., and R. T. Holmes. 1983. Foraging niches and the structure of forest bird communities in contrasting montane habitats. *Condor* 85:121–38.

Salt, G. W. 1953. An ecologic analysis of three California avifaunas. *Condor* 55: 258–73.

Salt, G. W. 1957. An analysis of avifaunas in the Teton Mountains and Jackson Hole, Wyoming. *Condor* 59:373–93.

Schemnitz, S. D. 1980. *Wildlife management techniques manual*. 4th ed. Washington, D.C.: Wildlife Society.

Schoener, T. W. 1974. Resource partitioning in ecological communities. *Science* 185:27–39.

Schultz, A. M., R. P. Gibbens, and L. Debano. 1961. Artificial populations for teaching and testing range techniques. *Journal of Range Management* 14: 236–42.

Scott, J. M., F. L. Ramsey, and C. B. Kepler. 1981. Distance estimation as a variable in estimating bird numbers from vocalizations. *Studies in Avian Biology* 6:334–40.

Selander, R. K. 1966. Sexual dimorphism and differential niche utilization in birds. *Condor* 68:113–51.

Severinghaus, W. D. 1981. Guild theory development as a mechanism for assessing environmental impact. *Environmental Management* 5:187–90.
Shannon, C. E., and W. Weaver. 1949. *The mathematical theory of communication*. Urbana: Univ. of Illinois Press.
Short, H. L., and K. P. Burnham. 1982. Technique for structuring wildlife guilds to evaluate impacts on wildlife communities. USDI Fish and Wildlife Service Special Scientific Report, Wildl. no. 244.
Simpson, E. H. 1949. Measurement of diversity. *Nature* 163:688.
Sokal, R. R., and F. J. Rohlf. 1981. *Biometry*. 2d ed. San Francisco: W. H. Freeman.
Sorenson, T. A. 1948. A method of establishing groups of equal amplitude in plant sociology based on similarity of species content, and its application to analyses of the vegetation on Danish commons. *Kongelige Danske Videnskabernes Selskab Biologiske Skrifter* 56:1–34.
States, J. B. 1976. Local adaptations in chipmunk (*Eutamias amoenus*) populations and evolutionary potential at species borders. *Ecological Monographs* 46:221–56.
Steel, R. G. D., and J. H. Torrie. 1960. *Principles and procedures of statistics*. New York: McGraw-Hill.
Stewart-Oaten, A., W. M. Murdoch, and K. R. Parker. 1986. Environmental impact assessment: "Pseudoreplication" in time. *Ecology* 67:929–40.
Sturman, W. A. 1968. Description and analysis of breeding habitats of the chickadees, *Parus atricapillus* and *P. rufescens*. *Ecology* 49:418–31.
Szaro, R. C. 1986. Guild management: An evaluation of avian guilds as a predictive tool. *Environmental Management* 10:681–88.
Thomas, J. W., ed. 1979. *Wildlife habitats in managed forests: The Blue Mountains of Oregon and Washington*. USDA Forest Service Agricultural Handbook no. 553.
Tramer, E. J. 1969. Bird species diversity: Components of Shannon's formula. *Ecology* 50:927–29.
U.S. Fish and Wildlife Service. 1980. *Habitat evaluation procedures (HEP)*. Ecological Services Manual no. 102. Washington, D.C.: GPO.
Van Deusen, P. C., and G. S. Biging. 1984. *Crown volume and dimension models for mixed conifers of the Sierra Nevada*. Northern California Forest Yield Cooperative Research Note no. 9, Dept. of Forestry and Resource Management, Univ. of California, Berkeley.
Van Horne, B. 1982. Niches of adult and juvenile deer mice (*Peromyscus maniculatus*) in seral stages of coniferous forests. *Ecology* 63:992–1003.
Van Horne, B. 1983. Density as a misleading indicator of habitat quality. *Journal of Wildlife Management* 47:893–901.
Verner, J. 1983. An integrated system for monitoring wildlife on the Sierra National Forest. *Transactions of the North American Wildlife and Natural Resources Conference* 48:355–66.
Verner, J. 1984. The guild concept applied to management of bird populations. *Environmental Management* 8:1–14.
Wagner, J. L. 1981. Seasonal change in guild structure: Oak woodland insectivorous birds. *Ecology* 62:973–81.
Whitmore, R. C. 1981. Applied aspects of choosing variables in studies of bird habitats. In *The use of multivariate statistics in studies of wildlife habitat*, ed. D. E. Capen, 38–41. USDA Forest Service General Technical Report RM–87.

Wiens, J. A. 1969. An approach to the study of ecological relationships among terrestrial birds. *Ornithological Monographs* no. 8.

Wiens, J. A. 1977. On competition and variable environments. *American Scientist* 65:590–97.

Wiens, J. A. 1984. The place of long-term studies in ornithology. *Auk* 101:202–203.

Wiens, J. A. 1989. Spatial scaling in ecology. *Functional Ecology* 3:385–97.

Williamson, P. 1971. Feeding ecology of the red-eyed vireo (*Vireo olivaceus*) and associated foliage-gleaning birds. *Ecological Monographs* 41:129–52.

Willson, M. F. 1974. Avian community organization and habitat structure. *Ecology* 55:1017–29.

Wolda, H. 1981. Similarity indices, sample size and diversity. *Oecologia* 50:296–302.

Yule, G. U. 1912. On the methods of measuring association between two attributes. *Journal of the Royal Statistical Society* 75:579–642.

5 Foraging Behavior and Habitat Resources

Introduction

"Each of the many foods potentially available to an animal has a different nutritional value, a different pattern of spacing and abundance, and a different cost of capture and processing. . . . Because the animal has only limited amounts of time and energy, its choices among different potential foods may critically affect its survival and reproductive success" (Morse 1980:51). This quotation aptly summarizes the critical role that the acquisition of food plays in the life of any animal. The foraging animal usually has numerous food resources in its environment. Of interest to habitat ecologists are what foods animals consume, when they consume them, how much food they take, and how they allocate time in searching for foods. The foraging animal does not exist in a vacuum, of course: various competitors—competing either directly for the food item(s) or indirectly for space (territory) or other resources—and predators impact the time, place, and even method of foraging used by the animal.

Many studies of habitat use have focused on foraging behavior, because of its importance in understanding the energy balance of the animal. Analyses of foraging behavior show how animals *actively* use their habitats. In contrast, simple presence-absence or relative abundance studies rely on indirect correlations between the animal and general habitat features to infer selection of habitat components, substrates, or food resources.

Although they are possibly more informative than purely correlative studies, observations of foraging behavior by themselves are fraught with uncertainties. What is the quality (energy content, digestibility, micronutrients) of the food consumed? What food does the animal prefer, how does it use the food, and what food is available? How is the consumed food related to the survival and, ultimately, to the reproductive success of the individual? Foraging studies involve far more complicated issues than simply confirming that an animal is foraging in a particular location in a particular fashion at a given time.

It should be evident, then, that the study of foraging behavior must begin with the development of a sound theoretical framework that elucidates how the animal perceives and then utilizes its environment. For example, what food does the animal perceive as available? Is the choice of food based on gross abundance of the item, its nutrient content, its distance from cover, or some combination of these or other factors? Determination of an animal's perception of its environment is difficult to achieve through simple behavioral observations. Thus, three general areas of study have attempted to provide insight into foraging behavior: development of theoretical models, laboratory experimentation, and field experimentation.

The development of foraging theory encompasses a wide array of concepts and models, including general prey, marginal value, and central-place foraging models. Researchers can evaluate the validity of predictions from these models through observational and experimental studies. Laboratory experimentation involves studies of food choice and competitor interactions. Field experimentation classically involves removal of both conspecifics and potential competitors or the removal or supplementation of food, in an attempt to determine the food-energy requirements of an animal and/or the incremental impact of food and competitors on the animal's behavior.

Regardless of the theoretical basis upon which we construct our foraging studies, we must employ proper methods of data collection and analysis. Such considerations include specific methods by which animals are visually observed and the biases associated with the methods, the proper application of statistical procedures, and the accumulation of adequate numbers of samples. Unfortunately, scant attention has historically been given to these concerns in studies of foraging behavior.

In this chapter we will first review the general theoretical framework upon which most foraging studies have been conducted. The principal methods used to observe foraging animals and to assess resource abundance and availability will be discussed. We will cover some of the more popular methods used to analyze foraging data, emphasizing problems inherent in their use. Detailed, taxon-specific treatments of foraging behavior (including diet analysis) are currently available (see Korschgen 1980; Riney 1982; Cooperrider, Boyd, and Stuart 1986; Morrison et al. 1990). A short section on energetics will briefly outline topics that could be included in more detailed studies of energy acquisition by animals. Finally, we discuss the interesting and ever-changing area of foraging theory. Here we emphasize drawing the theoretician and the empiricist toward a common understanding of each other's goals and methods. All too often these two groups appear to act as opposing armies, seemingly unwilling to work in a complementary fashion.

Theoretical Framework

Although this chapter deals with foraging behavior, the reader should remember that any study of wildlife habitat is essentially a study of animal behavior in the broadest sense. How an animal uses food resources is only one component of how it selects specific substrates, general habitat types, and geographic locations for a wide variety of biological needs. Selection of habitats, like that of food, reflects the evolutionary history of the species as well as the current ecological conditions of the environment and the effects of other animals. In the end, all behavior—whether for foraging, dispersal, migration, reproduction, or predator avoidance—is modified and guided by the selective advantage of increasing fitness of individuals and vitality of populations.

How an Animal Perceives Its Environment

The theoretical framework developed in Chapter 2 on the perception of habitat by an animal—its niche gestalt, stimulus summation, and the like—clearly applies to the ways in which an animal recognizes and differentiates food items. We can view the animal as perceiving its environment on a series of scales, from recognizing the general type of habitat (such as forest) to identifying specific features that make a particular location "suitable."

Animals are usually adapted to most efficiently exploit specific types of foraging substrates. A substrate is the specific location and surface at which the animal directs foraging. The animal must compare the risk associated with foraging in a particular manner with risks associated with other methods and/or other locations. The choice of foraging method and location is based not only on the number of prey present but also on the quality of that prey. That is, do numerous, low-quality items that are easy to obtain have a net energy benefit over scarce, high-quality items that are difficult to obtain? Schoener (1969) hypothesized that animals can achieve this net balance by one of two extreme strategies: energy maximization or time minimization. Energy maximizers try to obtain the greatest amount of energy possible within a given period of time. Time minimizers seek to minimize the time required to obtain a given amount of food. Time minimizers thus have more time available for other activities, such as grooming and parental care (see also Morse 1980:53–54). As Stephens and Krebs recognized, however, maximization of net energy is not necessarily a desirable goal. The confusion, they observed, concerns equating the net rate of energy consumption with the ratio of benefit to cost (or foraging "efficiency"). We repeat their example: "Maximizing efficiency . . . ignores the time required to harvest resources, and it fails to distinguish tiny gains made at a small cost and larger gains made at a larger cost: for example, 0.01 calories gained at a cost of 0.001 gives the same benefit/cost ratio as a gain of 10 calories costing 1. The 10-calorie alternative, however, yields 1000 times the net profit of the 0.01 alternative" (Stephens and Krebs 1986:8–9).

Optimal Foraging Theory

These and related considerations form the basis of a large area of scientific investigation collectively known as optimal foraging theory. The term *optimal foraging* is, however, a poor choice of words. Although it is reasonable to try to distinguish foraging models that use maximization, minimization, or stability arguments from those that do not, *optimal foraging* can be interpreted to imply some claim about the single best way to forage. Foraging theory per se makes no such claim (Stephens 1990). Stephens and Krebs' book (1986) provides excellent coverage of foraging theory. Here we will only examine these theories and their associated models as they relate to our descriptions of how animals might perceive their environment. Notions about perception lead, in turn, to our determination of how we should observe and record animals as they exploit food resources. Thus, we will explore how models of foraging behavior can help us design our studies.

In considering how animals determine whether or not to forage in an area, we have two rather distinct alternative views: Is the animal selecting among various prey distributed (in some fashion) throughout a generally suitable area, or is the animal distinguishing between various patches of prey? Foraging theorists thus distinguish between prey-choice models and patch-choice models. An animal can forage "optimally" within either of these two basic constructs. These models have clear implications for the design of foraging studies, both observational and experimental. That is, if prey are distributed patchily, then our assessment of prey abundance or availability must be of proper spatial and temporal scale to recognize these patches. An overall average of prey over a large area would fail to identify their patchy nature. Likewise, studies conducted at too small a scale would identify patches that the animal might not even recognize.

Both prey-choice and patch-choice models assume that a foraging animal sequentially searches for prey, encounters the prey, and then decides whether or not to consume (or to attempt to consume) the prey. As outlined by Stephens and Krebs (1986:13–14), however, the form of the decision taken by the animal upon encountering a prey item differs between these two models. In the prey-choice model an animal must decide whether it should take the item or continue searching for another item. One can state various rules upon which the animal could base its decision. By contrast, in the patch-choice model the animal decides how long it should forage in the patch encountered. Both models consider how an animal can best make these decisions with the goal of maximizing the long-term average rate of energy intake.

These models thus develop a general theoretical constraint upon which we can build our foraging studies. Clearly, the rate and method of searching, the frequency of encounter, and the type of attack used when the decision has been made to consume prey all tell us a great deal about the ecology of a species. Further, variations in search, encounter, and prey consumption may lend insight into the current physiological condition of an animal. Thus, measuring aspects of foraging rate, the frequency and types of encounter, and related aspects of foraging is important in understanding why animals occupy, or do not occupy, certain areas. Many of these aspects of foraging behavior will be discussed later in this chapter.

Foraging theory predicts, when the animal considers prey choice, that the decision to take a specific item will depend not only on its own abundance but also on the abundance of higher-ranking items. Food items are ranked by their ratio of food value to handling time (Morse 1980:54). Handling time is the amount of time necessary to pursue, capture, and consume the prey (Stephens and Krebs 1986:14). Thus, foraging theory suggests that we should be concerned not only with how an animal goes about foraging (that is, the search-encounter) but also with the types and relative ranking of abundance and quality of prey encountered by the animal. Interested readers would do well to peruse chapter 3 of Morse's 1980 text and especially Stephens and Krebs's 1986 monograph on foraging theory (and references cited by these authors).

Three terms often used in foraging theory are *use, selection,* and *requirement.* In this chapter we define *use* as a demonstrated presence of a particular prey item in an animal's diet. *Selection,* however, is use coupled with evidence that the frequency of occurrence in the diet (by number of prey items, biomass, foraging frequency, etc.) is significantly greater statistically than the frequency in the animal's environment. Selection is the parameter that is typically estimated by several indices. *Requirement* is the presence, or a particular minimal amount, of a food item that the animal must obtain so it can live or reproduce. Few studies of foraging behavior or diet have demonstrated requirements. Observational studies of diet and foraging behavior of animals in the wild usually cannot be designed so that the researcher can conclude whether an animal selects a food item because of a physiological or behavioral requirement. Observational studies generally lack control groups, and various food items in various combinations and densities are not randomly present.

Ultimately, most observational studies of wildlife foraging behavior cannot and should not address requirements. If many observational studies conducted across different time periods or locations consistently report selection of particular food items (as with selection of particular habitat types or components), however, then one can safely infer that the species is exhibiting a behavior of adaptive significance. This strongly implicates requirements (Ruggiero et al. 1988).

Measuring Behavior

The Process of Studying Foraging Behavior

Martin and Bateson (1986:11–15) outlined the general steps to follow in designing and implementing a study on behavior (fig. 5.1). Their outline applies to most areas of concern. We are specifically interested in the choice of variables, choice of recording methods, accumulation of sufficient sample sizes, and data analysis. More thorough overviews of research methods in behavioral studies have been detailed in a number of publications (see Altmann 1974; Kamil and Sargent 1981; Hazlett 1977; Colgan 1978; Lehner 1979; Martin and Bateson 1986; Kamil, Krebs, and Pulliam 1987; Morrison et al. 1990).

We can divide most behavioral observations into three general categories: structure, consequence, and relation (Martin and Bateson 1986: 38–39). Researchers must recognize these categories before defining specific variables to record; failure can lead to problems in the analysis and

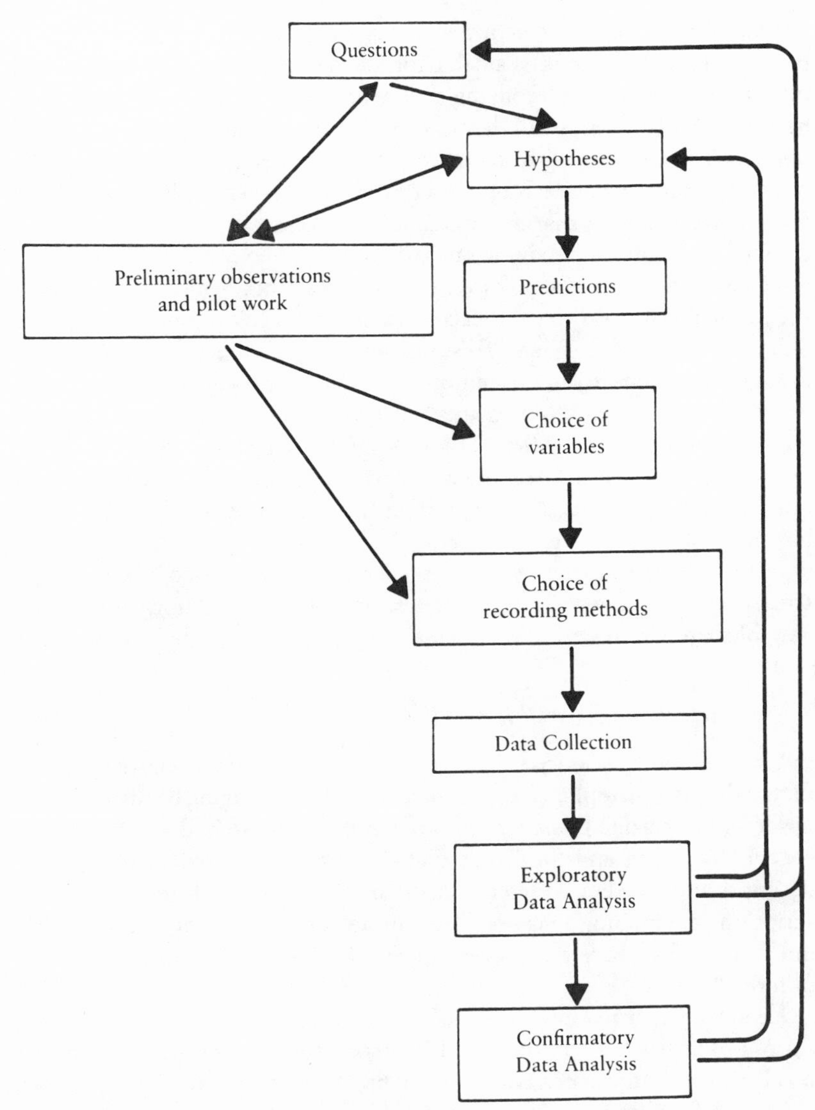

Figure 5.1. The processes involved in studying behavior. (From Martin and Bateson 1986, fig. 1.2.)

interpretation of data. *Structure* is the appearance, physical form, or temporal pattern of the animal's behavior, described in terms of posture and movements. For example, saying that an ungulate "reaches down and removes a leaf from a bush" describes the structure of the behavior. *Consequence,* in contrast with structure, describes the effects of the animal's behavior. Here, behavior is described without reference to how the effects are achieved. Saying that an ungulate is "browsing leaves" describes the consequences of the behavior but tells nothing about how the browsing took place (that is, the structure). *Relation* describes behavior in terms of the animal's spatial proximity to features of the environment, including other animals. Here, the animal's position or orientation relative to something is the salient feature. Relation is important in foraging studies, since it describes the orientation of an animal to its foraging substrate. For example, a bird can "hang" (its relation) from a perch while it "gleans" (takes prey, the consequence) from the underside of a leaf, or it can simply stand upright while it gleans. Clearly, noting the relation or orientation of the bird adds substantial information to the behavioral description; different types of prey could occupy different parts of a plant. Keep in mind the structure, consequence, and relation of an animal as we now discuss the variables used to describe foraging.

Selection of Variables

In Chapter 4 we discussed selection of variables in habitat analyses. Much of that discussion applies, in general, to analysis of foraging behavior (especially with regard to considerations of temporal and spatial variation, sex-specific behavior, and the like). Whereas habitat analysis focuses on the description of the environment surrounding the animal, however, the description of foraging behavior concentrates on the specific actions of an individual. Thus, here we discuss aspects of behavior specific to the taking of food.

Animals perform a myriad of activities during the course of a day. They sleep, groom, engage in intra- and interspecific interactions, feed, and so on. To quantify these behaviors in any practical way requires that we devise some form of record keeping. One could compile a running account of every action an individual makes. Clearly, such an effort would be extremely tedious and would require that the individual animal be observed constantly; observer fatigue would be great, and record keeping voluminous. To overcome such severe restrictions, researchers characterize virtually all behavioral data by categories. Regardless of the sophisticated techniques later used to analyze the data, all will be for naught if the worker did not properly determine the categories. Although no set rules exist for categorizing behavior, various guides are available. We are dealing primarily with foraging behavior in this chapter (and not complete "activity budgets" or "ethograms"), but the guides presented here apply to most aspects of animal behavior.

Slater (1978) outlined five basic guides for classifying behavioral acts. At the outset of a study, one must beware of simply massing behaviors because of presumed similarities of causation or function. This will enter observer-induced bias in the raw data. Slater's first guideline is that behavioral categories must be discrete. This means that acts within a category must have clear points of similarity that they do not share with acts outside the category. Second, behavioral categories must be homogeneous. All the acts in a category should be similar in form so that there is little danger of massing two different behaviors in the same category. Once acts are assigned to a category, there is little chance of separating them. Third, it is better to split than to lump. Within reason, two similar behaviors with possibly different consequences should be split into different categories; they can always be lumped later. Fourth, names of categories should avoid causal or functional implications. Applying meaning or function to categories often results in biased interpretation of data. For example, describing a call as a "hunger call" places anthropomorphic meaning onto an unknown behavior. Finally, the number of categories used must be manageable. Although we previously cautioned against using too few categories, too many categories will reduce an observer's reaction time and lower accuracy. Automated recording devices can help overcome this problem (see Martin and Bateson 1986, chap. 5; Raphael 1988). The work of Martin and Bateson (1986:39–42) and Miller (1988) should be consulted for further, related discussions of describing behavior. Remsen and Robinson (1990) also developed a detailed classification scheme for describing the foraging behavior of birds.

Recording Methods

Once researchers have developed behavioral categories, they must rigorously define methods for recording these behaviors. After all, well-developed categories mean little if methods used to fill them are poorly defined and biased. Martin and Bateson (1986, chap. 4) provided an excellent framework for recording behavioral observations by defining sets of sampling rules and recording rules (see fig. 5.2).

Sampling Rules

Before collecting data, the researcher must objectively decide which individuals to observe and when to observe them. In Chapter 4 we discussed

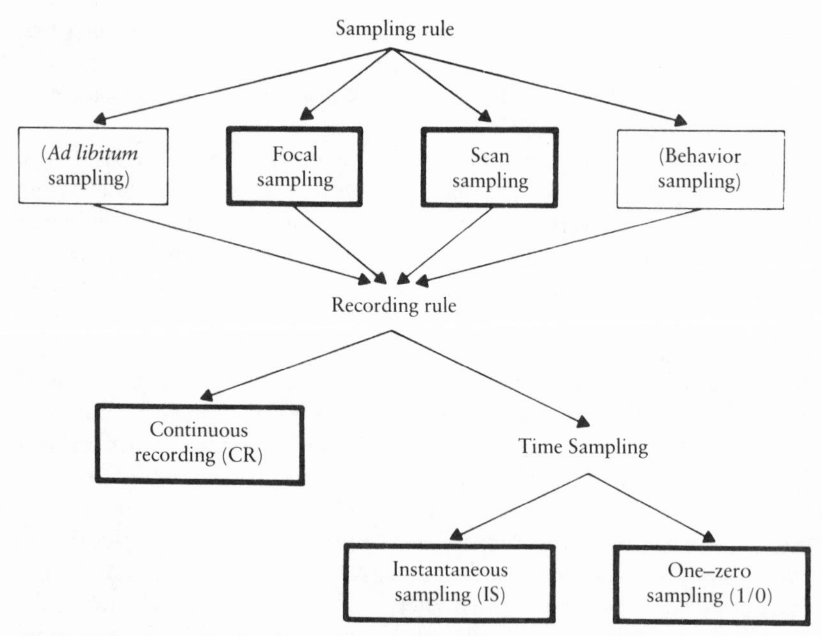

Figure 5.2. The hierarchy of sampling rules (determining who is watched and when) and recording rules (determining how behavior is recorded). (From Martin and Bateson 1986, fig. 4.1.)

methods for the selection of study species and the importance of temporal and spatial stratification of data. Here we concentrate on observations of individual animals within the chosen species (and population[s]) and within the context of proper spatial and temporal study design. The researcher should carefully read the work of Altmann (1974) before initiating a study of behavior.

Focal-Animal Sampling

Focal-animal sampling involves observing one individual, even if it is located within a group of animals, for a specified period of time. This method has the clear advantage of allowing adequate records to be collected on various classes (for example, sex and age class) of a species. Further, if animals are individually tagged or show unique markings, then researchers can accumulate information on individual variation as well as information on changes in behaviors over time.

A problem with focal sampling involves conspicuousness of individual animals. By focusing on individuals, researchers may produce records of behavior that vary considerably in length. For example, foraging birds are obvious when they are near branch tips, but they often disappear within the canopy. Thus, samples may be short and biased toward those individuals which are most obvious (see Slater 1978). Unfortunately, no sampling method employing only an observer's eyesight can overcome this problem. The advent of miniaturized radio transmitters (that is, radio telemetry) promises to solve this problem at least partially. Although one still cannot see specifically what the out-of-sight bird is doing, some transmitters have mercury switches or vital rate sensors to detect body position, heart rate, and other activities. Event recorders currently available for larger animals (such as many mammals and larger birds) send a signal whenever the animal lowers its head, presumably to forage. Williams (1990) provided a review of the use of telemetry in studies of avian foraging behavior, and Kenward (1987) reviewed in detail techniques for radio-tagging wildlife.

Scan Sampling

Using scan sampling, researchers collect data on an entire group of animals (a flock, a herd) at regular intervals, recording the behavior of each individual. The behavior of each individual scanned is, of course, an instantaneous sample. Scan sampling usually restricts the observer to recording only a few or simple categories. Scan sampling takes from only a few seconds to several minutes, depending upon the amount of information being recorded and the size of the group.

Bias in scan sampling is influenced largely by the period of time that elapses between "scans." Rare or inconspicuous behaviors will likely be observed at a lower frequency than they actually occur. For example, most animals spend the majority of their foraging time searching for prey, while relatively little time is spent in the actual pursuit or consumption of food items. Harcourt and Stewart (1984) showed that studies using scan sampling of foraging gorillas in the wild resulted in underestimates of actual foraging time. Focal-animal sampling resulted in a more accurate estimate of actual feeding time because the researchers could follow individuals continuously (gorillas are less visible while feeding). Thus, scan sampling, like focal-animal sampling, can be affected by the conspicuousness of animals performing various activities.

Scan sampling can be combined with focal-animal sampling during the same observation session. Researchers record the behavior of individuals in detail, but at specific intervals they also scan sample the entire group for simpler behavioral categories. This technique is especially useful when the location and activities of group members are thought to influence specific individuals in the group. Workers have shown, for flocking birds and herding mammals, that the time an individual spends foraging is related to the number and proximity of group members: an individual must watch more vigilantly for predators as group size decreases (see Hamilton 1971; Pulliam 1973; Alexander 1974; Caraco, Martindale, and Whittham 1980).

Focal-animal and scan sampling are the most common methods used to observe foraging animals. Readers should consult the work of Altmann (1974) and Martin and Bateson (1986) for discussion of other recording methods.

Recording Rules

The means of actually recording behavior, after the sampling methods are determined, are called recording rules. Figure 5.2 depicts the relationship between sampling rules and recording rules. We can divide recording rules into two basic categories: continuous sampling and time sampling.

In continuous sampling, researchers record each occurrence of a behavioral act, as well as the time of the activity and pertinent environmental information. Continuous sampling gives true frequencies, latencies, and durations of behavior *if* an animal can, indeed, be watched continuously for a sufficient period of time. Obviously, termination of a recording session will usually result in the worker's underestimating durations of behavior. Continuous recording is necessary when the sequence of behaviors and the time reference of the behaviors are of interest. Continuous sampling is tedious, and in practice, only a few categories can be measured reliably.

In studies of foraging behavior, observers often record the behavior of an animal continuously for a specific period of time. After some time has elapsed, the worker observes either the same or a different individual for the same period of time. Within continuous observation periods, the duration of each behavior is timed. For example, Block (1990) watched foraging white-breasted nuthatches (*Sitta canadensis*) for ten to fifteen seconds at a time, noting the duration of time spent searching for and procuring prey. But because a foraging bird—especially one in dense vegetation—can seldom be continuously followed for more than a minute, accurate estimates of latency (the time from a specific event, like the beginning of the recording session or the presentation of a stimulus, to the onset of a behavior) and especially duration are difficult to obtain. Researchers use continuous sampling methods to avoid recording only the most conspicuous behavior, since an observer can follow an individual through a series of behaviors.

Time sampling involves sampling behavior periodically and can be divided into instantaneous sampling and one-zero sampling (Martin and Bateson 1986; see fig. 5.2). In time sampling, each observation session is divided into successive short periods of time called sample intervals.

Instantaneous sampling, or point sampling, refers to the instant (point) at which the activity of an individual is recorded. The data obtained depends largely upon the length of the sample interval. If the interval is long relative to the average duration of the behavior, then one can obtain a measure of the proportion (not frequency) of all sample points at which the behavior occurred. If the sample interval is short in relation to the average duration of the behavioral pattern, however, then instantaneous sampling can approximate the results of continuous sampling.

Instantaneous sampling is a common method used to record foraging behavior. Typically, the observer follows an animal while it remains in view or places a limit on the observation period, while recording the specific foraging act being performed at set intervals. For example, Morrison (1982) followed foraging warblers for up to ten minutes at a time, recording the activity of an individual every sixty seconds.

One-zero sampling is similar to instantaneous sampling in that the recording session is subdivided into short sample intervals. In one-zero sampling, however, the observer merely records whether or not a particular behavior occurred during the sample period, noting no information on the frequency or duration of the act. One-zero sampling, unlike instantaneous sampling, consistently overestimates duration, because the behavior is recorded as though it occurred throughout the sample interval when it need have occurred at only one brief instant. In fact, Altmann (1974) argued against the use of one-zero sampling in all applications. Because of the controversy over one-zero sampling, we advise against its use. For a more detailed discussion of the pros and cons of this method, see the works of Altmann (1974), Slater (1978), and Martin and Bateson (1986:56–63), and references cited therein. Unfortunately, few researchers have examined quantitatively the issue of length of sample interval.

Independence

During the design of behavioral sampling, the researcher must consider the independence of the samples collected. Failure to give sufficient attention to this topic, as well as to observer bias, could doom even an intensively conducted study to obscurity.

Most statistical tests carry the assumption that the data analyzed represent random samples from populations and that each datum is statistically independent (see Zar 1984). Thus, the number of *individuals* for which data are taken characterizes or defines one's total sample. Machlis, Dodd, and Fentress (1985) noted that the objective of research is to obtain measurements from an adequate number of individuals, not to obtain large samples of measurements. Machlis, Dodd, and Fentress called this the pooling fallacy. Unfortunately, studies of behavior are replete with examples of such pooling errors (see Hurlbert 1984).

The literature on foraging, likewise, is filled with similar examples. Researchers commonly collect a series of instantaneous records on a single individual and then consider each point sample or each sample interval as an independent datum. Following an individual animal and recording behavioral acts every ten seconds for, say, sixty seconds results in six samples. A proper procedure would be to average the six points for the individual, resulting in $n = 1$, where n is sample size. Another sample interval taken on the same individual a short while later would, likewise, not represent another independent sample. This second interval (and those subsequent) would require averaging with the first; n remains 1.

The problem, then, becomes one of determining when samples become independent. Martin and Bateson (1986:50) noted that samples "must be adequately spaced out over time." No one to our knowledge, however, has defined *adequate*. If a sample is taken on an individual at 10:00, when can a second independent sample be taken on this individual? Perhaps not at 10:10, for the animal might be foraging in a similar location if not in a similar manner and might be affected by the same set of biotic and abiotic conditions. Ideally, samples should be taken only once on an individual during a biologically relevant period of time. Biologically relevant periods are clearly recognizable, albeit observer-defined, periods of time, such as morning or evening and spring or summer. But even if all individuals are individually recognizable, as with color-banded birds or marked mammals, it would indeed take a Herculean effort to relocate adequate numbers of animals. Further, foraging studies are usually confined to specific geographic areas for purposes of logistics, available habitat, site-specific management interests, and the like. Thus, compromises between sample size and sample independence are clearly warranted.

Within-individual variation is an important, although seldom analyzed, aspect of behavioral research. Researchers have tended to treat populations as homogeneous units, thus largely ignoring individual differences in behavior. In fact, in a monograph on this subject, Lomnicki (1988) contended that further advances in population ecology will require consideration of individual differences, such as unequal access to resources. Thus, repeatedly collecting samples on a number of different individuals gives important ecological data.

This does not, however, solve the problem of overall sample size and independence. Because foraging studies, especially those involving birds, usually involve unmarked individuals within confined areas, we must define recording rules that seek to maximize independence of observations, while realizing that full independence has probably not been achieved. A few such "rules" can be defined as follows: Only one individual within a group (flock, herd) will be recorded per recording session. Observers will systematically cover the study area (which is as large as possible), seeking out new groups or new individuals (where solitary); that is, they will not repeatedly sample the same group. Recording sessions will be stratified by time periods, both within and between days. Recording sessions will be distributed throughout the identified biological period (such as the breeding season) to avoid grouping samples within short periods of time.

Thus, no clearly defined procedure—other than nonrepeated sampling of individuals—is available to guarantee perfect independence of data. If researchers make systematic efforts to maximize sampling independence, however, then reasonable biological interpretation of observed behavior will likely result.

Observer Bias

Observers usually bias behavioral studies; avoiding bias is virtually impossible. This bias takes various forms: the influence of the observer on an animal, observer consistency in recording data (intraobserver reliability), and consistency among observers in recording data (interobserver reliability). Although these biases are difficult to eliminate, it is critical that they be recognized and their influence systematically reduced.

Researchers conducting field studies of foraging animals should be aware that the presence and activities of observers might influence an animal's activities. Researchers sometimes offer that their presence "did not markedly alter an individual's behavior" or that they allowed a period of several minutes to elapse for the animal to resume "normal activities" before data collection began. We should remember, however, that wild animals are constantly vigilant for predators and competitors. The presence of an observer likely heightens this awareness. It is also highly likely that a foraging animal is fully aware of, and reacting to, the observer's presence well before the observer even notices the animal. Predator vigilance and group size influence foraging activities of many animals. Further, animals that appear to become habituated to the presence of an observer may have simply adopted a modified pattern of foraging that allows them to keep the observer under surveillance.

Intraobserver reliability is a measure of the ability of a specific observer to obtain the same data when measuring the same behavior on different occasions (see Martin and Bateson 1986:88–89). It thus measures the ability of an observer to be precise in his or her measurements. *Precision* describes the repeatability of a measurement and is not synonymous with accuracy. Because we seldom know what the actual behavioral pattern is—that is, the "true" or "correct" pattern—we can not directly measure an observer's accuracy. Assessing intraobserver reliability in field-based behavioral studies is difficult; animals seldom repeat their behavior in exactly the same fashion. One test is to videotape foraging animals and then repeatedly to present (in some random fashion) individual sequences to the observers. Researchers can use the results of these trials to estimate the degree of observer reliability. Martin and Bateson (1986:88–94) discussed some simple ways to evaluate statistically the results of such trials.

Interobserver reliability measures the ability of two or more observers to obtain the same results on the same occasion (Martin and Bateson 1986: 89). To what extent is interobserver reliability a problem in field studies? Unfortunately, there have been few evaluations of the magnitude of this problem in foraging studies. Ford, Bridges, and Noske (1990), for example, found that comparisons of foraging behaviors of individuals of the same species in different areas or years, recorded by different observers, needed to be treated cautiously. Problems were particularly evident when observers had not previously agreed on standard methods of observations or classification of terms. Differences in experience among observers apparently accounted for much of the interobserver variability noted.

In Chapter 4 we mentioned the problem of interobserver reliability in recording habitat data (Block, With, and Morrison 1987). Others have noted this problem in studies involving bird counts (see Ralph and Scott 1981). Intra- and interobserver reliabilities have been dramatically improved by careful, rigorous, and repeated training. Each new observer is taught how data should be recorded, initially working with others in the field, comparing data and discussing reasons for decisions. Observer reliability increases, in our experience, when observers become informed about and comfortable with why a particular behavior is categorized in a certain manner. Training should continue throughout the study, with frequent sessions to "recalibrate" the observers. Each behavior should be carefully defined in writing. It usually helps to define a behavior by its structure (for example, *probe* means "insert bill beneath surface of substrate"). Commonly, in protracted studies definition and criteria tend to "drift" with the passage of time, as observers become more familiar with behaviors and possibly lazier in their evaluations. Careful and repeated calibration training will help solve this problem.

Determination of Sample Size

Researchers should incorporate evaluation of sample size into the design phase of each study. Such planning guides the collection of a sufficient number of samples and avoids "overkill," the collection of too many samples. Authors of many published studies have had to combine data between years and/or sexes in order to obtain "adequate" sample sizes. Such lumping of data may not be appropriate, however, for some study objectives.

In Chapters 2 and 4 we discussed how temporal and sexual differences affect differential habitat use by animals. Studies of foraging behavior show that, even within what we often consider a "season," combining data over even a few months can obscure important patterns of resource use. In designing studies, researchers must carefully determine the number and types of variables for which they will have time to collect data.

For example, Brennan and Morrison (1990) found that significant variation in the use of tree species by foraging chestnut-backed chickadees (*Parus rufescens*) occurred throughout the year (fig. 5.3). Using some of the same data, Morrison (1988) showed how lumping of data can result in inappropriate interpretations: lumping tends to "average-out" many possibly important ecological relationships (see fig. 5.4).

Sakai and Noon (1990) showed that western flycatchers (*Empidonax difficilis*) significantly altered their foraging behavior within the breeding period. We should not, of course, be surprised by these results. Animals must respond to changes in resource abundance and availability and the demands placed upon them by both abiotic and biotic factors. Many of the equivocal results obtained in behavioral studies likely result from lumping of data. Authors seldom justify the lumping process beyond noting that such a process was necessary to obtain sufficient samples over some specified period. Unfortunately, an evaluation of the effect of sample size on study results does not accompany such "justification." From our own experience, we believe that such a lack of evaluation results primarily from a feeling that "we collected all the data we could." While understandable, this does not justify the publication of results based on insufficient samples and/or samples combined into inappropriate periods. Thus, not only must researchers evaluate sample sizes needed for reliable descriptions and statistical studies, but they should also carefully evaluate these samples over short periods of time that define ecological periods. Chapter 4 describes methods used to analyze sample sizes. Studies of sample-size effects in foraging studies are found in the work of Morrison and his colleagues (1990; see also Morrison 1984).

Assessing Use and Availability of Resources

A host of biotic and abiotic factors influences the evolution of adaptations of animals. One of the major themes of this book is the identification and quantification of these factors. Food plays a crucial role in the dynamics of populations, and scientific literature exists on the diets of many species. Recently, attention has turned toward the quantification of food *abundance* or *availability*. These terms, often used synonymously, do in fact have different definitions with quite different ramifications for the way we interpret

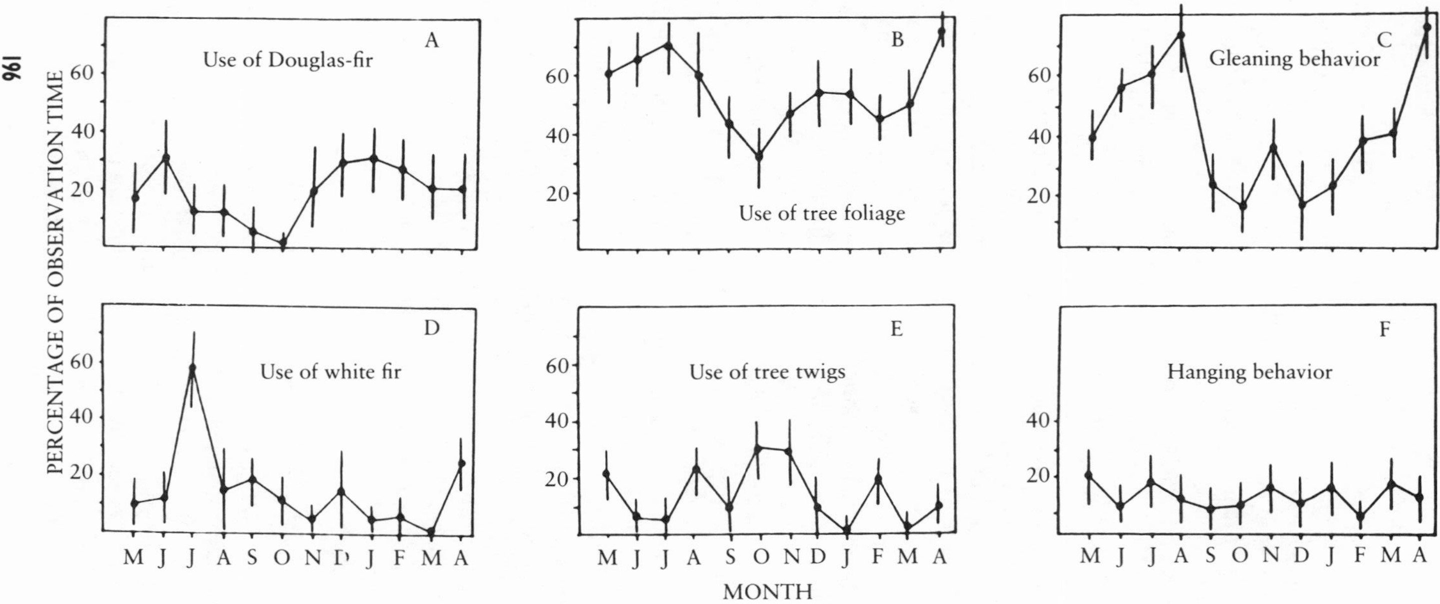

Figure 5.3. Seasonal variation in use of tree species, use of substrates, and foraging modes by chestnut-backed chickadees, using a one-month interval. Dots represent mean values, and vertical bars represent one standard deviation. (From Brennan and Morrison 1990, fig. 4.)

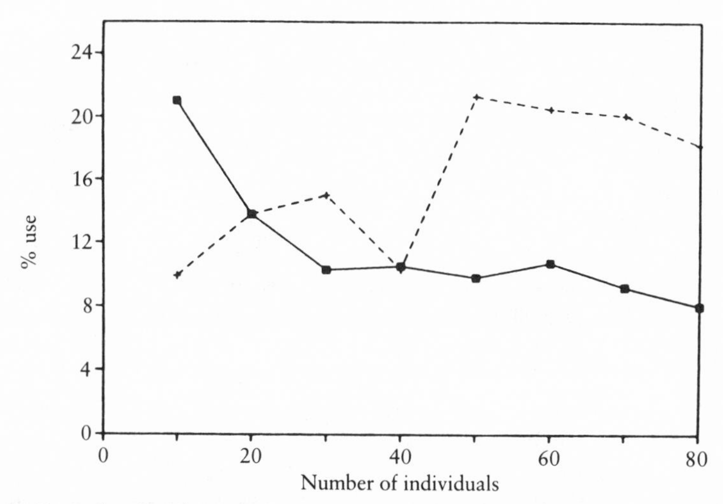

Figure 5.4. Influence of sample size on the observed use of terminal buds (coniferous and deciduous) by foraging chestnut-backed (solid line) and mountain (dashed line) chickadees in a mixed-conifer forest (Sierra Nevada, California) during September and October 1986. (From Morrison 1988, fig. 1.)

behavioral and ecological data. To point out the perceived importance of food in behavioral studies, Hutto (1990) examined a dozen ecological and ornithological journals between 1978 and 1986. He found 155 articles on landbirds alone that dealt specifically with the relationship between food supply and various ecological relationships, including timing of annual cycles of reproduction, molt, migration, territoriality, habitat use, diet, and population size.

In this section we discuss goals and considerations inherent in studies of food use and techniques for measuring abundance and availability of food for wild animals. This section does not catalog the many methods available for sampling food habits of animals; such methods are varied, as well as species- and situation-specific. A number of factors, however, underlie the proper use of most of these methods, and we emphasize those.

Use of Food

A dichotomy exists in the literature regarding the emphasis placed on quantification of animal diets. Wildlife biologists and economic entomologists have expended a good deal of effort to determine the actual food items consumed by wild animals. Korschgen (1980) observed that, in the late 1800s and early 1900s, studies of food habits examined the economic importance of avian feeding habits, concentrating on the plunder of agricultural crops, poultry, and livestock. The greatest concentration of activity in food habits studies took place in the 1930s and 1940s, emphasizing waterfowl and upland game birds. Regarding deer food habits, papers dealing with diets dominated the literature prior to 1950. The proportion of research reporting on food availability, food digestibility, and food requirements has grown steadily since that time.

In contrast, scientists studying ecological relationships of animals seldom attempt to quantify the occurrence of prey items in the diet. Rather, studies have concentrated on indirect measures of food use, such as foraging location (Hutto 1990; Rosenberg and Cooper 1990). Although morphological differences between species undoubtedly reflect some degree of evolutionary response to varying resources, they may not necessarily be good predictors of species' diets, especially under local environmental conditions (Rosenberg and Cooper 1990).

Although studies of food habits abound, most are single-species studies from single locations. Thus, little generalization is possible. Further, as noted by Rosenberg and Cooper (1990), one of the reasons that studies of avian diets have been neglected by modern ornithologists is that researchers fear the detail, tedium, and technological expertise thought to be necessary for such studies. Regardless of the reasons, little literature exists on the diets of most species of animals in the world.

Sampling Techniques

The variety of methods used to study diets of vertebrates can be divided into three basic categories: those involving collection (sacrificing) of individual animals, those involving capture or other temporary disturbances of individual animals, and those requiring little or no disturbance of individuals (Rosenberg and Cooper 1990).

Several reviews of dietary assessment are readily available. Although many of these studies directly concern specific groups of animals (such as seabirds), many of the methods also apply to other groups of animals. Rosenberg and Cooper (1990) provided a recent and thorough review of methods for birds. Ratti, Flake, and Wentz (1982:765–913) reprinted papers on food habits and feeding ecology of waterfowl; they included a bibliography of other important references. Each new edition of the Wildlife Society's *Wildlife Management Techniques Manual* includes reviews of methodologies for birds and mammals (see Korschgen 1980). Riney (1982:124–37) summarized mammalian food habits studies, and many authors in the work of Cooperrider, Boyd, and Stuart (1986) covered diet studies.

The most frequently used method of sampling diets is direct examination of stomach contents. The primary advantage of such sampling is that adequate numbers of samples (stomachs and prey items) are relatively easy to obtain. That is, animals can be collected through shooting or other capture techniques, including mist nets for birds and bats, live traps for mammals, and pitfalls for amphibians, reptiles, and small mammals. With shooting, an individual animal can be collected after the researcher observes its specific foraging behaviors; one can then attempt to relate the specific food items in the stomach to those sampled from the foraging substrate. For game animals, researchers often take gut samples from animals at hunter check stations. Another advantage of gut sampling is that the entire contents of the gut are obtained. Unfortunately, such direct sampling necessitates the killing of the animals; obviously, such a procedure does not allow repeated sampling of the same individual or the subsequent recording of other observations (such as reproductive success or movements).

Nondestructive methods of sampling stomach contents are available for many species. Live-caught animals can be forced to regurgitate using a variety of chemical emetics (see Rosenberg and Cooper 1990). Some mortality occurs from the use of emetics, however. The rate of such mortality depends on the species involved, the dosage of the emetic, and the general condition of the animal. Any researcher planning to use emetics should carefully review the literature associated with this method. Regurgitation in birds has been induced using only water, with a good deal of success and low mortality (Rosenberg and Cooper 1990). The entire digestive tract can also be flushed in some animals, using, for example, a warm saline solution.

Most nonintrusive methods of determining food habits involve observation of animal foraging behavior, analysis of food removal rates, and analysis of feces. Direct observation of food items eaten is possible with some species. Diurnal birds of prey, such as eagles and large hawks, can be observed when they forage in open areas. Herbivorous animals, especially large ungulates, can sometimes be observed browsing or grazing: "bite counts," calculated as the number of bites per plant species, are recorded (see Willms et al. 1980; Thill 1985).

Scientists have used many methods to indirectly determine food consumed by animals. Most of these methods were designed for studying foraging ungulates and largely involve quantification of plant material removed by herbivores, using various survey methods on fixed plots. Researchers assess the height, weight, and condition of plants over periods of time and then relate the results to the type and amount of food consumed (see Dasmann 1949; Severson and May 1967; Willms et al. 1980). Such methods can be combined with bite counts and other direct observations to further describe the food consumption of animals. Other workers have used species-specific evidence of feeding on plants to determine the types of foods consumed by herbivores (Mackie 1970; Dusek 1975).

Evaluation of stomach contents, either through collection of the animal or through use of emetics (Gavett and Wakely 1986), is the most common method of determining food habits of wild animals. There are, however, many biases associated with such analyses, including the fullness of the stomach, differential digestive rates of food items, and difficulty in identification and quantification of small and/or fragmented food items (Severson and Medina 1983; Rosenberg and Cooper 1990).

Differential digestion rates of food items impose the largest potential bias in any study of gut contents. Different kinds of foods consumed at about the same time often are digested at different rates. Further, steps must be taken to prevent excessive postmortem digestion of food. For example, small-bodied insects may be gone from the gizzard within five minutes, whereas hard seeds may persist for several days (Swanson and Bartonek 1970). Several authors have developed correction factors for the differential rates of digestion shown by animals (see Mook and Marshall 1965). Differential digestion of food items is not confined to the intertaxonomic level. For example, Rosenberg and Cooper (1990) discussed data that showed that birds digested second and third instar gypsy moth larvae (*Lymantria dispar*) in less than half the time it took them to digest fourth and fifth instars.

Although bite counts are conceptually simple, this method is difficult to use because it is hard for a researcher to properly identify plants during observations of foraging. Biases toward taller, more easily identified species are inherent in this method. Further, differences between food habits of tame animals—commonly used in this method—and those of wild individuals has not been adequately analyzed.

Methods for direct measurement or estimation of forage utilization, such as point frames, caged plot forage weights, or volume removed, were developed originally for use in livestock studies or range utilization surveys. It is more difficult, however, to detect use by native ungulates than use by livestock with such methods. Severson and May (1967) showed the infeasibility of estimating sagebrush browse by tagging twigs and weighing browsed and unbrowsed leader groups, because of the tight, knotty growth habit of this plant species. Additionally, native ungulate species tend to move more and take less per plant than domestic species while feeding. Further, it is difficult to detect and quantify use of some food items, such as individual leaves, fruits, mosses, and lichens, or of plants that are entirely consumed (like mushrooms). It is often impossible to assign use to any one animal species when several herbivores are present (A. J. Shlisky, unpublished paper, 1988).

Perhaps the most easily obtainable samples of diets come from feces. These can be collected from any species that can be captured alive or at any location where fecal samples are obtainable. Such sampling can be integrated into any study that uses mist nets, mark-recapture, or population monitoring methods. Furthermore, droppings can be collected year-round from animals of any age or any reproductive state, and repeated sampling from known individuals is possible. Researchers can also collect droppings from dens, communal roosts, feeding sites, and under nests.

Ralph, Nagata, and Ralph (1985) describe in detail a technique for collecting and analyzing bird droppings. They discuss materials and equipment required, time and personnel constraints, and tips for identifying various food items. This and similar techniques have been used successfully in studies of many bird species (Davies 1976, 1977a,b; Waugh 1979; Waugh and Hails 1983; Tatner 1983; Ormerod 1985) and small mammals (Meserve 1976; Dickman and Huang 1988). Many papers have detailed methods of fecal collection and analysis in ungulates (see Korschgen 1980; Riney 1982:129–31).

Level of Identification

Our review above indicates that researchers have given some attention to biases involved in diet analyses. The topic of the proper taxonomic level of food identification, however, has received little attention. The taxonomic level selected for diet analysis can have substantial impact on the ecological interpretation of the results. This problem is analogous to the selection of variables for inclusion in habitat models. The level at which foods are identified is likely to affect similarity measures and conclusions drawn from them.

Greene and Jaksic (1983) studied the influence of prey identification level on measures of niche breadth and niche overlap in several vertebrate groups, including raptors, carnivores, and snakes. They calculated niche metrics for the finest prey identification levels (usually specific or generic) reported in diet studies and then recalculated the metrics after combining the prey lists to the ordinal level. They found that niche breadth was consistently larger at the finer prey identification levels than for the ordinal level of classification for all vertebrate groups examined (table 5.1). Calculations using the ordinal levels underestimated niche breadth at higher-resolution levels by 17–242 percent and underestimated niche-breadth scores for single species even more extremely. Food-niche overlaps based on the ordinal level overestimated higher-resolution overlap. Overestimates ranged to infinity when two species did not coincide in the use of any prey species but appeared to do so because of an ordinal level of prey identification. Greene and Jaksic clearly showed, then, that using ordinal levels of prey classifi-

Table 5.1. Biases in estimating food-niche relations at high (H) levels of prey identification, and when ordinal (O) resolution levels are used

Predator assemblages	Breadth underestimate	Overlap overestimate	Nearest neighbor correlation	Guild size	
				H	O
Falconiforms					
So. Michigan	17.3 ± 17.1 (6)	16.5 ± 29.1 (15)	1.000 ± 0.000 (6)*	4	4
So. Wisconsin	75.8 ± 61.7 (8)	301.1 ± 572.4 (24)	0.632 ± 0.258 (8)*	2	4
No. Utah	49.9 ± 25.0 (7)	171.3 ± 573.0 (21)	0.886 ± 0.128 (7)*	3	3
No. California	86.2 ± 127.8 (8)	78.7 ± 158.4 (27)	0.786 ± 0.359 (8)*	3	3
Ce. Chile	104.0 ± 88.6 (5)	236.9 ± 178.4 (10)	0.480 ± 0.532 (5)	3	3
So. Spain	49.1 ± 82.6 (9)	135.0 ± 230.6 (33)	0.919 ± 0.087 (9)*	2	3
Strigiforms					
So. Michigan	63.7 ± 46.3 (4)	28.3 ± 22.8 (6)	0.375 ± 0.629 (4)	4	3
So. Wisconsin	54.7 ± 28.8 (7)	82.8 ± 111.2 (21)	−0.002 ± 0.198 (7)	7	3
No. Utah	82.6 ± 71.3 (4)	108.2 ± 66.1 (6)	0.844 ± 0.237 (4)*		
Ce. Chile	199.5 ± 171.4 (3)	327.6 ± 300.3 (3)	—	2	2
So. Spain	53.8 ± 93.2 (3)	49.8 ± 186.9 (3)	—	2	2
Carnivores					
So. Spain	17.7 ± 17.0 (2)	0.0 ± — (1)	—	—	—
Ce. California	241.9 ± 187.4 (2)	16.6 ± — (1)	—	—	—
So. California	132.7 ± 97.3 (4)	382.2 ± 603.8 (6)	0.500 ± 0.707 (4)*	2	2
Snakes					
So. Spain	124.2 ± 139.6 (5)	1516.3 ± 4667.7 (10)	0.920 ± 0.110 (5)*	2	3
Ce. California	79.0 ± 76.6 (5)	755.3 ± 1300.8 (10)	0.840 ± 0.261 (5)*	2	2

Source: Greene and Jaksic 1983, table 1.
Note: Figures are for $\overline{x}$ ± SD (sample size).
*Significant ($P < 0.05$) correlations.

cation can result in serious misinterpretations of some of the potentially most important food-niche and community parameters in assemblages of many animals. They noted, however, that food items do not necessarily need to be identified using Linnaean nomenclature. Phenotypically distinguishable taxa (such as acridid grasshopper A, B, and so on) can substitute for Linnaean identification (see also Wolda 1990).

When should one use the level of order, and when should one use a finer level of identification? While no hard and fast rules exist for this situation, Cooper, Martinat, and Whitmore (1990) offered several suggestions. Taxonomic levels should be identified that contain enough observations to make analysis meaningful. There are several practical considerations here. First, variables (prey categories) with high numbers of zero counts will not be normally distributed and usually cannot be transformed to normality. Thus, multivariate statistical procedures such as principal components and discriminant analysis (explained in Chapter 7) lose validity. Also, in a procedure such as cluster analysis, large numbers of prey categories produce results that are frequently difficult to interpret. Thus, dividing arthropod orders into families may prove impractical for some relatively uncommon orders. More important, one should consider whether dividing a particular order into finer levels will result in any benefit. That is, if the ecological and behavioral characteristics of two families or groups of families within an order are not very different, then it is unlikely that subdivision of the order will provide much additional information. Conversely, if several subgroups within an order exhibit very different characteristics, such as size, location, or behavior, then subdivision will likely yield useful information. Such suggestions apply, in general, to all studies dealing with the problem of level of taxonomic identification, whether of plant or animal material.

Availability of Food

Foraging studies often use *presence, abundance, density,* and other related terms interchangeably with *availability*. Availability has a distinct definition, however, that encompasses ecological concepts that separate it from other terms.

Availability is defined as the quality or state of being available, whereas *available* means usable, obtainable, accessible. Thus, available items are those to which an animal has access. This is in clear contrast with food density, the total amount of food per unit area (or some index thereof). Food density is usually termed food abundance. That is, if an animal does not have access to a food item, then the item imparts to the animal no direct benefit. (Indirect benefit may be obtained if potential competitors or predators ignore an animal because of these alternate foods.) Access might be denied for many reasons and can change for even the same prey item over short periods of time. Even food present in the area might temporarily become unavailable due to weather conditions, prey behavior, and presence and activity of other animals, including the observer!

Food availability for browsing ungulates is related to the height, location, and/or density of plants. The browsing height limit is set by the easy-reach height of the animal studied. This varies with the size of individuals within the same species (for example, adult versus young deer). Further, some of the interior growth of a plant may not be available for browsing (Riney 1982:208–9). Riney developed a relative scale of plant availability that acknowledged the difficulty in determining exactly what food is available to the animal; he described, for example, how to classify the availability of interior plants in a dense thicket. The availability of such food will change as exterior parts of plants are browsed and as the natural phenology of the plant changes its accessibility to animals. Riney classified plants in this situation as "all available," "largely available," "mostly available," or "unavailable."

It is critical to remember that a sampling design might inaccurately represent both food density and food availability. Many different techniques exist for measuring the same resource, be it potential animal or plant food, and all of these methods have biases associated with their use. In discussing arthropod sampling methods in ornithology, Morrison, Brennan, and Block (1989) called for the use of several techniques to measure the same resource. Multiple techniques at least reduce the risk of inappropriate conclusions. This general idea can be applied to any plant or animal food.

Researchers should not ignore potential foraging locations and prey items when assessing availability, even if the animal is not using them to any degree at the moment. In this way, if any shift—gradual or sudden—to an item occurs, the observer will know the status of that item prior to the shift. Earlier in this chapter we discussed the rapid changes in foraging behavior noted in many species of animals. All potential foraging locations should be sampled as well as those currently in use. General knowledge of the animal's behavior, especially temporally, will help guide such sampling. The problem remains, however, that the sampling methods might not reflect availability.

Analysis

Statistical analysis of foraging data should compose an integral part of an initial study design. A large and varied number of methods of statistical analysis—uni- or multivariate, parametric or nonparametric—have been used with foraging data, based largely on the objectives of the researcher and the form of the data collected. Important relationships exist between the number of variables recorded and the sample sizes necessary for analysis. Clearly, sufficient and appropriate planning should guide the proper use of the preferred analyses.

We will not review the myriad of statistical techniques that can be applied to foraging data. Readers can consult virtually any general statistics text for direction on analyzing various types of behavioral data (such as continuous or categorical). Especially useful texts are those of Siegel (1956), Conover (1980), Snedecor and Cochran (1980), Sokal and Rohlf (1981), and Zar (1984). Texts dealing more specifically with quantitative methods in ethology include Colgan (1978) and Lehner (1979). Martin and Bateson (1986:116–39) presented a concise but thorough review of fundamental univariate statistical techniques for the study of animal behavior. Although not dealing specifically with statistical methods, Kamil (1988) discussed the application of experimental methods to ornithology from the perspective of a behaviorist; researchers should consult his article regardless of the animal group studied.

Foraging data are most frequently recorded as categorical variables, such as foraging mode. Continuous data are often later classified into categories for analysis, such as foraging height. When data are so classified, the result is a contingency table, the cells of which contain frequencies of the various category combinations, such as foraging mode by sex. The null hypothesis of homogeneity of the categories is then tested using contingency analysis, such as the *chi*-square or *G* log-likelihood statistic. When three or more categories (dimensions) are compared, multidimensional contingency table analysis is used (see Colgan and Smith 1978). Most foraging studies involving categorical variables have employed contingency table analysis.

Researchers usually record foraging behaviors of animals in some sequential fashion; this is especially true for birds. Analyses of such data using contingency tables and *chi*-square (and related) analyses may not be valid, however; sequential observations are likely not independent and, as such, violate the critical assumption of independence of most statistical contingency tests, including *chi*-square (Raphael 1990). Many papers using contingency table analysis violate the independence assumption. When reviewing foraging studies, readers should carefully compare the data collection methods with the assumptions underlying the analytical techniques used. This does not mean that the mere violation of assumptions negates a study. It behooves the researcher, however, to address these violations and discuss their possible influence on results.

The often sequential nature of data collection can, in fact, be a benefit in the elucidation of animal foraging behavior. That is, examining the sequence of behaviors of an individual provides potentially much more information on how the animal exploits its environment than does an overall lumping of observations. For example, a bird might perform the following sequence of behaviors: glean, glean, probe, probe, glean, probe, peck, hover-glean, glean, probe. Here, the bird gleaned for 40 percent of the sequence, probed for 40 percent, and pecked and hover-gleaned for 10 percent each. If another individual were found to forage using these behaviors with the same frequency, we would conclude that the two foraged in an identical manner. But how does the sequence of behaviors affect our conclusion? That is, if the second individual probed four times in a row, then pecked and hover-gleaned, and then gleaned for the final four motions, the way in which it exploited its environment would be quite different from that of the first bird. Note also that the first bird only probed after it gleaned. Does this mean it was obtaining different food than the second bird was? Although we cannot answer this question from the data provided here, we can at least conclude that there is some individual variation in foraging behavior and that researchers should analyze the specific substrates being used. Given that much of our interest in optimal foraging theory involves the decisions animals make in choosing and moving among potential resources, we need to apply greater emphasis to analysis of behavioral sequences.

A sequence in which the behavioral pattern always occurs in the same order is considered deterministic. Classical behaviorists refer to such behavioral sequences as fixed action patterns. Vertebrates seldom if ever repeat foraging behaviors in the same order, but they exhibit some amount of variability that may be predictable. These sequences are considered *stochastic* (or probabilistic). Sequences which show no temporal structure, in which the component behavior patterns are sequentially independent, are considered random sequences. In a random sequence, one behavior or set of behaviors can be followed by any other behavior with equal probability. The conditional probability that one behavior follows another—that is, the probability that B follows A, given that A has occurred—denoted *P* (B/A), is called the transition probability (Martin and Bateson 1986:63–64).

There are several methods available for analyzing sequential data, such as time-series analyses. Of particular interest to us are analyses involving Markov chains. Markov analysis is a method for distinguishing whether a sequence is random or whether it contains some degree of temporal order. A first-order Markov process is one in which the probability of occurrence for the next event depends only on the immediately preceding event. If the probability depends on the two preceding events, then the process is considered a second-order one. Higher-order processes are involved as additional events are considered. One analyzes sequences by comparing

Sequence: ABABABBABABAABABABABA

1st behavior pattern

2nd behavior pattern	A	B
A	0.1 (0.5)	0.9 (0.5)
B	0.9 (0.5)	0.1 (0.5)

Figure 5.5. A highly simplified transition matrix, analyzing the sequence shown above it, which comprises only two different behavior patterns (A and B). The matrix shows the empirical transition probabilities for the four different types of transition ($A|A, B|A, A|B, B|B$). For example, the lower left cell shows that the conditional probability of B given that A has occurred ($B|A$) is 0.9 (nine out of ten transitions after A has occurred). For comparison, the transition probabilities under a random model (0.5 for each type in this example) are shown in parentheses. The matrix confirms that A and B tend to alternate (the probabilities of $B|A$ and $A|B$ are high), while repeats are rare (the probabilities of $B|B$ and $A|A$ are low). (From Martin and Bateson 1986, fig. 4.7.)

the observed frequency of each transition with the frequency of transitions that would be expected if the sequences were random (Martin and Bateson 1986:64). A simple example of the Markov analysis was given by Martin and Bateson (1986:64–65) and is repeated here in figure 5.5. Raphael (1990) presented a detailed example of the application of Markov chains to foraging data. Other papers using Markov analysis of foraging include those of Colwell (1973) and Riley (1986). Raphael (1990) reviewed these and other behavioral studies that involved Markov chains.

Indices

Scientists have developed a type of methodology to quantify the use of food in relation to its availability. Widely known as preference or electivity indices, these measures seek to compare frequency of food items in the diet with the availability of those items in the animal's environment by representing these data in a single index value. Researchers have also applied some of these indices to habitat analysis (see Morrison 1982; Chapter 4). Many of these indices have received a good deal of attention in the wildlife literature, especially with regard to ungulates. Because of this attention and the potential application of the indices to a wide variety of situations, we will discuss some important considerations in their use.

The general approach to using electivity indices is to establish a ratio between frequency of food consumed by an animal and the amount of food available for consumption. For example, Ivlev's electivity index (Ivlev 1961) compares relative availability of food types in the environment (p) with their relative use in the diet (r): $E_i = (r_i - p_i)/(r_i + p_i)$. Other indices have similar forms (for a review, see Lechowicz 1982). If r and p are equal for all food types, then the animal is choosing food types at random, that is, in direct proportion to the relative availability of the food. If r and p differ, then one usually concludes that the animal is either avoiding (a negative index value) or selecting (positive value) an item.

The most straightforward indices simply consist of the estimated percentage of a food item in an animal's diet divided by the total amount of food in the animal's environment. Values range from -1 to 0 for avoidance (negative selection), to 0 to infinity for positive selection. Such indices are often called the forage ratio (Jacobs 1974) and have been used in a number of studies (see Heady and Van Dyne 1965; Chamrad and Box 1968; Petrides 1975; Hobbs and Bowden 1982). The major drawback to these simple indices, however, is their intrinsic asymmetry, that is, unbounded positive values. A log transformation is necessary to utilize forage ratios (Jacobs 1974; Strauss 1979; Lechowicz 1982). Unfortunately, the forage ratio also changes with the relative abundance (p) of food in the environment. Thus, this index cannot be used if one wishes to examine the relationship between the relative abundance of food and food selection (Jacobs 1974; see also Lechowicz 1982).

Readers need only refer to our earlier discussions on the great difficulty in accurately quantifying use, and especially availability, of food (or habitat) to identify the overriding factor influencing index values. Clearly, proper determination of the variables to compare in such analyses will largely determine the results. An example modified from Johnson (1980) will illustrate this point. Suppose an investigator collects an animal and finds that its gut contains food items A, B, and C in the percentages shown in table 5.2(A) under "Usage." A sample of the animal's feeding area reveals that the items were present in the proportions shown under "Availability." Many investigators would conclude that item A was avoided because use

Table 5.2. Results of comparing usage and availability data when a commonly available but seldom-used item is included (A) and when excluded (B) from consideration

Item	Usage (%)	Availability (%)	Conclusion	Rank		
				Usage	Availability	Difference
			(A)			
A	2	60	Avoided	3	1	+2
B	43	30	Preferred	2	2	0
C	55	10	Preferred	1	3	−2
			(B)			
B	44	75	Avoided	2	1	+1
C	56	25	Preferred	1	2	−1

Source: Johnson 1980, table 1.

was less than availability, while items B and C were preferred. But suppose other investigators do not believe that item A is a valid food item; perhaps it was ingested only accidentally while the animal consumed other foods. They would then consider the data in table 5.2(B), obtained by deleting item A from the analysis. Now, although item C is still deemed preferred, the assessment of item B has changed from preferred to avoided. Thus, we see that conclusions drawn from such analyses will depend markedly upon the array of components thought by the investigator to be available to the animal. Remember also our earlier discussions on the effects of the level of prey identification on quantification of diet. Not only does the data put into these indices influence conclusions drawn from them, but the indices themselves will also yield differing results because of peculiarities of their mathematical structure.

The work of Ivlev (1961), Jacobs (1974), Chesson (1978, 1983), Strauss (1979), Vanderploeg and Scavia (1979a,b), and Johnson (1980) describes the development of the more widely used electivity indices. Lechowicz (1982) compared the characteristics of seven of the electivity indices proposed by these researchers; most of the indices are permutations of Ivlev's original index (see table 5.3). Lechowicz found that Ivlev's, Strauss's, and Jacobs's indices could not potentially obtain the full range of index values for all values of r and p. The index values for intermediate values of r and p depend on the relative abundance of other items in the environment or the diet. A critical problem with most of the indices is that direct comparisons between indices derived from samples differing in relative abundances are

Table 5.3. Various indices of electivity or feeding preference based on the proportions of food in the diet (r_i) and in the environment (p_i)

Algorithm	Comment	Reference
1. $E_i = (r_i - p_i)/(r_i + p_i)$	Ivlev's electivity index	Ivlev 1961
2. $E_i' = r_i/p_i$	Ivlev's forage ratio	Ivlev 1961
3. $D_i = \dfrac{r_i - p_i}{(r_i + p_i) - 2r_ip_i}$		Jacobs 1974
4. $Q_i = \dfrac{r_i(1 - p_i)}{p_i(1 - r_i)}$	Use of $\log_{10} Q$ recommended	Jacobs 1974
5. $L_i = r_i - p_i$		Strauss 1979
6. $\alpha_i = W_i = \dfrac{r_i/p_i}{\Sigma_i\, r_i/p_i}$	Chesson's *alpha*, Vanderploeg and Scavia's selectivity coefficient	Chesson 1978; Vanderploeg and Scavia 1979a
7. $E_i^* = \dfrac{W_i - (1/n)}{W_i + (1/n)}$	n = number of kinds of food items	Vanderploeg and Scavia 1979b

Source: Lechowicz 1982, table 1.

inappropriate. (Exceptions are available, however, such as Chesson's *alpha* and Vanderploeg and Scavia's selectivity coefficient; see Chesson 1983.)

To avoid the problems associated with inclusion or exclusion of specific food items in the calculation and evaluation of indices, Johnson (1980) developed a procedure based on ranks. He proposed using the difference between the rank of usage and the rank of availability. Using the earlier example with item A included (table 5.2), the differences in ranks of usage and availability are +2, 0, and −2 for items A, B, and C, respectively. Excluding item A from the analysis, we find that B and C have values of +1 and −1, respectively. Although the values themselves change, the difference between B and C remains 2; C is preferred to B, regardless of whether A is included or excluded. The loss of information concerning the absolute differences between food items realized when using ranks is likely of little consequence. Statistical methods based on ranks are nearly as efficient as methods based on the original data; this is especially true if the assumptions necessary to treat the original data are not met (for example, if measurements are not exact or the distribution is not normal). As we have noted previously, we have good reason to doubt the accuracy of most usage and, especially, availability measures. Johnson (1980) also presented methods for determining statistical significance among components of the data.

Energetics

Assessments of foraging energetics commonly assume that the goal of a foraging animal is to maximize its net energy balance. Many theoretical and empirical studies have examined how specifically the animal achieves this balance. Townsend and Hughes stated, "We can conclude that there is validity in the basic assumption that the rate of energy gain from food has been a significant limiting factor and has constituted an important selection pressure in the evolution of foraging behavior" (1981:107). Thus the neurological and physiological capabilities of the animal itself link an animal to its environment. One must think of the environment in its functional relation to the animal, rather than merely as the geographic area and physical structure of the habitat in which the animal lives, in order to understand the dynamic relationships between an animal and its environment (Moen 1973:21). In this brief section we simply wish to stress these points.

Thermal energy is exchanged between the animal and its environment by radiation, conduction, convection, and evaporation (see fig. 5.6). Each of the methods of thermal energy exchange change relative to one another as the animal's environment alters; rain, wind, and other abiotic factors affect energy exchange over the short term. Changes in such factors as plant cover, for example, influence energy exchange over a long period of time.

Animals are, of course, subject to the first law of thermodynamics: energy can be neither created nor destroyed, but only changed in form. The conversion of energy from one form to another in a living organism is called metabolism. A wild animal has an ecological metabolic rate that is an expression of the energy "cost of living" for the purpose of daily activities and other physiological processes. Ecological metabolism varies from one activity to another and from one species to another. There are various direct and indirect methods for determining metabolic rates of animals under varying environmental conditions, both in the laboratory and the field (see Moen 1973; Van Soest 1982). Studies of free-ranging animals are, however, quite difficult relative to their laboratory-based counterparts (see Bell 1990; Goldstein 1990). It seems clear that we need closer, direct studies of the energetic gains and costs of animal behavior, including foraging.

Foods are only the carriers of the nutrients and potential energy in an animal's diet. Thus, food habits lists are helpful for only the most general analysis of the animal-environment relationship (Moen 1973:135). Further, studies of feeding rates, food utilization, and the like may be unrelated

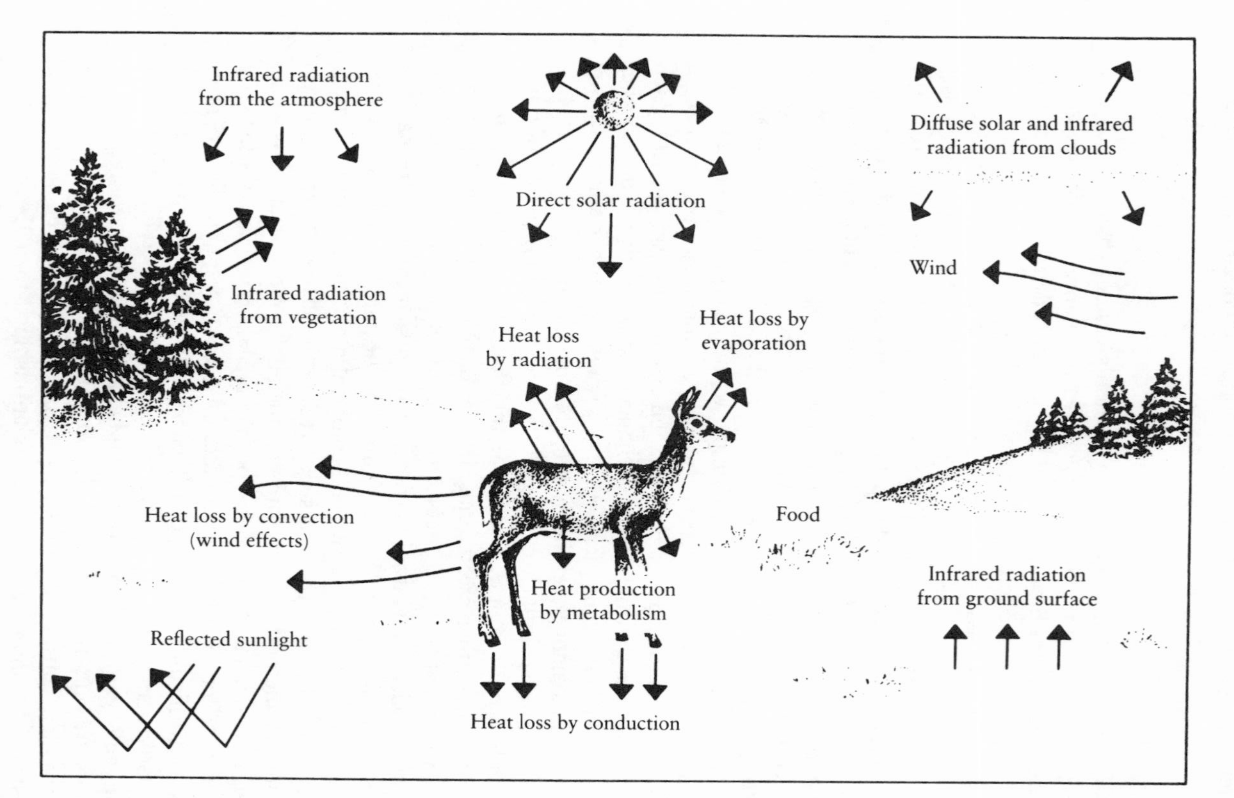

Figure 5.6. The thermal energy exchange between an animal and its environment includes radiation, conduction, convection, and evaporation as the four basic modes of heat transfer. (From *Wildlife Ecology*. By Aaron N. Moen. Copyright © 1973 by W. H. Freeman and Company. Reprinted with permission.)

to the functional nutritive relationship between the animal and the environment. For example, the chemical composition of two plant species may be very similar, and an animal may be able to digest and assimilate each plant species in a similar manner; the animal thus makes no biochemical distinction between the two species. The biologist, however, lacking knowledge of the chemical nature of the food, separates the plants along taxonomic lines; such a separation might be unnecessary from a nutritional viewpoint (Moen 1973:135–36).

Implications for Foraging Ecology

A foraging animal must choose among behaviors that impose upon it different energy costs. Simultaneously, the animal must acquire sufficient energy to cover both these costs and the costs of the remaining daily maintenance, activity, thermoregulation, storage, and production. Studies of foraging behavior have demonstrated that a changing physical environment may strongly influence an animal's pattern of use of time. Changing weather may alter food availability, thereby necessitating a change in foraging strategies. In more extreme situations, the physical environment imposes such a strenuous thermal load that foraging is either impossible (in excessive heat) or energetically too expensive to be profitable (in extreme cold). In these cases foraging may cease altogether. An understanding of the energy costs of different activities provides a means for evaluating the costs and benefits of changing behavioral strategies. In addition, studies of daily energy expenditure have demonstrated that more subtle influences, such as convective heat loss and acclimatization of resting metabolic rate, can also have important impacts on overall energy balance. It is this overall level of energy expenditure which must be balanced by the energy gained during foraging. Hence, from the perspective of foraging ecology, studies of daily energy expenditure have provided a number of important tools and insights. The currency of energetics may be fruitfully applied to understanding costs and benefits of different foraging strategies. These costs depend not only on the types of activities employed but also on subtle patterns of acclimatization of metabolic rates and thermal balance with the environment. Careful studies of the energetics of individual animals in different circumstances, and of species with similar diets but different behavioral patterns, should help to illuminate the forces governing the patterns of time use by animals (Goldstein 1990).

A complete and useful assessment of foraging ecology must thus consider nutritional and energetic relationships in addition to the behavioral attributes of the animal. As noted by Moen, such a holistic approach to foraging ecology has clear implications for habitat management: "The success of management practices depends on the degree to which the relationships between the animal and its habitat are optimized. The successful management of the habitat is dependent on a knowledge of what the animal requires behaviorally as well as physiologically" (1973:187). We hope that future studies in foraging behavior will seek to more directly link behavioral and physiological factors influencing animals; a clear theoretical framework for such studies is currently being developed.

Synthesis: Empiricism and Theory

With few exceptions, empirical studies of foraging behavior—whether they involve dietary analysis or are purely behavioral—rarely even mention what, if any, theory underlies their development. This may in large part result from researchers' failure to fully understand the theory itself. Stephens commented, "Many would-be testers do not understand the theory well enough to perform meaningful tests" (1990:454). Further, there is a general feeling among many, if not most, biologists that foraging theory is "a complete waste of time" (Pierce and Ollason 1987; see Stephens and Krebs 1986 for a strong defense of foraging theory). Although this debate will likely continue, we should remember that the fundamentals of the scientific method call for clear construction and evaluation of hypotheses based on well-developed theory. Gray (1987) and Schoener (1987) offer related discussions on the success and failure of foraging theory.

Many of the quantitative predictions of the major models of foraging theory—those concerning patch and prey choice—have not stood up well under empirical examination. For example, Stephens and Krebs (1986) found only eleven unambiguous quantitative fits in their tabulation of 125 studies. Stephens (1990) found, however, two consistent qualitative trends. First, the prediction that the time a forager spends exploiting a depleting patch should increase as the time required to travel between patches increases has held in most situations in which it has been examined. Second, as predicted by the prey model, foragers selectively attack prey items that are most profitable, that is, those that have the highest ratio of energy available to time required for handling and consumption.

Alternatively, the models have also had their qualitative failures. The prey-choice model's prediction of absolute preference—that is, a given prey type should either always be accepted or always ignored—has been consistently rejected. Overall, however, much data from many taxa are consistent with the simple notion of rate maximization. Long-term, average rate maximization is a specific way to combine less search and handling time with more energy intake. Thus, as a minimal and conservative conclusion, foraging animals seem to act "economically" in that they tend to choose alternatives that yield more food in less time (Stephens 1990). That animals forage economically, or profitably, represents an important generalization about the capabilities of foraging animals. Thus, foraging theory has clearly given us new and more sophisticated paths to follow in our studies of animal foraging ecology. Dietary studies are still needed, but they take on added meaning and importance only when made an integral part of the study of the overall feeding strategy of the animal. Our individual studies of foraging behavior will lead to more rapid understanding and synthesis of overall patterns of foraging ecology if we begin with a clear development of the theory upon which the studies are based.

Foraging theory directly relates to wildlife management. As developed by Nudds (1980) for ungulate herbivores, the type of foraging model most closely followed by an animal (energy maximizing, equal food value, nutrient optimizing, unequal food value) can guide land management decisions concerning the amounts and types of food to emphasize. Nudds concluded that deer, as well as other temperate-latitude ungulates, are primarily habitat specialists but become diet generalists in winter. The foraging behavior of deer in winter adhered most closely to the predictions of the energy-maximizing models; it seemed energetically less costly to remain in sheltered habitats and fast than to forage in exposed habitats. Translating these conclusions to a management scenario, Nudds suggested that manipulating winter habitats of deer by increasing only the abundance of "preferred" food would not be warranted. Management would more beneficially be directed toward the physical structure of the winter habitat. Although some of Nudds's suggestions have been criticized (Jenkins 1982; but see Nudds 1982), his 1980 paper is important in attempting to relate foraging theory to practical wildlife management. That is, how can foraging theory guide our studies of food selection? As Jenkins noted, "Foraging theory, combined with good empirical work on food selection, may lead to valuable new insights about certain wildlife management problems" (1982:256).

Literature Cited

Alexander, R. D. 1974. The evolution of social behavior. *Annual Review of Ecology and Systematics* 5:325–83.

Altmann, J. 1974. Observational study of behavior: Sampling methods. *Behaviour* 49:227–67.

Bell, G. P. 1990. Birds and mammals on an insect diet: A primer on diet composition analysis in relation to ecological energetics. *Studies in Avian Biology* 13:416–22.

Block, W. M. 1990. Geographic variation in foraging ecologies of breeding and nonbreeding birds in oak woodlands. *Studies in Avian Biology* 13:264–69.

Block, W. M., K. A. With, and M. L. Morrison. 1987. On measuring bird habitat: Influence of observer variability and sample size. *Condor* 89:241–51.

Brennan, L. A., and M. L. Morrison. 1990. Influence of sample size on interpretation of foraging patterns by chestnut-backed chickadees. *Studies in Avian Biology* 13:187–92.

Caraco, T., S. Martindale, and T. S. Whittham. 1980. An empirical demonstration of risk-sensitive foraging preferences. *Animal Behaviour* 28:820–30.

Chamrad, A. D., and T. W. Box. 1968. Food habits of white-tailed deer in south Texas. *Journal of Range Management* 21:158–64.

Chesson, J. 1978. Measuring preference in selective predation. *Ecology* 59:211–15.

Chesson, J. 1983. The estimation and analysis of preference and its relationship to foraging models. *Ecology* 64:1297–1304.

Colgan, P. W., ed. 1978. *Quantitative ethology*. New York: John Wiley and Sons.

Colgan, P. W., and J. T. Smith. 1978. Multidimensional contingency table analysis. In *Quantitative ethology*, ed. P. W. Colgan, 145–74. New York: John Wiley and Sons.

Colwell, R. K. 1973. Competition and coexistence in a simple tropical community. *American Naturalist* 107:737–60.

Conover, W. J. 1980. *Practical nonparametric statistics*. 2d ed. New York: John Wiley and Sons.

Cooper, R. J., P. J. Martinat, and R. C. Whitmore. 1990. Dietary similarity among insectivorous birds: Influence of taxonomic versus ecological categorization of prey. *Studies in Avian Biology* 13:104–9.

Cooperrider, A. Y., R. J. Boyd, and H. R. Stuart, eds. 1986. *Inventory and monitoring of wildlife habitat*. USDI Bureau of Land Management Service Center, Denver, Colo.

Dasmann, W. P. 1949. Deer-livestock forage studies in the interstate winter deer range in California. *Journal of Range Management* 2:206–12.

Davies, N. B. 1976. Food, flicking, and territorial behavior of the pied wagtail (*Motacilla alba yarelli* Gould). *Journal of Animal Ecology* 45:235–52.

Davies, N. B. 1977a. Prey selection and social behavior in wagtails (Aves: Motacillidae). *Journal of Animal Ecology* 46:37–57.

Davies, N. B. 1977b. Prey selection and the search strategy of the spotted flycatcher (*Muscicapa striata*): A field study on optimal foraging. *Animal Behaviour* 25:1016–33.

Dickman, C. R., and C. Huang. 1988. The reliability of fecal analysis as a method for determining the diet of insectivorous mammals. *Journal of Mammalogy* 69:108–13.

Dusek, G. L. 1975. Range relations of mule deer and cattle in prairie habitat. *Journal of Wildlife Management* 39:605–16.

Ford, H. A., L. Bridges, and S. Noske. 1990. Interobserver differences in recording foraging behavior of fuscous honeyeaters. *Studies in Avian Biology* 13:199–201.

Gavett, A. B., and J. S. Wakely. 1986. Diets of house sparrows in urban and rural habitats. *Wilson Bulletin* 98:137–44.

Goldstein, D. L. 1990. Energetics of activity and free living in birds. *Studies in Avian Biology* 13:423–26.

Gray, R. D. 1987. Faith and foraging: A critique of the "paradigm argument from design." In *Foraging behavior*, ed. A. C. Kamil, J. R. Krebs, and H. R. Pulliam, 69–140. New York: Plenum Press.

Greene, H. W., and F. M. Jaksic. 1983. Food-niche relationships among sympatric predators: Effects of level of prey identification. *Oikos* 40:151–54.

Hamilton, W. D. 1971. Geometry for the selfish herd. *Journal of Theoretical Biology* 31:293–311.

Harcourt, A. H., and K. J. Stewart. 1984. Gorillas' time feeding: Aspects of methodology, body size, competition and diet. *African Journal of Ecology* 22: 207–15.

Hazlett, B. A., ed. 1977. *Quantitative methods in the study of animal behavior*. New York: Academic Press.

Heady, H. F., and G. M. Van Dyne. 1965. Botanical composition of sheep and cattle diets on a mature animal range. *Hilgardia* 36:465–92.

Hobbs, N. T., and D. C. Bowden. 1982. Confidence intervals on food preference indices. *Journal of Wildlife Management* 46:505–7.

Hurlbert, S. H. 1984. Pseudoreplication and the design of ecological field experiments. *Ecology* 54:187–211.

Hutto, R. L. 1990. On measuring the availability of food resources. *Studies in Avian Biology* 13:20–28.

Ivlev, V. S. 1961. *Experimental ecology of the feeding of fishes*. New Haven: Yale Univ. Press.

Jacobs, J. 1974. Quantitative measurement of food selection. *Oecologia* 14:413–17.

Jenkins, S. H. 1982. Management implications of optimal foraging theory: A critique. *Journal of Wildlife Management* 46:255–57.

Johnson, D. H. 1980. The comparison of usage and availability measurements for evaluating resource preference. *Ecology* 61:65–71.

Kamil, A. C. 1988. Experimental design in ornithology. In *Current Ornithology*. Vol. 5, ed. R. F. Johnston, 312–46. New York: Plenum Press.

Kamil, A. C., and T. D. Sargent, eds. 1981. Foraging behavior: Ecological, ethological, and psychological approaches. New York: Garland STPM Press.

Kamil, A. C., J. R. Krebs, and H. R. Pulliam, eds. 1987. *Foraging behavior*. New York: Plenum Press.

Kenward, R. 1987. *Wildlife radio tagging*. New York: Academic Press.

Korschgen, L. J. 1980. Procedures for food-habits analyses. In *Wildlife management techniques manual*, ed. S. D. Schemnitz, 113–27. 4th ed. Washington, D.C.: The Wildlife Society.

Lechowicz, M. J. 1982. The sampling characteristics of electivity indices. *Oecologia* 52:22–30.

Lehner, P. N. 1979. *Handbook of ethological methods*. New York: Garland STPM Press.

Lomnicki, A. 1988. *Population ecology of individuals*. Princeton: Princeton Univ. Press.

Machlis, L., P. W. D. Dodd, and J. C. Fentress. 1985. The pooling fallacy: Problems arising when individuals contribute more than one observation to the data set. *Zeitschrift fuer Tierpsychologie* 68:201–14.

Mackie, R. J. 1970. Range ecology and relations of mule deer, elk and cattle in the Missouri River Breaks, Montana. *Wildlife Monographs* 20:1–79.

Martin, P., and P. Bateson. 1986. *Measuring behaviour*. Cambridge: Cambridge Univ. Press.

Meserve, P. L. 1976. Food relationships of a rodent fauna in a California coastal sage scrub community. *Journal of Mammalogy* 57:300–319.

Miller, E. H. 1988. Description of bird behavior for comparative purposes. In *Current Ornithology*. Vol. 5, ed. R. F. Johnston, 347–94. New York: Plenum Press.

Moen, A. N. 1973. *Wildlife ecology*. San Francisco: W. H. Freeman.

Mook, L. J., and H. W. Marshall. 1965. Digestion of spruce budworm larvae and pupae in the olive-backed thrush, *Hylocichla ustulata swainsoni* (Tschudi). *Canadian Entomologist* 97:1144–49.

Morrison, M. L. 1982. The structure of western warbler assemblages: Ecomorphological analysis of the black-throated gray and hermit warblers. *Auk* 99: 503–13.

Morrison, M. L. 1984. Influence of sample size and sampling design on analysis of avian foraging behavior. *Condor* 86:146–50.

Morrison, M. L. 1988. On sample sizes and reliable information. *Condor* 90: 275–78.

Morrison, M. L., C. J. Ralph, J. Verner, and J. R. Jehl, Jr., eds. 1990. Avian foraging: Theory, methodology, and applications. *Studies in Avian Biology* no. 13.

Morrison, M. L., L. A. Brennan, and W. M. Block. 1989. Arthropod sampling methods in ornithology: Goals and pitfalls. In *Estimation and analysis of insect populations*, ed. L. McDonald, B. Manly, T. Lockwood, and J. Logan, 484–92. New York: Springer-Verlag.

Morse, D. H. 1980. *Behavioral mechanisms in ecology*. Cambridge, Mass.: Harvard Univ. Press.

Nudds, T. D. 1980. Forage "preference": Theoretical considerations of diet selection by deer. *Journal of Wildlife Management* 44:735–40.

Nudds, T. D. 1982. Theoretical considerations of diet selection by deer: A reply. *Journal of Wildlife Management* 46:257–58.

Ormerod, S. J. 1985. The diet of dippers *Cinclus cinclus* and their nestlings in the catchment of the River Wye, Mid-Wales: A preliminary study of faecal analysis. *Ibis* 127:316–31.

Petrides, G. A. 1975. Principal foods versus preferred foods and their relation to stocking rate and range condition. *Biological Conservation* 7:161–69.

Pierce, G. J., and J. G. Ollason. 1987. Eight reasons why optimal foraging theory is a complete waste of time. *Oikos* 49:111–18.

Pulliam, H. R. 1973. On the advantages of flocking. *Journal of Theoretical Biology* 38:419–22.

Ralph, C. J., and J. M. Scott, eds. 1981. Estimating numbers of terrestrial birds. *Studies in Avian Biology* no. 6.

Ralph, C. P., S. E. Nagata, and C. J. Ralph. 1985. Analysis of droppings to describe diets of small birds. *Journal of Field Ornithology* 56:165–74.

Raphael, M. G. 1988. A portable computer-compatible system for collecting bird count data. *Journal of Field Ornithology* 59:280–85.

Raphael, M. G. 1990. Use of Markov chains in analyses of foraging behavior. *Studies in Avian Biology* 13:288–94.

Ratti, J. T., L. D. Flake, and W. A. Wentz. 1982. Waterfowl ecology and management: Selected readings. Bethesda, Md.: The Wildlife Society.

Remsen, J. V., Jr., and S. K. Robinson. 1990. A classification scheme for foraging behavior of birds in terrestrial habitats. *Studies in Avian Biology* 13: 144–60.

Riley, C. M. 1986. Foraging behavior and sexual dimorphism in emerald toucanets

(*Aulacorhynchus prasinus*) in Costa Rica. M.S. thesis, Univ. of Arkansas, Fayetteville.

Riney, T. 1982. *Study and management of large mammals*. New York: John Wiley and Sons.

Rosenberg, K. V., and R. J. Cooper. 1990. Approaches to avian diet analysis. *Studies in Avian Biology* 13:80–90.

Ruggiero, L. F., R. S. Holthausen, B. G. Marcot, K. B. Aubry, J. W. Thomas, and E. C. Meslow. 1988. Ecological dependency: The concept and its implications for research and management. *Transactions of the North American Wildlife and Natural Resources Conference* 53:115–26.

Sakai, H. F., and B. R. Noon. 1990. Variation in the foraging behaviors of two flycatchers: Associations with stage of the breeding cycle. *Studies in Avian Biology* 13:237–44.

Schoener, T. W. 1969. Optimal size and specialization in constant and fluctuating environments: An energy time approach. *Brookhaven Symposium in Biology* 22:103–14.

Schoener, T. W. 1987. A brief history of optimal foraging ecology. In *Foraging behavior*, ed. A. C. Kamil, J. R. Krebs, and H. R. Pulliam, 5–67. New York: Plenum Press.

Severson, K. E., and M. May. 1967. Food preferences of antelope and domestic sheep in Wyoming's Red Desert. *Journal of Range Management* 20: 21–25.

Severson, K. E., and A. L. Medina. 1983. Deer and elk habitat management in the Southwest. *Journal of Range Management Monographs* no. 2.

Siegel, S. 1956. *Nonparametric statistics for the behavioral sciences*. New York: McGraw-Hill.

Slater, P. J. B. 1978. Data collection. In *Quantitative ethology*, ed. P. W. Colgan, 7–24. New York: John Wiley and Sons.

Snedecor, G. W., and W. G. Cochran. 1980. *Statistical methods*. 7th ed. Ames: Iowa State Univ. Press.

Sokal, R. R., and F. J. Rohlf. 1981. *Biometry*. 2d ed. San Francisco: W. H. Freeman.

Stephens, D. W. 1990. Foraging theory: Up, down, and sideways. *Studies in Avian Biology* 13:444–54.

Stephens, D. W., and J. R. Krebs. 1986. *Foraging theory*. Princeton: Princeton Univ. Press.

Strauss, R. E. 1979. Reliability estimates for Ivlev's electivity index, the forage ratio, and a proposed linear index of food selection. *Transactions of the American Fisheries Society* 108:344–52.

Swanson, G. A., and J. C. Bartonek. 1970. Bias associated with food analysis in gizzards of blue-winged teal. *Journal of Wildlife Management* 34:739–46.

Tatner, P. 1983. The diet of urban magpies *Pica pica*. *Ibis* 125:90–107.

Thill, R. A. 1985. Cattle and deer compatibility on southern forest range. In *Proceedings of a conference on multispecies grazing*, ed. F. H. Baker and R. K. Jones, 159–77. Morrilton, Ark.: Winrock International.

Townsend, C. R., and R. N. Hughes. 1981. Maximizing net energy returns from foraging. In *Physiological ecology: An evolutionary approach to resource use*, ed. C. R. Townsend and P. Calow, 86–108. Sunderland, Mass.: Sinauer Associates.

Vanderploeg, H. A., and D. Scavia. 1979a. Two electivity indices for feeding with special reference to zooplankton grazing. *Journal of the Fisheries Research Board of Canada* 36:362–65.

Vanderploeg, H. A., and D. Scavia. 1979b. Calculation and use of selectivity coefficients of feeding: Zooplankton grazing. *Ecological Modelling* 7:135–49.

Van Soest, P. J. 1982. *Nutritional ecology of the ruminant.* Ithaca, N.Y.: Cornell Univ. Press.

Waugh, D. R. 1979. The diet of sand martins in the breeding season. *Bird Study* 26:123–28.

Waugh, D. R., and C. J. Hails. 1983. Foraging ecology of a tropical aerial feeding bird guild. *Ibis* 125:200–17.

Williams, P. L. 1990. Use of radiotracking to study foraging in small terrestrial birds. *Studies in Avian Biology* 13:181–86.

Willms, W., A. McLean, R. Tucker, and R. Ritchey. 1980. Deer and cattle diets on summer range in British Columbia. *Journal of Range Management* 33: 55–59.

Wolda, H. 1990. Food availability for an insectivore and how to measure it. *Studies in Avian Biology* 13:38–43.

Zar, J. H. 1984. *Biostatistical analysis.* 2d ed. Englewood Cliffs, N.J.: Prentice-Hall.

6 Development of Predictive Models

Introduction

The main objectives for developing models of wildlife-habitat relationships are to formalize our current understanding about a species or an ecological system, understand which environmental factors affect distribution and abundance of a species, predict future distribution and abundance of a species, identify weaknesses in our understanding, and generate hypotheses about the species or system of interest. The goals of modeling wildlife-habitat relationships are usually associated with prediction. In this book, *predictive modeling* refers to estimating the presence, distribution, or abundance of a wildlife species or group of species, given information on actual or possible habitat conditions.

Making Predictions

There are two main classes of predictions that can be made from models. Hindcasting identifies key environmental variables, typically those of vegetation structure, that account for observed variation in species variables. Hindcasting is used to explain patterns observed in species occurrence and abundance and is pertinent, strictly speaking, only to the time and place at which the original data were gathered. The other class of prediction, forecasting, involves an explicit attempt to predict species conditions, given environmental conditions at a time or place not represented by the field data originally used to generate the model. Many workers use results of hindcasting, such as those obtained by the use of correlation, regression, or multivariate statistics, to predict future species conditions. Without proper design of the initial investigation and without validation studies, however, predictions from hindcasting may prove quite unreliable, because environmental, demographic, and ecological conditions may vary significantly among locations or over time. At best, using hindcasting models for prediction in new situations entails assuming that factors not accounted for in the prediction model are insignificant or unchanged.

What is the best means of predicting species responses to environmental conditions? Proper forecasting techniques account for autocorrelation of a variable over some time series or over spatial gradients (such as environmental ones). Researchers using forecasting should consider causes of the distribution and abundance of a species, rather than the simple correlations employed in hindcasting. The degree to which observational data should be used to tell something about habitat selection and resource requirements of a species, for example, should be examined very carefully before such extrapolations are made.

This chapter explores the use of predictive modeling in assessing wildlife-habitat relationships. First, we offer definitions and a classification of wildlife-habitat models. We discuss how scientific uncertainty affects wildlife management and how model development and application should be used in light of uncertainties. We then review current model forms used in predicting wildlife-habitat relationships and describe how vegetation and wildlife models can be linked. Knowledge-based and decision-aiding models are then introduced. Finally, we discuss the often-overlooked topic of model validation.

Definitions and Uses of Modeling

As a prelude to delving into ecological models of various types, Bunnell's *Alchemy and Uncertainty* (1989) is recommended reading. Bunnell provides a common-sense interpretation of the value of models for making decisions about land allocations and habitat management. Also, the recommendations of Farmer, Armbruster, Terrell, and Schroeder (1982) may prove helpful for articulating key assumptions of a modeling approach and for guiding model development. Overton's strategy for model construction (1977) and Schroeder's review of concepts and community models for use in evaluating wildlife impact (1987) are of value as well.

A model is any representation of some part of the real world. Hall and Day (1977) offered that a model may take one of four forms. Listed in order of increasing specificity, these are: conceptual, diagrammatic, mathematical, and computational. One can also view these forms as stages in a logical model-building process.

Developing a conceptual model may entail synthesizing current scientific understanding, field observations, and professional judgment of a particular species or habitat and then proposing a few hypotheses to explain the species' distribution and abundance. Even at the conceptual stage, one must explicitly state assumptions and simplifications necessary for the model to be true or useful. The diagrammatic stage takes a conceptual model one step further by showing interrelationships among various environmental parameters and species' behaviors. The mathematical stage quantifies these relationships by applying coefficients of change and formulas of correlation or causality. Finally, in the computational stage one explores or solves the mathematical relationships by analyzing the formulas with computers.

Levins (1968) argued that an ecological theory is only a combination of all models used to describe the theory. He failed to address the origin of these models, however. Our view is that the mathematical models to which Levins referred are themselves derived from conceptual models of how we understand the world. We stress this point because the conceptual and diagrammatic stages of modeling are often the most difficult, and the most revealing, stages of building ecological theories and enhancing understanding.

We have already mentioned that predictive models of wildlife-habitat relationships are used for forecasting. A fuller classification of wildlife-habitat models follows (refer to fig. 6.1).

Theoretical and Empirical Models

Wildlife-habitat models may be classified first as theoretical or empirical. Theoretical models stress relationships between species and their environment as we think the relationships ought to be represented and quantified. Theoretical models may be descriptive or mathematical. Descriptive theoretical models reflect the conceptual and diagrammatic stages of model development of Hall and Day (1977).

An example of a descriptive theoretical model is optimal foraging theory

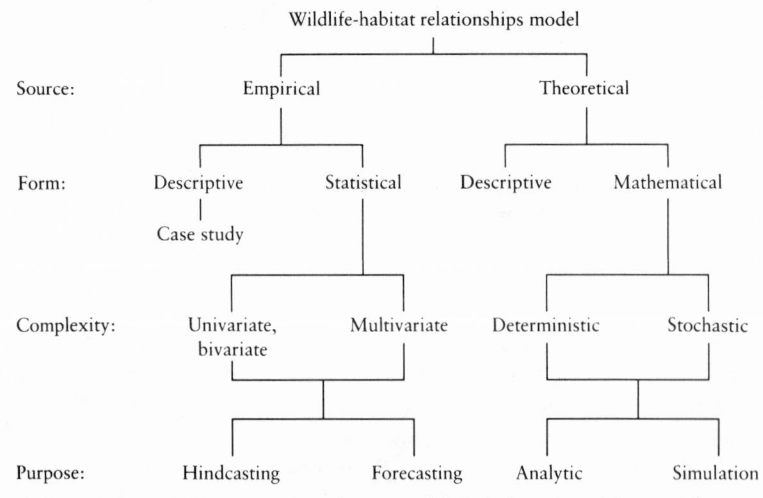

Figure 6.1. Classification of predictive wildlife-habitat models according to source of data and utility of the model

(Pyke 1984), which is a synthesis of observations and hypotheses about a particular behavior of habitat use. The basic tenet is that animals will seek food in order to optimize their use of resources and foraging environments in light of the energy required to obtain the resources (Garton 1979; Krebs, Ryan, and Charnov 1974). That is, the net balance between energy expended for searching for food will be minimized with respect to the net energy gained from the food. Thus, the foundation is the theory of optimality. What makes this particular body of concepts descriptive is that each case is analyzed differently; no one specific, unifying mathematical formula represents all observations and hypotheses.

Theoretical models of wildlife-habitat relationships also may be more fully mathematical in nature. Mathematical theoretical models may take many forms, such as graph theory models of food webs (Cohen 1978; Levins 1975) and resource competition models (Tilman 1982). What distinguishes mathematical models from descriptive models is reliance on strict symbolism and mathematical structure to express the behavior of animals and the relationships between species and environments.

Theorists have advanced many complex mathematical models as ways of understanding the behavior of populations. Such models include those of catastrophe theory (Cobb and Zacks 1985), which assumes that changes in environmental conditions and population responses might occur as thresholds which can take sudden, major shifts, rather than as gradual and continuous changes. Fractal geometry (Mandelbrot 1977) has been used in evaluating old forest patch conditions for spotted owls (*Strix occidentalis*) (Dixon et al. 1990) and bald eagles (*Haliaeetus leucocephalus*)(Pennycuick and Kline 1986).

We build mathematical models mostly because we believe that mathematical properties represent the operation of the real world. Manipulating and exploring mathematical relationships supposedly deepens our insight into the functioning of real systems. This approach relies on two main and usually tacit assumptions. The first assumption is that the real-world system has been adequately represented in the mathematical formulation; that is, no vital parameter or relationship affecting the outcome was omitted. The second assumption is that the behavior of the mathematical formulas represents the true behavior of the real system. These preconditions may or may not hold true in any given mathematical model of wildlife-habitat relationships.

Despite our efforts, we may end up learning more about the behavior of the mathematics used in the model than about the behavior of the biological entities that the model supposedly represents. Risks of this kind of error are high if few empirical field studies are available by which we can validate the mathematical relationships. The seriousness of the error is a function of how the model is to be used. It may be serious if, for example, we rely on the model to teach us about biological behavior. It may be serious also if the model is used to irreversibly allocate resources or habitats with high opportunity values; an example might be modeling species' requirements of estaurine environments in an urban setting. Such errors may be less serious, however, if land allocations or effects of management actions are easily reversed.

Mathematical models may be couched in deterministic or stochastic forms as well. In deterministic mathematical models, parameters and outputs take on single values, such as mean litter size or total forage quantity. Such single values are often called point estimates. In stochastic mathematical models, however, parameters may be represented by a specified distribution of values, such as the range of litter size or the standard deviation of forage quantity. Solutions of stochastic models may be more variable than those of deterministic models. That is, with stochastic models, small variations in the values of the parameters may result in greater variation in the results than with deterministic models.

Finally, mathematical models may be distinguished as analytic or simulationary in form. An analytic mathematical model has one set of relations—formulas and their variables—and has one or a fixed set of exact solutions. Once the researchers have defined the parameters and coefficients, they obtain the fixed results by solving the formulas. On the other hand, simulation mathematical models are usually stochastic in nature, they are run on computers, and their results typically vary from run to run. Analytic models are most useful in exploring well-defined problems in which mathematics cleanly represents the species and environmental relationships. Simulation models are most useful when the problem involves parameters whose values vary or are not well known.

Examples of theoretical wildlife-habitat relationship models are habitat suitability index (HSI) models as formulated by the USDI Fish and Wildlife Service (Schamberger, Farmer, and Terrell 1982). In HSI models, an overall habitat suitability index is calculated from the geometric mean of the values of three environmental parameters. The geometric mean presumably represents the best "balance" and interaction of all three factors. With a geometric mean, if the value of any one variable is zero, then the composite HSI score is zero. This property is desirable in models for which all environmental variables must be nonzero for any positive response by the species.

HSI models are theoretical in nature because the underlying relationship between habitat capability and each environmental parameter is assumed from the outset, rather than derived from hindcasting data from field observations. Actual coefficients used in the relationships are often synthesized from field observations and expert opinion. Further, HSI models are mathematical because of the explicit formula (the geometric mean) used. They are also deterministic, because only single values for environmental parameters are used, resulting in only single values of solutions. Finally, they are analytic in form because the mathematical formulas can be solved exactly for particular values of environmental parameters.

In contrast with theoretically derived models, empirically derived models are founded on a set of field observations. Relationships between species and environments are discerned by exploring the patterns found in data. Thus, empirical models of wildlife-habitat relationships should entail proper use of sampling theory, field techniques, and descriptive and predictive statistics.

Empirical models may be classed as generally descriptive or more strictly statistical in nature. Descriptive empirical models may take the form of anecdotal information on such topics as species ranges, use of various environments, social behavior, and food habits. Quantitative models in wildlife biology are often based on descriptive empirical models of this sort.

A caution in using descriptive empirical models is in order. Often, results of a case study are taken to represent conditions with a broader geographic area or range of ecological conditions than represented in the study per se. In case studies, sample sizes of locations or animals often fall short of numbers required for using strict statistical tests. Also, sites or animals are selected nonrandomly because of their particular interest to the researchers. For example, study of behavior of white-tailed deer (*Odocoileus virginianus*) in a bottomland selected as a potential reservoir site does not provide information on use of other environments and potential effects on all white-tailed deer from similar projects in a broader geographic area. Case studies are conducted because they are easier and cheaper than fuller, statistically based studies; information is required on only particular sites or animals; and there is a misunderstanding that results of case studies can be extended to a broader geographic area or set of conditions. When broader information is needed, a proper statistical study should be conducted.

Statistical empirical models are more formal in structure than descriptive empirical models. They typically entail demonstrating statistical patterns of species-habitat associations from a set of field observations. Statistical models may be univariate, where only a single factor is considered at a time; bivariate, where a species's response to a single environmental variable is considered; or multivariate, where one or more species parameters are related to two or more environmental variables. Scientists have developed many forms of bivariate, multivariate, and other statistical approaches to analyzing ecological data in recent decades. For example, James (1971) used principal components ordination to evaluate environmental relationships among breeding birds.

Most of these approaches should be used to hindcast. Such statistical models, however, are typically used for explaining ecosystem function or for predicting future values of ecological parameters. This is risky unless the models have been validated against an independent data set.

One should also understand the assumptions of statistical procedures—even simple ones—used to create models that predict ecological conditions. For example, Garsd (1984) warned about accounting for false correlations in ecological modeling. Morrison, Timossi, and With (1987) discussed developing and testing linear regression models for predicting bird-habitat relationships. For monitoring, a sound statistical basis for sampling and data interpretation is paramount (see Schindler 1987; Shettleworth et al. 1988; States et al. 1978; and Verner 1983).

The anecdotal word models of Errington (1963) and Lack (1954, 1966) are examples of empirical descriptive models. Examples of empirical statistical models are the regression models of bird-habitat relationships by Balda, Gaud, and Brawn (1983). Balda, Gaud, and Brawn surveyed the frequency of occurrence of five common secondary cavity–nesting species of birds at replicate study sites over eight years. They gathered data on environmental conditions at each site and formulated linear multiple regression models with which to predict future densities of bird species given potential environmental conditions. This approach used hindcasting of observational data to forecast future conditions. The researchers, however, used the first five years' data to construct the models and the last three years' data to validate them. Their models explained between 50 and 89 percent of the variation in breeding bird density.

Another aspect of evaluating statistical ecological models is the use of so-called null models in ecology (Harvey et al. 1983). A null model is a depiction of an ecological system that functions much like a null hypothesis; it is constructed for the purposes of falsification. Proponents of null modeling in ecology (see Connor and Simberloff 1979; Strong, Szyska, and Simberloff 1979) argue strongly that ecological modeling cannot proceed in a scientific fashion without the formalization of a falsifiable hypothesis, such as that afforded by a null model. There has been considerable controversy, however, about what constitutes a null model, how to build one (Pleasants 1990), and whether it has any value in the scientific process (Roughgarden 1983). At any rate, the number of discussions of null modeling in recent ecological literature has diminished.

Scientific Uncertainty

Biological data and models are rarely adequate or precise enough for predicting species distribution and abundance with little or no error. Rather, analysts making resource management decisions affecting environmental conditions typically encounter the following obstacles: imprecise data; uncertain inferences; limiting and fallacious assumptions; unforeseen environmental, administrative, and social circumstances; and risks of failure. Imperfections are often present in habitat analyses and management decisions, but they are especially important when a risk of reducing a wildlife population or eliminating a species exists. One may encounter uncertainty when analyzing biological data, when making inferences about species' responses to environmental conditions, and when selecting and instituting a management plan.

Uncertainty may be classified as scientific uncertainty or decision-making uncertainty. Analyzing species and habitats entails a different process from that used to make decisions on resource management. The kinds and implications of uncertainties from the analysis process and from the decision-making process are also quite distinct. Results of a technical study such as a risk analysis of population viability, however, may be part of the information used by a decision-maker in developing a habitat management plan.

Scientific uncertainty in habitat modeling results from the nature of the data and the ways in which information on species and habitats is represented and applied. Scientific uncertainty essentially reflects the fact that our predictions of how species respond to environmental conditions are not perfect. Uncertainty may occur because the system itself is naturally variable and very complex, and thus difficult to predict; the process of estimating values of parameters in the habitat model entails a high degree of error; models used to generate predictions are in some sense invalid; or the scientific question being asked is ambiguous or incorrect.

Variability of Natural Systems: Noise in the Message

Many aspects of natural systems vary over time. Predicting attributes of the system—the "message" we are trying to interpret—may often involve observing and modeling traits that are influenced by outside factors—the "noise" inherent in the message. Such noise introduces variation in measurements and uncertainty in estimating and predicting attributes of the system. In statistical models of habitat relationships, noise is typically depicted as unexplained variation in the occurrence or abundance of a species. One kind of unexplained statistical variation is the value of residuals in linear regression models.

Uncertainty of Empirical Information: Errors of Estimation

Values of environmental parameters are typically estimated from a sample set of observations. A parameter, for example, may be the average number of tree stems per hectare or the variance of litter sizes of black-footed ferrets (*Mustela nigripes*) that can be attributed to individual and environmental variation. When a parameter is estimated from a sample set of observations, from a statistical viewpoint, uncertainty or errors in estimation may occur. The estimation may be biased if the values of the observations are consistently lesser or greater than actual (unknown) values; inaccurate if the estimated value of the parameter of interest (such as a mean or a variance) is substantially different from the true value; or imprecise if values of individual observations vary widely. Each of these errors in estimating the value of a parameter constitutes a different kind of scientific or statistical uncertainty.

Such errors of estimation can arise from a number of sampling problems. These may include inadequate sample size, observations taken from disparate times or places, and samples taken nonrandomly or nonsystematically, depending on the assumptions of the estimator being used. Errors of estimating the value of parameters may also arise from applying the wrong kind of estimator. An example is applying a formula for calculating variance. If correct use of the formula assumes that observations were made independently and randomly when they were actually made over a time series or systematically, such as at even intervals over a transect, then the analyst has made the error of applying the wrong kind of estimator.

Uncertainty of Model Structure: Model Validation

Model validity refers to a broad spectrum of performance standards and criteria, including model credibility, realism, generality, precision, breadth, and depth (Marcot, Raphael, and Berry 1983). The various criteria refer to such attributes of models as the number of parameters in a model and their interactions, the context within which a model was developed or should be used, and the underlying and simplifying assumptions of the model structure. A parameter that is estimated precisely, accurately, and without bias may still be used inappropriately; a model might be applied to the wrong environment, location, season, or species.

Appropriateness of the Problem: Asking the Right Question

The context in which one applies a theory or uses a model may introduce yet another source of scientific uncertainty. Even if a model has been validated—that is, shown to be a useful tool and to generate acceptable predictions according to particular criteria—it still may be applied to the wrong problem.

For example, a life table model that assumes equal sex ratios and assumes that adults breed each year may generate acceptable predictions for use with Dall sheep (*Ovis dalli*) but may generate grossly inaccurate predictions when used for species with variable or quite different social breeding organizations, such as pronghorn (*Antilocapra americana*). This calls into question the reliability of the model when used with some species or under some circumstances.

Further, the original hypothesis or problem may be ambiguous or even unanswerable. For example, a model of species-habitat relationship that describes vegetation types used for feeding and breeding may be validated against field data and provide a relatively realistic depiction of occurrence of species given vegetation structure and composition of individual stands. It may not, however, provide a useful foundation for answering questions about landscape patterns necessary for maintaining viable populations of the species.

A Review of Models of Habitat Relationships

Wildlife biologists have a number of models of wildlife-habitat relationships available to them. A review follows listing models classified by function (table 6.1). The list omits a number of models taken from theoretical or highly mathematical literature.

Models of Vegetation Structure

Stand Growth Models

Researchers have developed a number of models that display and predict composition and structure of vegetation stands. These include silvicultural stand growth and yield models (Avery and Burkhart 1983; Clutter et al. 1983). Stand growth and yield models include CLIMACS (Dale and Hemstrom 1984); Douglas-Fir Simulator (DF–SIM; Curtis, Clendenen, and DeMars 1981; Fight, Chittester, and Clendenen 1984); FORCYTE (Kimmins 1987); FOREST (Ek and Monserud 1974); FREP (Leary 1979); Prognosis (Wykoff, Crookston, and Stage 1982); Stand Projection System (SPS; Arney 1985); STEMS (Belcher, Holdaway, and Brand 1982); VARP (Tappeiner, Gourley, and Emmingham 1985); and WOODPLAN (Williamson 1983). Growth and yield information is generally available for a variety of forest types on commercial forest land. For example, Oliver and Powers (1978) presented growth models for unthinned ponderosa pine (*Pinus ponderosa*) plantations in northern California, and Ramm and Miner (1986) reviewed fourteen growth and yield programs in the north central region of the United States. Meldahl (1986) compared and critiqued alternative modeling methodologies for growth and yield prediction models, including yield tables, multiple regression models, diameter distribution models, differential-difference equation models, and individual tree models. He concluded that a manager must first describe available inputs and the specific kinds of information needed from growth and yield projection systems before selecting the correct model.

Many stand growth and yield models of forests assume even-age silvicultural management of a single-species stand. Such models typically depict stand growth as beginning at final harvest, such as clear-cutting. Data are available for other methods of forest regeneration, however, such as shelterwood cutting. The user may specify such parameters as stand origin (such as artificial planting or natural seeding), seedling spacing, fertilization, presence and degree of precommercial thinning, number and intensity of commercial thinnings, sanitation entries, other intermediate treatments, and timing of final harvest. Such models typically describe the expected structure of a forest stand in terms of stem density per acre, stem volume

Table 6.1. Classification of wildlife-habitat relationships models

- Models of vegetation structure
  - Stand growth models
    - Stand growth and yield models
  - Succession models
- Models of species' response to vegetation structure
  - Single species models
    - Life history models
    - Habitat preference models
    - Optimal foraging models
    - Correlation models
    - Multivariatestatistical models
    - Habitat suitability index (HSI) models
    - Habitat capability (HC) models
    - Habitat evaluation procedures (HEP)
    - Bayesian and pattern recognition (PATREC) models
    - Population trend and viability models
    - Indicator species models
  - Multiple species models
    - Species-habitat matrices
    - Guild and life-form models
    - Community structure models
    - Community and ecosystem simulation models
    - Hierarchy models
    - Semantic models
- Models of landscapes
  - Vegetation disturbance models
  - Fragmentation models
  - Cumulative effects models
  - Models of insular biogeography
- Models for monitoring species and habitats
  - Adaptive resource management
- Knowledge-based and decision-aiding models
  - Decision support models
  - Expert systems

(board feet or cubic feet), stem basal area per acre, quadratic mean diameter at breast height (dbh), and tree height. A few stand growth models, such as SPS, provide stem dbh distributions as well as quadratic mean dbh.

Unfortunately, researchers often use stand growth and yield models for stands outside the geographic scope, age, and conditions in which the empirical growth and yield data were gathered. Thus, such models may produce unreliable, and possibly optimistic, predictions of stand conditions.

Some stand growth and yield models provide estimates of suppression mortality, which may be useful for predicting the density and size of future snags or down wood in the stand. For example, Neitro and his coworkers (1985) used DF-SIM to model the occurrence of snags in even-age stands of Douglas-fir (*Pseudotsuga menziesii*) in western Washington and Oregon. SPS also predicts the rate of suppression mortality and can be used similarly.

Dennis and his colleagues (1985) reviewed problems associated with mathematically modeling growth and yield of renewable resources. In particular, they pointed out that variation in yield estimates, aside from sampling variations inherent in inventories, are usually large and uncontrollable. Techniques to help plan for use levels, such as linear programming or forest stand growth models, cannot account for such variation. Even if they could, we seldom have data on the frequency and effects of factors such as storms, insect epidemics, and fires that affect resource amounts and growth rates. Such random events can greatly influence optimal solutions for the best rate of use of resources. Also, growth and yield models seldom account for spatial variation, such as specific tree stem spacing affecting mortality functions.

Succession Models

Another class of vegetation growth and change models tracks changes in acreages of successional or structural vegetation stages. Examples include the DYNamically Analytic Silviculture Technique (DYNAST; Boyce 1980) and FORPLAN, which are used by the USDA Forest Service for habitat and general resource planning (see Kirkman et al. 1984; Benson and Laudenslayer 1984; Sweeney 1984). These models operate on current acreages of various forest types and their growth stages, as well as rates of succession for each type. The output displays the acreage of each forest type and growth stage over time. Figure 6.2 displays the results of one such model.

Often, researchers use such models to calculate habitat capability for a variety of wildlife species. Such models, however, typically lack sensitivity to spatial patterns of forest stages; the same array of acreages of various forest developmental stages will produce the same habitat capability estimates regardless of habitat patch sizes or arrangements. This may be of little consequence to wildlife species that use early successional to midsuccessional stages, if patches are widely distributed, but it may produce great errors when managers are predicting responses of species requiring scarce

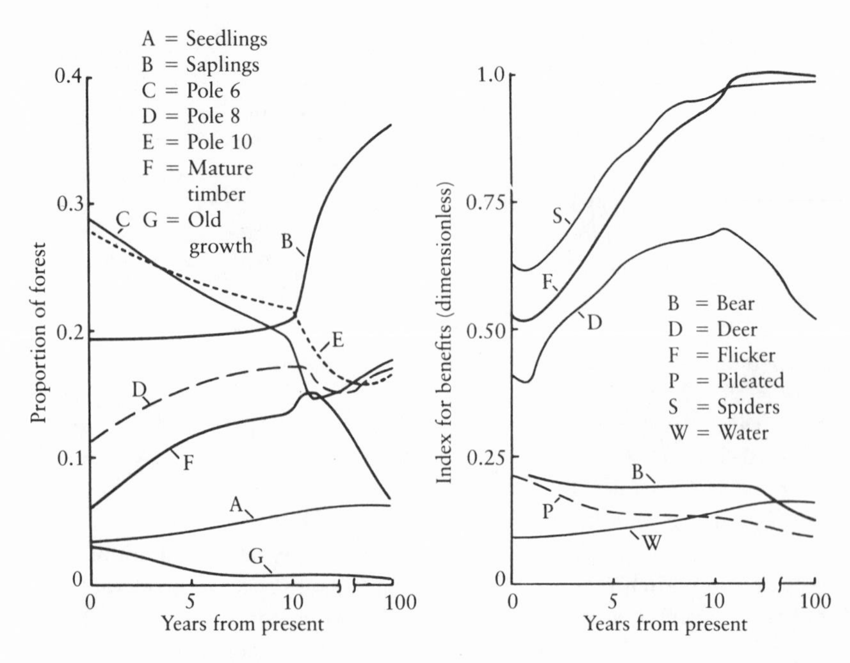

Figure 6.2. Output from a DYNAST simulation of, at left, seven forest growth conditions and, at right, five species and one environmental variable in an upland oak-hickory forest in North Carolina. (From Boyce 1985.)

or declining habitats, such as specific edge or forest-interior conditions in some locations.

Models of Species' Response to Vegetation Structure

Modeling species' response to vegetation structure ultimately entails linking models of vegetation growth, structure, and succession with predictions of occurrence and abundance of wildlife species. A variety of model forms serves this purpose. We can categorize such models generally as single species models or multiple species models.

Single Species Models

Life history models. Ecological models depicting and predicting life history characteristics of species have been proposed for describing the evolution of behavioral and phenotypic traits (Partridge 1978; Barclay and Gregory 1981; Holm 1988). Although not directly useful for depicting wildlife-habitat relationships, such models help us understand the ecological roles of species and species groups.

Habitat preference models. The study of animals' selection and preference for food and habitats is essentially a study of the adaptive advantages of particular behaviors. Recently, scientists have critically evaluated the concepts of ecological dependency and preference (Carey 1984; Ruggiero et al. 1988). Porter and Church (1987) assessed effects of environmental patterns on analysis of habitat preference. Rosenzweig (1987) reviewed how habitat selection contributes to biological diversity.

Optimal foraging models. Related to life history and habitat preference models are models of foraging behavior (Townsend and Hughes 1981; Pyke 1984). For example, Belovsky (1987) related foraging requirements to body size to help explain the life history traits of home range area.

Other models of animal behavior and response to environmental factors also prove instructive. For example, Caraco (1979) reviewed the ecological response of animal group size to environmental conditions. Such a model may be useful for interpreting herding behavior of ungulates and understanding how environmental conditions might affect herd size, composition, and distribution.

Correlation models. Correlation models display the degree to which environmental parameters explain species parameters. Correlations in such models are typically based on empirical data and are best used for hindcasting rather than for forecasting species' responses. Too often, however, analysts use unvalidated correlation models as predictive models. Garsd (1984) reviewed various pitfalls of depicting and using correlation models, including the common mistake of interpreting spurious correlation as a causal relation.

Multivariate statistical models. Modeling species-habitat relationships with multivariate statistics is a common practice (see Folse 1979; Capen 1981). Multivariate approaches include multiple regression, various forms of principal components analysis, discriminant function analysis, and canonical correlation (see Chapter 7). Recently, logistic regression has come into favor, as it accounts for some of the nonlinearity between predictor and response variables (see Hassler, Sinclair, and Kallio 1986). In general, multivariate models help identify significant environmental parameters that account for observed variation in the distribution and abundance of wildlife species.

One shortcoming of a multivariate approach is that results can be difficult to interpret. Mathematically, many multivariate statistical techniques combine several environmental parameters into one collapsed function, which is then correlated with species' distribution and abundance. Biologically, however, it is not always clear what the collapsed functions or the correlations mean, particularly if axes are rotated and if data are standardized and transformed to increase the degree of correlation. Nevertheless, multivariate models are indispensable for exploring patterns in large empirical data sets (that is, for hindcasting) and for understanding relations between wildlife species and environmental variables.

Examples of multivariate approaches to modeling species-habitat relationships include assessments of nest site selection by kingbirds (MacKenzie and Sealy 1981) and habitat selection by northern spotted owls (Solis and Gutierrez 1982). Multivariate statistics can be used on vegetation data alone; for example, Radloff and Betters (1978) used a multivariate approach to classify physical site data for wildlife management.

Habitat suitability index (HSI) models. One of the more popular approaches to modeling environmental conditions is the use of habitat suitability index (HSI) models. The USDI Fish and Wildlife Service and other federal resource management agencies use HSI models extensively (Schamberger, Farmer, and Terrell 1982). These models typically denote habitat suitability of a species as the geometric mean of three environmental variables deemed to most affect species presence, distribution, or abundance. The general model form of the HSI is

$$HSI = (V_1 \times V_2 \times V_3)^{1/3}$$

where V_1, V_2, and V_3 are three key environmental variables. Each variable and the resulting HSI values are scaled from 0 to 1. The overall HSI value represents the final response of the species to the combination of the values of the environmental parameters.

For example, the three environmental variables denoted in an HSI model for yellow warbler (*Dendroica petechia*) are percentage of deciduous shrub crown cover, average height of deciduous shrub canopy, and percentage of shrub canopy comprising hydrophytic shrubs (Schroeder 1983). The resulting suitability index in the yellow warbler model represents relative habitat values for reproduction. Figure 6.3 depicts this model and displays values of the HSI for several forest conditions. Researchers have constructed HSI models for a wide variety of species in the United States, including bobcat (Boyle and Fendley 1987), black bear (Rogers and Allen 1987), hairy woodpecker (Sousa 1987), Rocky Mountain elk (Wisdom et al. 1986), fish (Terrell 1984), and yellow warbler (Schroeder 1983). O'Neil, Roberts, Wakeley, and Teaford (1988) provided a procedure for modifying HSI models.

HSI models are useful for representing in a simple and understandable form the major environmental factors thought to most influence occurrence and abundance of a wildlife species. HSI models are best viewed, however, as hypotheses of species-habitat relationships rather than as causal functions (Schamberger 1982). Their value lies in documenting a repeatable

V_1 Percentage of deciduous shrub crown cover

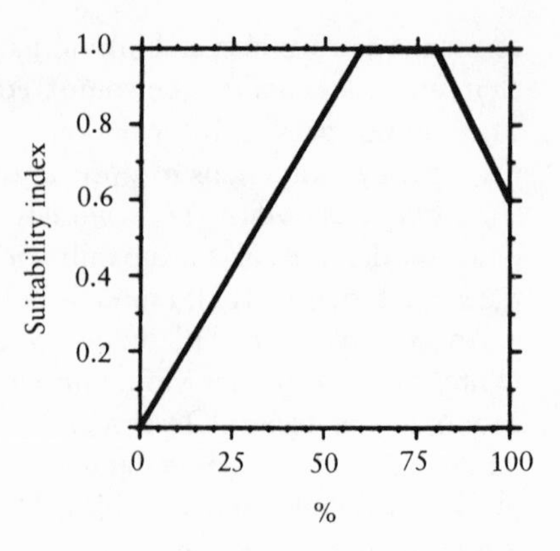

V_2 Average height of deciduous shrub canopy

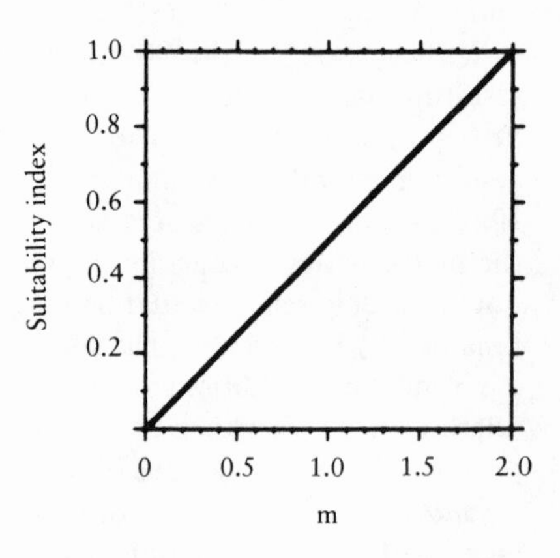

V_3 Percentage of deciduous shrub canopy comprised of hydrophytic shrubs

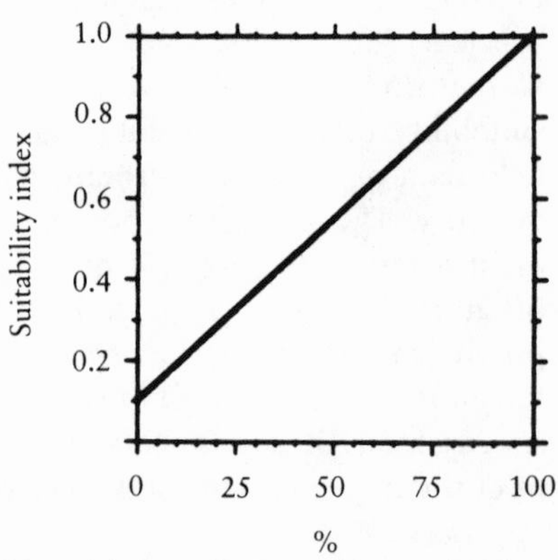

Figure 6.3. Habitat suitability index (HSI) model for yellow warbler and index values for forest conditions. (From Schroeder 1983.)

assessment procedure and providing an index to particular environmental characteristics that can be compared with alternative management plans. They do not provide information on population size, trend, or behavioral response by individuals to shifts in resource conditions.

Habitat capability (HC) models. Closely allied to habitat suitability index models are habitat capability models. HC models perform essentially the same function as HSI models but may vary slightly in structure.

An example is the HC model for assessing habitat effectiveness for the winter range of Rocky Mountain elk (*Cervus elaphus nelsonii*) in the Blue Mountains of eastern Oregon and Washington (Wisdom et al. 1986). This model calculates an elk habitat effectiveness index as the geometric mean of four environmental variables. The model is currently undergoing field testing by the USDA Forest Service. Figure 6.4 shows the model structure and its use.

With both HC and HSI models, it is difficult to interpret whether the resulting index value is intended to represent environmental conditions or population response. Also, in both model forms, the sensitivity of the resulting habitat index values to any one environmental variable is diminished as more variables are added to the model. This occurs as a result of the mathematics of a geometric mean model, and hence, the model may not accurately reflect actual habitat effectiveness or population response. Finally, like HSI models, HC models should be used to represent relative environmental conditions and to generate hypotheses about species-habitat relationships rather than as definitive statements of cause-and-effect relations or reliable predictions of species response.

Habitat Evaluation Procedures (HEP). The USDI Fish and Wildlife Service uses Habitat Evaluation Procedures extensively to assess environmental conditions at the species level (Flood et al. 1977; U.S. Fish and Wildlife Service 1980). The procedure is based on the habitat unit (HU), defined as the product of habitat quality (on a 0 to 1 index, as from a habitat suitability index) and habitat quantity. HEP models may require much field data on specific environmental attributes, such as forage quality or quantity. The procedure, however, provides a structured way to document a repeatable assessment of environmental conditions. Researchers often use HEP models to evaluate impacts of, and mitigations for, proposed projects on environmental conditions for species of special interest. Recently, Roberts and O'Neil (1985) provided a procedure for selecting species for HEP assessments. Rewa and Michael (1984) described a way of evaluating environmental quality for ecological guilds by using an HEP approach.

Wakeley and O'Neil (1988) presented methods to increase efficiency in applying HEP. Their suggestions included delineating cover types by using

Figure 6.4a–d. Habitat capability (HC) model for assessing Rocky Mountain elk winter range in eastern Oregon or Washington. (From Wisdom et al. 1986.)

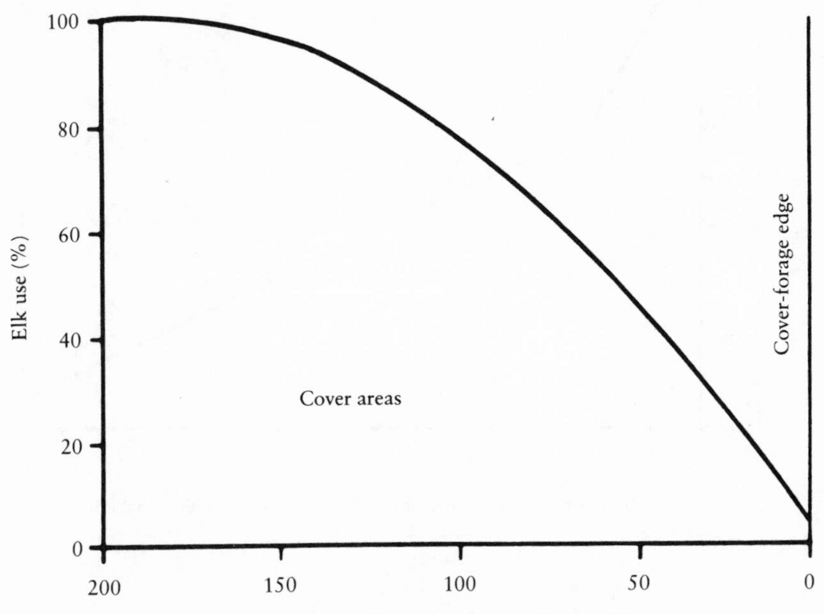

	Distance from cover-forage edge (yards)	Percent use from figure 3	Rating of habitat effectiveness
	>400	5	0.05
	301--400	10	0.10
Forage	201--300	25	0.25
area	101--200	70	0.70
Edge	0--100	100	1.0
	0--300 [1]	100	1.0
Cover	301--400	80	0.80
area	401--500	60	0.60
	501--600	40	0.40
	>600	20	0.20

[1]Cover areas not large enough to exceed 100 yards from edge (or 200 yards in width)

Distance from cover-forage edge (yards)	Percent use from figure 5	Rating of habitat effectiveness
0--100	50	0.50

Figure 6.4a.

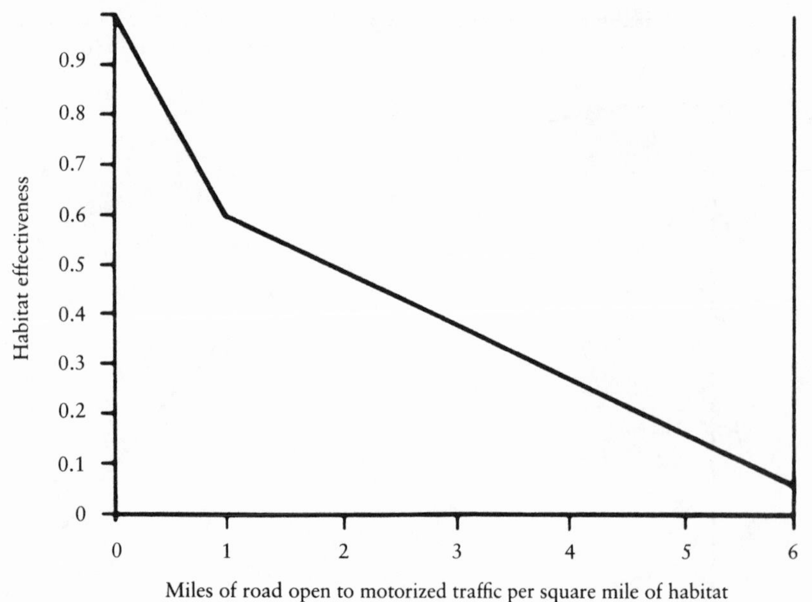

Figure 6.4b.

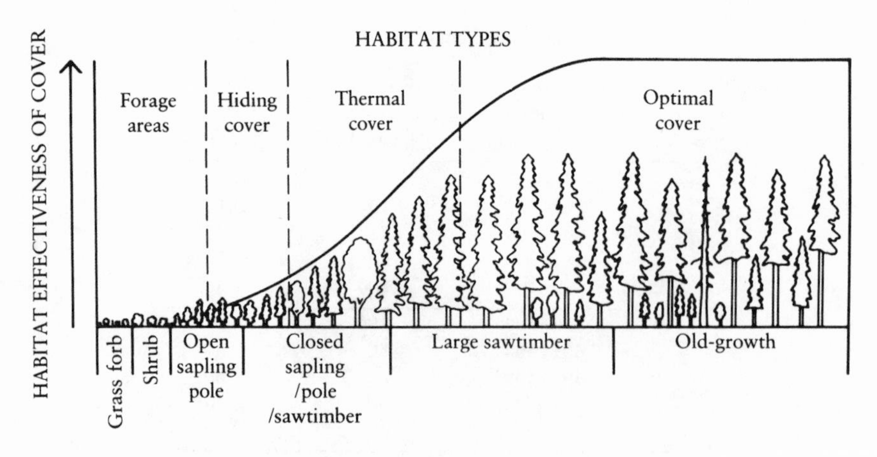

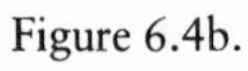
Figure 6.4c.

Land type before treatment application	Treatment category	Clearcutting	Burning	Seeding	Fertilization	Commercial thinning	Shelterwood cutting	No treatments	Rating of effectiveness
		Treatments							
Cover[1]	1	■	■	■	■				1.0
	2	■	■	■					0.75
	3	■	■						0.50
	4	■							0.25
	5					■			0.10
	6						■		0.10
Meadow/ pasture[2]	7				■				1.0
	8							■	0.75
Talus[3]	9							■	0.05

[1]Any forest stand with overstory canopy closure of 60 percent or higher before clear-cutting, commercial thinning, or shelterwood cutting. After treatment application, areas are assumed to be forage areas.

[2]Any **permanent** forage area not classified as cover or talus before treatment application.

[3]Any **permanent** forage area dominated by rock outcrops or talus with minimal soil present.

Figure 6.4d.

remote imagery and combining types; choosing wildlife species to model for which there is available inventory information; choosing model forms that make the best use of available inventory data and that focus on the most important life history components; designing field sampling for environmental conditions to be cost-effective and tailored to the range of modeled conditions; and using computers to aid in collecting and analyzing field inventory data and conducting model analysis.

Bayesian and pattern recognition (PATREC) models. Pattern recognition models are useful for predicting effects of changes in environmental conditions on wildlife in the form of a risk analysis. PATREC, based on the use of Bayes's Theorem (Williams, Russell, and Seitz 1977), calculates the probability of a species' response, such as its population density, given the probabilities of various environmental conditions being present. The general form of the model is

$$P(S|E) = \frac{P(E|S)P(S)}{P(E)}$$

where $P(S|E)$ is the posterior probability of species density (S), given that specific environmental conditions (E) are present; $P(E|S)$ is the prior probability of environmental conditions being present, given a specific species density; P(S) is the unconditional probability of the species having a specific density; and P(E) is the unconditional probability of the environment being in specific conditions. Researchers have used PATREC models in forest planning by integrating them with vegetation response models (Kirkman et al. 1984). An example of a PATREC model in use is the evaluation of the habitat of bighorn sheep (*Ovis canadensis*) by Holl (1982), shown in table 6.2.

Values of prior probabilities, $P(E|S)$, are often difficult to determine empirically and usually represent best professional judgment. Also, posterior probabilities, $P(S|E)$, are often quite sensitive to values of the prior probabilities. Thus, if priors are not known with much certainty, posteriors may be wrought with a large, and unknown, degree of error.

Researchers may also find a Bayesian approach useful in modifying sample wildlife surveys by analyzing interim results. This so-called empirical Bayes approach (Marks and Woodruff 1979; Morris 1983) was illustrated by Johnson (1985, 1989) for use with waterfowl samples. In his study, he modified the design of the USDI Fish and Wildlife Service sampling of waterfowl by analyzing results of surveys to date, so that estimators of waterfowl density were more accurate. The increase in accuracy was gauged by a reduction of an average 30 percent in the mean square errors of one density estimate (a Bayesian ratio estimator). The bias of the density estimates, however, also increased. Estimates were also larger or smaller than the usual estimators, but the author deemed the decrease in precision insignificant.

Population trend and viability models. In this new and rapidly growing area of ecology, concepts and models are being advanced to help scientists understand and predict potential risks of extinction of populations. Burgman, Akcakaya, and Loew (1988) recently reviewed the use of some extinction models for conserving species. Roozen (1987) and Strebel (1985) explored how variability in population parameters and environmental conditions affects extinction likelihoods. Quinn and Hastings (1987) explored

Table 6.2. Pattern recognition (PATREC) model of bighorn sheep

Habitat attributes	Conditional probabilities High	Low
1. Area is below 6,000 ft. elevation	0.60	0.40
2. More than 60% of the area has aspects between 130 and 230° from North	0.60	0.35
3. Water is available within 150 yards of escape terrain	0.80	0.65
4. More than 35% of the area is a birch leaf mountain mahogany association	0.60	0.20
5. Percent of the area within 150 yards of escape terrain having a shrub canopy cover less than 35%		
a) less than 25%	0.20	0.40
b) 25% to 50%	0.30	0.35
c) more than 50%	0.50	0.25
6. Percent of the area occupied by trees		
a) less than 30%	0.60	0.20
b) 30% to 50%	0.30	0.25
c) more than 50%	0.10	0.55
7. Grasses and forbs compose more than 5% of cover	0.70	0.35

Source: Holl 1982.
Population Density Standards: High = 17/km^2; Low = 6/km^2.
Prior Probabilities: $P(H)$ = High = 0.60; $P(L)$ = Low = 0.40.

Computations to provide an output are quite simple and can be done by hand, using a small calculator or a computer. Once the required inventory data are gathered from an area, they are used in the following equation, Bayes's Theorem:

$$P(H_{ID}) = \frac{P(H) \times P(ID_H)}{[P(H) \times P(ID_H)] + [P(L) \times P(ID_L)]}$$

where $P(H_{ID})$ is the probability that the area supports a high-density population, based on the inventory data; $P(H)$ and $P(L)$ are the probabilities of having a high- or low-density area, respectively (prior probabilities); and $P(ID_H)$ and $P(ID_L)$ are the probabilities that the inventory data have a high- or low-density potential, respectively (conditional probabilities).

The following will serve as an example. An imaginary area, Bighorn Canyon, has been inventoried, and attributes are met for 1, 3, 5b, 6b, and 7. Attributes 2 and 4 were not met. First, the probabilities that the inventory data for the given area have high- or low-density potentials are calculated. Notice that when an attribute is not found, both conditional probabilities are subtracted from 1.

$$P(ID_H) = (0.6)(1 - 0.6)(0.8)(1 - 0.6)(0.3)(0.3)(0.7)$$
$$= 0.0048$$

(*table continued on following page*)

Table 6.2. (*Continued*)

$$P(ID_L) = (0.4)(1 - 0.35)(0.65)(1 - 0.2)(0.35)(0.25)(0.35)$$
$$= 0.0041$$

Substituting these data into the equation, we calculate

$$P(H_{ID}) = \frac{0.0048(0.6)}{0.0048(0.6) + 0.0041(0.4)}$$
$$= 0.64$$

From these data we would conclude that the probability that Bighorn Canyon supported a high-density population (17 sheep/km^2) was 0.64. The model also provides insight into some management options which are available to improve the area. Although habitat attributes 2 and 4 were not met, it is unlikely that management could change these. Attributes 5b and 6b can be managed, however; thus, the probability that Bighorn Canyon would support a high-density population could be increased.

how isolated patches of habitat enhance extinction probabilities. Taylor (1990) reviewed concepts describing the effects of metapopulation dynamics, dispersal, and predation on population viability.

Dennis, Munholland, and Scott (in press) provided procedures for analyzing the susceptibility of populations to randomly varying environmental conditions. Their approach allows estimating several parameters pertinent to declines and extinctions of populations, including the continuous rate of increase; the finite rate of increase; the geometric finite rate of increase; the probability of reaching a lower threshold population size; the mean, median, and most likely time of attaining the threshold; and the projected population size.

Conner (1988) argued for conservation of populations at high levels to ensure long-term viability, rather than at minimally adequate levels. Other recent reviews of quantitative population models and their uses have been offered by Fagen (1988) and Rabinovich, Hernandex, and Cajal (1985).

Microcomputers are being used extensively for simulating population trends and viability effects (see Ferson 1988; Fowles 1988; Maguire, Shaffer, and Tipton 1988). In an early approach of this type, Walters and Gross (1972) simulated big game populations for the purpose of developing management plans.

Indicator species models. Federal law in the United States mandates the use of management indicator species in national forest planning, in regulations pursuant to the National Forest Management Act of 1976. Wildlife species of management significance constitute categories of management indicator species: these include species that are threatened, endangered, of economic or social value, and that ecologically represent particular environments or other wildlife species also associated with those environments.

Recent controversy and confusion has erupted regarding the last category—ecological indicator species. The controversy has focused on whether the population attributes of one wildlife species can validly represent those of other wildlife species (Landres, Verner, and Thomas 1988; Patton 1987). The confusion has arisen from equating the general concepts of management indicator species with the specific category of ecological indicators used in the precise context of one species indicating other species. Scientific arguments soundly caution that few true ecological indicator species can be identified (see Chapter 4).

Multiple Species Models

Researchers have developed several approaches to assess environmental conditions for multiple wildlife species. Principal methods include use of species-habitat matrices, guild and life-form models, and community structure models. In general, multiple species models have the advantage over single species models that they simultaneously assess potentially conflicting species' requirements. Schroeder (1987) reviewed many ecological community models and discussed their structures and utility.

Species-habitat matrices. Species-habitat matrices provide one simple form of representing relationships between wildlife species and their environments. These matrices are tables listing vegetation types and environmental conditions associated with particular wildlife species. Often the data are qualitative and derive from a combination of field studies and professional judgment. Examples from the Wildlife Habitat Relationships Program of the USDA Forest Service include species-habitat matrices for amphibians, reptiles, birds, and mammals in the Pacific Northwest (Thomas 1979; Brown 1985), California (Marcot 1979; Verner and Boss 1980), Colorado (Hoover and Wills 1984), and New England (DeGraaf and Chadwick 1987). See Verner and Boss (1980) for examples of species-habitat matrices.

Such information bases may be useful for predicting sets of wildlife species associated with specific environmental conditions, such as old-growth forests in the Pacific Northwest (Marcot 1980). They can also be used in designing crude optimal patterns of vegetation conditions to meet requirements of many species simultaneously (Toth, Solis, and Marcot 1986), although spatial effects are usually not explicitly represented in the relationship information.

Validation of a wildlife-habitat relationships model by Raphael and Marcot (1986) suggested that researchers can probably best use such information bases to predict the occurrence of species in general vegetation types and in environmental conditions across broad regions rather than at the scale of an individual stand. A shortcoming of such models is that they generally do not quantify population response. Thus, such models cannot help to gauge population density or to quantify population trend.

Guild and life-form models. Guild or life-form models denote the response of a set of species with similar characteristics to changes in environmental conditions (see Thomas 1979; Severinghaus 1981; Short 1982; DeGraaf, Tilghman, and Anderson 1985). Such models simplify the assessment of many species by referring to a few sets of species. Guilds or life-forms may be defined *a priori* as sets of species with common attributes, as in the models of Thomas (1979) and Short (1983). They may also be defined by multivariate analysis of empirical data on species abundance and distribution, as in the assessment of Hubbard Brook bird guilds by Holmes, Bonney, and Pacala (1979). Figure 6.5 shows a guild representation of some land-dwelling bird species in the United States.

The guild approach may prove useful when environmental conditions and target species are well defined (see Landres and MacMahon 1980; Knopf, Sedgwick, and Cannon 1988; Maurer, McArthur, and Whitmore 1981; Block, Brennan, and Gutierrez 1987). Individual species of a guild, however, may vary disparately in response to environmental conditions while the guild as a whole shows little or no variation (Hariston 1981; Mannan, Morrison, and Meslow 1984; Block, Brennan, and Gutierrez 1987). Thus, whereas grouping species by guild might be useful for depicting groups of species with similar functions, the guild approach is not useful for predicting specific responses of each species to environmental conditions and changes. To adequately perform the latter task, a researcher could model each species within the guild individually and then combine results.

Community structure models. Community structure models describe wildlife species distribution, abundance, and diversity as a function of the structure of the environment. Multivariate statistics are commonly used to assess these relationships (see Erdelen 1984; Scott et al. 1987; Swift, Larson, and DeGraaf 1984). Raphael and Barrett (1983), for example, assessed the diversity of wildlife in late successional forests. A shortcoming of this approach, as with the guild approach, is that individual species' responses may vary considerably while the composite measures of community structure remain more or less constant. Also, researchers seldom express habitat management objectives in terms of wildlife species diversity indices (see Chapter 4).

A number of models discussed in this chapter can help resource managers plan for conserving biological diversity. The work by Scott, Csuti, Jacobi,

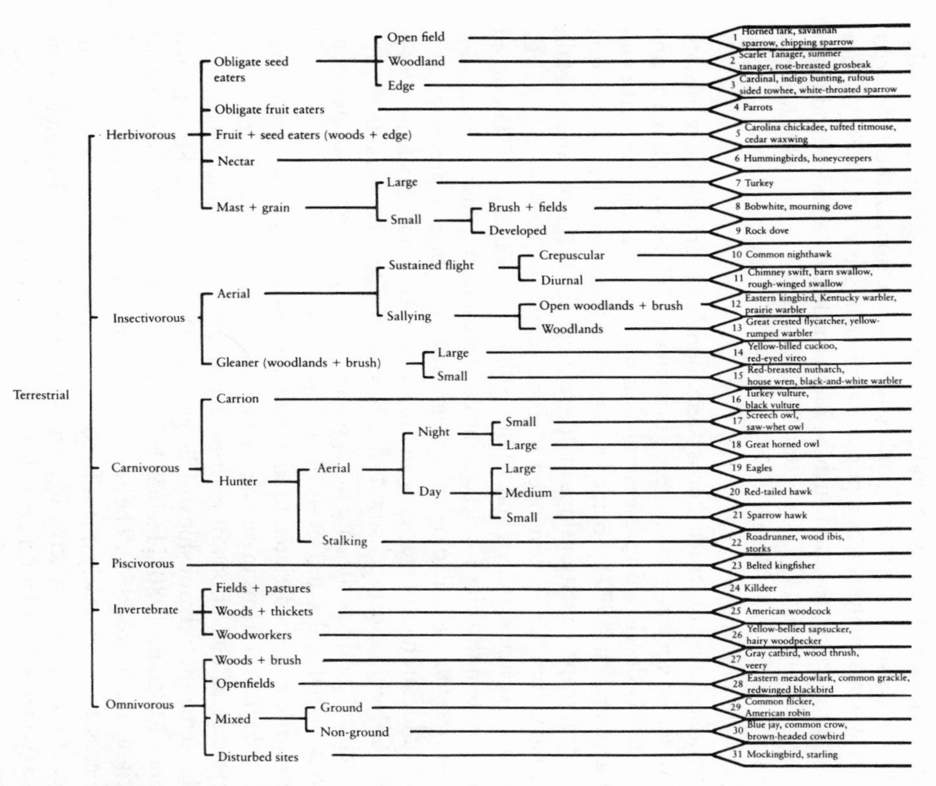

Figure 6.5. A guild classification of some land-dwelling bird species in the southwestern United States. (From Severinghaus 1981.)

and Estes (1987), however, specifically aimed at mapping geographic areas of high species richness for this purpose. Also, Udvardy (1984) presented a biogeographic classification system for similar planning purposes.

Community and ecosystem simulation models. Researchers use community and ecosystem simulation models to evaluate stochastic population and system responses to variable environmental conditions, effects of catastrophic changes in vegetation and environments on populations, and other aspects of species, community, and ecosystem function. In the 1960s ecological systems analysis was a popular approach (see Watt 1966). A number of more recent reviews (such as Biesterfeldt 1984) and introductory treatises (Eberhardt 1977; Gordon 1978; Law and Kelton 1982; Mihram 1976; Shannon 1975; Weiger 1975) on simulation modeling and systems ecology are available.

Scientists have constructed ecosystem models for grasslands, wetlands, and other ecological systems (e.g., Browder 1978). Smith, Shugart, and West (1981) developed forest simulation models for nongame bird management. French (1977) provided a conceptual framework for considering ecosystem trophic relationships in developing habitat classifications. Hall, Day, and Odum (1977) developed a "circuit language" for depicting energy flow in ecosystems. Patten and Auble (1981) and Walters (1971) offered general system theories for describing ecological niches and systems.

Hierarchy models. Ecological systems also have been represented as a series of hierarchies of elements and functions (Allen and Hoekstra 1984; Allen, O'Neill, and Hoekstra 1984; Kolasa 1989). The advantage of a hierarchical model is the ease of representing environments and species at various spatial scales and levels of biological organization.

Semantic models. Related to hierarchy models, semantic models represent ecological systems in semantic categories. That is, these models depict ecosystems as collections of entities that interrelate by functions similar to the semantic relationships of languages (Deutsch 1966; Ebeling and Jimenez-Montano 1980; Haefner 1974; Rashevsky 1965; Rescigno 1964; Rescigno and Segre 1965). The advantage of this approach is that it is often not necessary to precisely quantify population sizes and their interrelationships, as one must do when using strictly mathematical models such as the Lotka-Volterra population growth models or the Lotka differential competition equations. Use of semantic theory for depicting ecosystems and ecological communities, however, has lost favor in recent literature.

Models of Landscapes

Models included in this category display spacing and patterns of habitat patches explicitly at the landscape scale. Models of landscapes include vegetation disturbance models, fragmentation models, and cumulative effects models.

Vegetation disturbance models. Vegetation disturbance models simulate the extent and distribution of vegetation stages across a landscape given the frequency and intensity of disturbances such as fire and timber harvesting (see Shugart 1984; Shugart and Seagle 1985; Pickett and White 1985). Scientists can use such models to plan habitat management (Smith, Shugart, and West 1981; Karr and Freemark 1985).

Researchers have advanced many kinds of models for predicting changes in vegetation conditions. These include models of vegetation cover in forest stands (Moeur 1985); succession (Horn 1975; Culver 1981); disturbance and patch dynamics (Pickett and White 1985); development of old forest stand conditions (Dale and Hemstrom 1984); effects of fire on environmental conditions (Keane, Arno, and Brown 1990); and integration of forest succession and wildlife habitat capability models (Holthausen and Dobbs 1985).

Fragmentation models. Models displaying species response to fragmentation and isolation of habitat patches include incidence functions, which denote the probability of a species occurring in a particular patch given its size. The greater the degree of fragmentation, the higher the likelihood of extinction of species associated with forest interiors becomes (Wilcove 1987).

Fragmentation models address a variety of factors that may influence species persistence in a landscape: patch size (Askins, Philbrick, and Sugeno 1987; Blake and Karr 1987; Cole 1981; Lynch and Whigham 1984); isolation of patches (Bock 1987; Faanes 1984; Fahrig and Merriam 1985; Fahrig and Paloheimo 1988); patch patterns (Toth, Solis, and Marcot 1986; Lynch and Whigham 1981; Stamps, Buechner, and Krishnan 1987); and competition and structure of environments (Dueser and Porter 1986). These models also address the effect on population regulation caused by the interaction of environmental fragmentation with parasites (Dobson 1988) and predators (George 1987; Martin 1988; Savidge 1987; Yahner and Scott 1988). The use of insular biogeography theory for predicting effects of forest fragmentation on population persistence and quality of forest environments has served as the basis for some management guidelines (Franklin and Forman 1987). Noss and Harris (1986), Simberloff and Cox (1987), and others have addressed the utility of planning for corridors in forested landscapes.

Cumulative effects models. Researchers use cumulative effects models to assess effects on distribution and abundance of wildlife species across a landscape. The term *cumulative effects model,* a generic one, can apply to any assessment of the effects on wildlife species from management activities or natural disturbances across a geographic area or over time. Salwasser and Samson (1985) discussed the development of cumulative effects models, noting the major steps: stating management goals and standards, representing major environmental factors, projecting changes in environments, and estimating wildlife effects.

Weaver and his colleagues (1985) presented a cumulative effects simulation model for assessing grizzly bear habitat in the Yellowstone ecosystem. Their model consists of three submodels used to assess suitability of environments, bear displacement from human activities, and bear mortality under one or more planning scenarios. Each submodel in turn consists of variables deemed important to the presence of grizzly bears. For example, the habitat submodel accounts for food and thermal cover, habitat diversity, seasonal equity, and denning suitability. The cumulative effects of all environmental conditions, displacement, and mortality influences on the presence of grizzly bears are combined for a particular location and set of management conditions to produce habitat effectiveness values and a bear mortality risk index. Running the grizzly bear cumulative effects model requires at least a large minicomputer, as it is sizable and computation-intensive.

Habitat capability models can also aid in assessing cumulative effects when tied to models that predict future amounts of environments in sub-drainage scale areas. Examples include habitat capability models linked with FORHAB (Smith 1984; Smith, Shugart, and West 1981); HABSIM (Raedeke and Lehmkuhl 1984); ECOSYM (Henderson, Davis, and Ryberg 1978; Davis 1980; Davis and DeLain 1984); DYNAST (Benson and Laudenslayer 1984; Sweeney 1984; Holthausen 1984); FORPLAN (Davis and DeLain 1984; Holthausen 1984); TWIGS (Belcher 1982; Brand, Shifley, and Ohmann 1984); and STEMS (Belcher, Holdaway, and Brand 1982). Powers (1979) used linear programming to develop an optimal mix of agricultural and wildlife land use.

Holthausen and Dobbs (1985) presented the cumulative effects model FSSIM, which was used to assess management activities in management units of one thousand to twenty thousand acres and to project the consequences of those actions over time. Salwasser and Tappeiner (1981) developed a model designed to assess effects of timber harvest scheduling on spatial patterns of habitats.

Models of insular biogeography. These models, born of theory or empirical data, depict effects of habitat fragmentation and isolation on species presence and diversity on islands or in landscapes. For example, Blake and Karr (1987) studied the effects of the isolation of woodlots on breeding birds.

Related to models of biogeography are models depicting species' responses to spatial configurations of vegetation and environments (see Perry 1988). These models currently seem the most promising for helping us understand and ultimately prescribe habitat allocations, including size of habitat patches and connections or corridors between patches. Buechner (1987) and Stamps, Buechner, and Krishnan (1987) modeled vertebrate dispersal across boundaries of vegetation patches. Fahrig and Merriam (1985) explored the effects of patch connectivity on population survival. Hassell (1987) explored how patchiness of an animal's distribution would contribute to overall population regulation. Lynch and Whigham (1981) also evaluated patterns of habitat patches. See Hastings (1990) and Murray (1988) for recent reviews of concepts of spatial heterogeneity in ecological modeling and spatial dispersal of species.

Models for Monitoring Species and Habitats

Models of species-habitat relationships may prove useful for the expensive task of monitoring species and habitats over time. Single species models are probably best for monitoring management indicator species and other species of singular concern. Models depicting environmental associations, such as HSI, HC, PATREC, and HEP models, are useful when directly monitoring the species under consideration is too expensive. It is vital, however, to first demonstrate the degree of reliability and validity of the model. Also, population trends of species of high concern, especially species on state and federal threatened or endangered lists, should be monitored directly in the field rather than inferred through habitat relationships models.

In monitoring, a useful model is the adaptive management paradigm (Walters 1986; Walters and Hilborn 1978; Holling 1984). Adaptive management entails viewing a habitat management plan and its expected effects on wildlife species as a hypothesis; monitoring environments and species to ascertain how they respond to the management; and revising the direction of management if results so warrant. Monitoring, an essential step in the adaptive management feedback process, helps managers deal directly with limitations of biological uncertainty (Lee and Lawrence 1986).

Knowledge-Based and Decision-Aiding Models

Models that aid habitat evaluation and decision-making provide another class of tools for assessing species-habitat relationships. An explosion in computer-aided decision-making has occurred in recent years, and many fields of resource management have participated in the trend. Entomology offers a recent example (Coulson and Saunders 1987). Such models help organize and document factors associated with habitat evaluation and planning. These models include decision support models and models for helping to monitor species and environments.

Decision Support Models

Decision support models aid in weighing and prioritizing decisions and specific actions for habitat planning. These models may use optimization algorithms, as with the linear program FORPLAN, used by the USDA Forest Service to predict the future availability of commercially harvestable forests. The models can also help the manager document and organize information pertinent to developing a habitat plan.

A flurry of models derived from operations research became popular in the 1960s and 1970s (Hillier and Lieberman 1980; Milsum 1966; Rosen 1967). Such models included linear programming (Field 1977); goal programming (Dane, Meador, and White 1977; Schuler, Webster, and Meadows 1977); control theory (Williams 1985); and other approaches. Rudis and Ek (1981) provided a procedure for optimally prescribing forest island spatial patterns, although Salwasser and Tappeiner (1981) took a more programmatic approach to the same problem, using information on wildlife-habitat relationships.

One peculiar offshoot of decision theory models, game theory, has been used to explain and evaluate species interactions (Riechert and Hammerstein 1983). In game theory, typically two species, two individual animals, or even two competing or alternative management activities are represented in an interaction in which each has something to lose and something to gain. A game model helps evaluate how frequently various behaviors would occur, or what the optimal action should be, given the potential benefits and detriments to each. The optimal outcome maximizes the gain and minimizes the loss for one "player"; thus the optimal strategy is called a maximin solution. Using this approach, Hansen (1986) evaluated fighting behavior in bald eagles.

Expert Systems

Related to decision support models, advisory models help increase a manager's expertise in habitat evaluation. Such models include expert systems, computer programs that reason like human experts to solve a problem in classification, evaluation, and prescription by capturing the expertise of a specialist in if-then rules. Creating expert systems is a growing area of computer modeling (Marcot 1986b; Stock 1987; Rauscher 1988; Marcot, McNay, and Page 1988). These models can be used to diagnose, evaluate, and classify environmental conditions, prescribe management activities, and monitor environments and activities over time.

Marcot (1986b, 1988) discussed expert systems that aided evaluation of forest conditions for birds and identification of wildlife species. Expert systems also have assisted in fire management (Cohen et al. 1989; Davis, Hoare, and Nanninga 1986; Kourtz 1984, 1987); entomology (Stone et al.

1986); remote sensing of vegetation (Estes, Sailer, and Tinney 1986; Goldberg et al. 1985; McKeown 1987); forest road planning (Thieme 1987; Thieme, Jones, and Gibson 1986; Thieme et al. 1987); soils classification and management (McCracken and Cate 1986; Yost et al. 1986); scheduling of timber harvest (Hokans 1984); forestry and the forest products industry (Schmoldt and Martin 1986; Mills 1987); management of big game habitat (McNay, Page, and Campbell 1987); and conservation planning of special habitats (O'Keefe, Danilewitz, and Bradshaw 1987).

A rapidly growing area of expert system application is in interpretation of remote sensing data and development of geographic information systems (GIS) (Goldberg et al. 1985; Graklandoff 1985; Robinson, Frank, and Blaze 1986; Vogel 1989). The utility of GIS lies not only in compiling and graphically displaying geographic, cultural, and environmental data but also in manipulating those data. Most GIS have inherent functions for assessing spatial conditions such as area relationships and contiguity of mapped areas (polygons). More advanced functions include assessments of rates of change (such as slope depicted by rate of change in elevation) and spread functions that help assess border conditions surrounding polygons.

Many GIS allow such assessments to be automated into macro commands which can be used to help interpret and design landscape conditions and to schedule activities. Such automated assessments take on the form of an expert system when they account for the landscape context of a particular site being analyzed.

One example is ASPENEX, an expert system/GIS tool that determines site conditions of aspen stands (White and Morse 1987). ASPENEX assesses stand type, stand size, soil type, and distance to roads in helping to recommend management actions. ASPENEX combines the GIS model MOSS (public domain) with a commercially available "production rule" expert system shell. In one of its uses, ASPENEX identifies mature aspen stands of particular sizes on well-drained sites in close proximity to roads as having a high potential for timber management.

TEAMS (Terrestrial Ecosystem Analysis and Modeling System) is a multiresource model used for assessing scheduling of projects in southwestern ponderosa pine forests (Covington et al. 1988). The model, used at a subbasin scale of 250–8,000 ha, is one of the first operational models to link decision support functions, such as optimization (goal programming) routines, with geographic information systems (GIS) that display schedules and effects. TEAMS aids tactical planning of projects that comply with specific objectives and general standards and guidelines. TEAMS actually comprises eleven program packages, including a commercially available relational data base, a geographic information system, a linear program, and a graphics charting program. TEAMS uses ECOSIM, a model that simulates the effects of multiresource forest management on selected wildlife species and on timber. Planned enhancements include a module to help monitor project implementation and effects. One version of TEAMS predicts resource conditions and outputs over twenty years from annual treatment schedules applied over ten years. Another helps develop a thirty-year project schedule by decade and displays outputs over fifty years by five-year intervals. The system is being used for teaching at the School of Forestry, Northern Arizona University, Flagstaff, and for project analysis in ponderosa pine forests on Coconino National Forest and Navajo Tribal Trust lands in Arizona.

The U.S. Geological Survey is integrating an expert system knowledge base into a geobased information system to help interpret and assess earth science data (Vogel 1989). The hybrid system uses the geobased information system ARC/INFO from Environmental Systems Research Institute and the expert system development tool NEXPERT from Neuron Data. The marriage of GIS and expert systems tools is intended to help aid expert interpretations of areas with potential groundwater contamination in Thurston County, Washington. One output will be a map showing potentially contaminated areas.

Expert systems can also be used in developing GIS models through interpretation of remotely sensed data (McKeown 1987). Goodenough, Goldberg, Plunkett, and Zelek (1987) reported on expert systems used for interpreting remote sensing data for use in GIS at the Canadian Centre for Remote Sensing in Ottawa. In their system, the expert system model MICE (Map Image Congruency Evaluation) compares and reconciles maps and images, and the program module entitled Analyst Advisor helps with extraction of resource information from remotely sensed data.

Robinson, Frank, and Blaze (1986) outlined other potential benefits of an expert systems approach. An integrated expert system can help design maps, analyze terrain features, enhance the interface between the GIS digital data base and the user, and provide information critical for geographic decision-making. Robinson, Frank, and Blaze also reviewed examples of programs designed for each of these tasks.

What are the future prospects for a marriage of expert systems and GIS? One major area for further development is map production. Use of scale, color, symbols, legends, and the like can be automated by integrating expert system rules to help draw the final product (Robinson, Frank, and Blaze 1986). Future development of expert systems can aid in our understanding of general land use patterns, given knowledge about climate, soil, and terrain. Such a system is MAPS, which contains and interprets image data on the Washington, D.C., area (McKeown 1987).

Another major area for future development is landscape design and locating and scheduling activities. Given constraints of the land base, feasibility of operations, characteristics of activities, and planning goals and objectives, an expert system can help recommend potential map designs and activity schedules. There may typically be more than one solution to a map design problem. For example, in designing landscapes to meet wildlife management objectives, a number of configurations of habitat patch patterns may meet overall population and habitat management goals. The TEAMS model is beginning to address such design functions. From cartographic and topological perspectives, automating the design of landscapes is a difficult problem. Much work remains in this area.

Validating Wildlife-Habitat Models

Researchers should validate wildlife-habitat models as part of every step in building and using such tools (Marcot, Raphael, and Berry 1983). Model validation, best thought of as a general approach to developing as well as testing models, should be conducted in a variety of ways throughout the model development and application process (Caswell 1976; Gass 1977; Gentiol and Blake 1981; Law 1981; Leggett and Williams 1981; Mankin et al. 1977; Marcot, Raphael, and Berry 1983; Naylor and Finger 1967; Schellenberger 1974; VanHorn 1969, 1971).

Aspects of validating a model include model verification, verifying that mathematical equations are correct or that the computer program code has been written without bugs; testing the audience, ensuring that the audience for whom the model is intended will accept and use the tool; running the model, confirming that the model can be run with available or obtainable data; assessing purpose and context, ensuring that the purpose of the model and the conditions in which it is to be used have been clearly stated and adhered to in its use; and testing the output, assessing whether the output of the model matches real-world biological conditions. Researchers typically associate model validation with just the first and last items on this list. Each item, however, contributes to successful development and application of a wildlife-habitat model (see table 6.3).

Model Verification

Ensuring that formulas and computer codes are correctly written down is a simple but important aspect of model validation. A similar task is documentation. Documentation refers to explicitly explaining the development procedure used to create the model; writing down major assumptions and uncertainties inherent in the model; disclosing sources of information and analyses used to develop variables and their relationships in the model;

Table 6.3. Criteria useful for validating wildlife-habitat relationships models

Criterion	Explanation
Precision	Capability of a model to replicate particular system parameters
Generality	Capability of a model to represent a broad range of similar systems
Realism	Accounting for relevant variables and relations
Precision	Number of significant figures in a prediction or simulation
Accuracy	How well a simulation reflects reality
Robustness	Conclusions that are not particularly sensitive to model structure
Validity	A model's capability of producing all empirically correct predictions
Usefulness	If at least some model predictions are empirically correct
Reliability	The fraction of model predictions that are empirically correct
Adequacy	The fraction of pertinent empirical observations that can be simulated
Resolution	The number of parameters of a system which the model attempts to mimic
Wholeness	The number of biological processes and interactions reflected in the model
Heurism	The degree to which the model usefully furthers empirical and theoretical investigations
Adaptability	Possibilities for future development and application
Availability	Existence of other, simpler, validated models that perform the same function
Appeal	Matching our intuition and stimulating thought, and practicability
Breadth	Proportional to the number and kinds of variables chosen to describe each (habitat) component
Depth	Proportional to the number and kinds of variables chosen to describe each (habitat) component
Face validity	Model credibility

Table 6.3. (*Continued*)

Criterion	Explanation
Sensitivity	Model variables and parameters matching real-world counterparts, their variation causing outputs that match historical data; also, dependence of model output on specific variations of variables
Hypothesis validity	The realism with which subsystem models interact
Technical and operational validity	Identification and importance of all divergence in model assumptions from perceived reality, as well as the identification and importance of the validity of the data
Dynamic validity application	Analysis of provisions for application to be modified in light of new circumstances

Source: Marcot, Raphael, and Berry 1983.

and annotating any computer code. The more a model is verified, the more open it is to understanding and critique.

Verification is an especially important aspect of modeling when meeting legal mandates is a concern, as in developing models for use in preparing documents in accordance with the National Environmental Policy Act, such as environmental impact statements. In this case, keeping careful records is of paramount importance.

Testing the Audience

The best model in the world may fail to be used if it is too complex or esoteric. It will also be ignored if existing administrative organizations or policies do not provide for its use, or if it for some reason is not credible. In an operational sense, a model is valid, in part, if it is accepted (has face validity) and is usable in the intended work setting. Thus, developing models with teams combining managers and researchers helps enhance the utility of such tools (Bunnell 1989). A team approach helps ensure that the model addresses the correct question, is based on data available from existing data bases, is credible, and can be used in the everyday course of work.

Running the Model

In a typical management situation, a wildlife-habitat model is used to predict the response of wildlife species to potential environmental conditions created from alternative management activities. Such a model must run with data available from existing or easily obtainable inventories of vegetation and environments. Inventories that are dated or incomplete or fail to include pertinent variables may limit accuracy of model predictions, however. In such a case, the models may be more useful for suggesting changes to inventory procedures. At best, one might use proxy variables to represent missing variables; the size of a forest opening, for example, might substitute for a more complex representation of habitat patch juxtaposition. If proxy variables are used, then their degree of correlation with the intended variable should certainly be evaluated.

Purpose and Context

The purpose of a model, often incompletely stated, should guide the model's use. If a model is intended to be used to predict real-world environments and populations, then it should be evaluated against a different set of criteria than if it is intended to formalize our knowledge and understanding.

Also, the model-builder should specify the context of a model. Context includes the range of environmental conditions (such as weather), types of environments, and seasons in which the model was built and tested. The model-user must adhere to those conditions. When a model is used outside those conditions, its accuracy and reliability are essentially unknown.

Testing the Output

Too often, models created for prediction are untested in real-world situations. This occurs for a variety of reasons. Model-builders often do not consider validation until after the models are built, and postconstruction validation costs too much time and money. Sometimes models are built mostly from theory and are difficult or impossible to test, even if they are used for prediction. It may be unclear or unspecified what the model output represents, as with habitat capability index models.

Not all models require rigorous field testing. The validity of models used for helping make decisions about irreversible or expensive losses of environments and populations should be known, however. In this case, the researcher should evaluate the accuracy, bias, precision, and reliability of the model (see Golbeck 1986).

A habitat relationships model can predict the presence or absence of a species (table 6.4). (This simple framework for identifying correct and incorrect model predictions based on presence or absence of a species can be extended to evaluate predictions of population density or various levels of presence or abundance; see Raphael and Marcot 1986.) These predictions

Table 6.4. Comparing model predictions with true species' distributions results in two correct combinations and two kinds of error

	True distribution	
Model prediction	Species present	Species absent
Present	Correct	Type I error
Absent	Type II error	Correct

combined with true distributions yield two potentially correct outcomes and two erroneous ones. A Type I error in prediction occurs when the model predicts presence and the species is actually absent. This error could occur because of inadequate or incorrect sampling for the species, because the field study was conducted during the wrong season, because the species is inherently rare and does not maximally occupy all suitable environments, or because the model overstated the value of environmental parameters or failed to account for an environmental condition that deduces the presence of the species. The degree to which a model avoids Type I errors is given by the confidence coefficient P (where $P = 1 - \alpha$).

On the contrary, a Type II error in prediction occurs when the model predicts absence and the species is actually present. This error could occur because the animals detected were wandering or because their presence did not indicate actual environmental quality, because of sampling design, or because the model did not include a vital parameter that affects presence of the species. The degree to which a model avoids Type II errors is given by the power of the model, $1 - \beta$ (Toft and Shea 1984). Few habitat models or tests of their validity have addressed power.

The ramifications of each type of error of model prediction depend on the model's intended use (Marcot 1986a). If the objective is to identify needs for mitigation, in order to purchase or trade habitats with high opportunity costs or to restore or enhance environmental conditions, then the model must accurately predict species presence. That is, frequencies of Type I errors should be minimized because costs of actions based on model predictions are high. On the other hand, if the model is to be used for predicting impacts, especially on rare or vulnerable species, then errors in predicting species presence or positive responses may be tolerable, but false predictions of species absence or negative responses might be of greater concern than in the case of mitigation. Thus, the power of a model and its ability to avoid Type II errors are critical.

Few wildlife-habitat models intended to be used to inform managers and decision-makers have undergone formal statistical testing. Such studies are costly, tedious, unglamorous, and often viewed as unnecessary once a working model has been built. Notable exceptions include the work by Cook and Irwin (1985), who tested and revised an HSI model for pronghorn (*Antilocapra americana*). The model included five variables deemed important components for pronghorn winter range: shrub canopy cover, shrub height, shrub diversity, availability of winter wheat, and topographical cover. The HSI model converted field values of each of these variables to a 0 to 1 scale and then produced a composite value of all five rescaled values. Cook and Irwin collected environmental data from twenty-eight winter ranges in Montana, Idaho, Colorado, and Wyoming, calculated corresponding HSI values, and then correlated known pronghorn densities. The HSI model explained 39 percent of the variation in pronghorn density. By using simple linear regression, Cook and Irwin identified shrub cover as the most controlling variable. They then restructured the HSI model to maximize the correlation, which then increased to 50 percent (see Chapter 4 for additional discussion of this study).

In general, most HSI or habitat models can be expected to account for roughly half the variation in species density or abundance. On-site environmental conditions generally account for even less variation in population density when one is considering migratory species, especially neotropical bird migrants. At first it might seem that the low correlations from Cook and Irwin's tests suggest that the model is not very useful, but considering the large array of other factors that affect population density, the model—especially their modified version—performed well. The authors, too, recognized that density can be misleading as an index to suitability of an environment or to fitness of the individuals in the population.

Here also is a lesson for the manager who will use the model for maintaining environmental conditions. The manager must understand that most models that predict species presence, population density, or species richness from environmental characteristics will likely capture only a portion—typically half or less—of the variation in those species' parameters. This does not mean that habitat is unimportant; it is usually critical. It means that one cannot manage for environmental conditions alone and expect with high confidence that the population will show a direct response. Also, by managing for readily measurable environmental conditions, we control only a portion of the factors that affect the occurrence and abundance of species.

Given these validation results, the appropriate use of habitat models appears to be in helping us recognize the degree (correlation) to which we can provide for species presence and abundance and thus which environmental parameters under consideration are the most critical. Such models can also help us assess potential effects on species from alternative management scenarios. If such models are used specifically to predict population size, however, the predictions should be treated as hypotheses. Such predictions would assume that all factors not considered by the model—the 50 percent or more unexplained variation in occurrence or abundance—are unimportant or exist at optimal values. This assumption is invariably false.

Lancia, Miller, Adams, and Hazel (1982) tested a HEP model for predicting distribution of bobcats (*Felis rufus*) in southeastern evergreen forests of North Carolina. They tested how well the model predicted the frequency of use of vegetation patches by six radio-telemetered bobcats. To do this, they constructed a map showing HEP values for each cell in the study area. Their results suggested a significant correlation between habitat quality index values and bobcat frequency. Although these trends suggested that the model was useful for prognosticating occurrence of bobcats among locations in the study area, however, only 21 percent of predicted use levels exactly matched observed frequencies of use. By combining some use levels, the model predicted 56 percent of bobcat frequencies correctly. Some 12 percent of all cells showed a high frequency of bobcat occurrence but had low predicted habitat quality values. This error suggested the inability of the model to capture all components of good habitat. On the other hand, 32 percent of the cells had a low frequency of bobcat occurrence but a high predicted habitat quality. According to the authors, this error may have occurred because such sites usually existed adjacent to activity centers, and the bobcats may have used the sites less because of behavioral, population, or geometric effects, or because such sites may have been occupied by untagged cats.

The study by Lancia, Miller, Adams, and Hazel was essentially a test of how well their HEP model predicted habitat selection by individual bobcats within a study area, whereas the Cook and Irwin validation tested how well the pronghorn HSI model predicted population density between study areas.

Other validation studies of habitat relationships models include those of Laymon and Barrett (1986) and Laymon and Reid (1986), who tested a model predicting occurrence of spotted owls based on occurrence and distribution of old forests (Laymon, Salwasser, and Barrett 1985). Raphael and Marcot (1986) validated a wildlife-habitat relationships model for amphibians, reptiles, birds, and mammals in a Douglas-fir sere in California. In each of these tests, researchers use different criteria to test various aspects of model prediction, including the robustness of the models, the sensitivity of predictions, the precision of input variables, and the accuracy of predictions of species' abundances.

Overall, validating models is a multifaceted problem and should be done routinely as models are built and used. Validation should address the appropriateness of the objectives and structure of the model, the utility, reliability, accuracy, and completeness of the model, and its credibility.

Literature Cited

Allen, T. F. H., and T. W. Hoekstra. 1984. Nested and non-nested hierarchies: A significant distinction for ecological systems. In *Proceedings of the Society for General Systems Research Conference* Vol. 1, *Systems methodologies and isomorphies*, ed. A. W. Smith, 175–180. Salinas, Calif.: Intersystems Publications.

Allen, T. F. H., R. V. O'Neill, and T. W. Hoekstra. 1984. *Interlevel relations in ecological research and management: Some working principles from hierarchy theory*. USDA Forest Service General Technical Report RM–110.

Arney, J. D. 1985. *User's guide for the Stand Projection System (SPS)*. Report no. 1. Spokane, Wash.: Applied Biometrics.

Askins, R. A., M. J. Philbrick, and D. S. Sugeno. 1987. Relationship between the regional abundance of forest and the composition of forest bird communities. *Biological Conservation* 39:129–52.

Avery, T. E., and H. E. Burkhart. 1983. *Forest measurements*. 3d ed. New York: McGraw-Hill.

Balda, R. P., W. S. Gaud, and J. D. Brawn. 1983. Predictive models for snag nesting birds. In *Snag habitat management: Proceedings of the symposium*. J. W. Davis, G. A. Goodwin, and R. A. Ockenfels, technical coordinators, 216–22. USDA Forest Service General Technical Report RM–99.

Barclay, H. J., and P. T. Gregory. 1981. An experimental test of models predicting life-history characteristics. *American Naturalist* 117:944–61.

Belcher, D. M. 1982. TWIGS: The Woodsman's Ideal Growth Projection System. In *Microcomputers: A new tool for foresters*, ed. J. W. Moser, Jr., 70–95. West Lafayette, Ind.: Purdue Univ. Press.

Belcher, D. M., M. R. Holdaway, and G. J. Brand. 1982. *A description of STEMS—the Stand and Tree Evaluation Modeling System*. USDA Forest Service General Technical Report NC–79.

Belovsky, G. E. 1987. Foraging and optimal body size: An overview, new data and a test of alternative models. *Journal of Theoretical Biology* 129(3):275–87.

Benson, G. L., and W. F. Laudenslayer, Jr. 1984. DYNAST: Simulating wildlife responses to forest-management strategies. In *Wildlife 2000: Modeling habitat relationships of terrestrial vertebrates*, ed. J. Verner, M. L. Morrison, and C. J. Ralph, 351–55. Madison: Univ. of Wisconsin Press.

Biesterfeldt, R. C. 1984. System dynamics. *Forest Farmer* 43:10–11.

Blake, J. G., and J. R. Karr. 1987. Breeding birds of isolated woodlots: Area and habitat relationships. *Ecology* 68:1724–34.

Block, W. M., L. A. Brennan, and R. J. Gutierrez. 1987. Evaluation of guild-indicator species for use in resource management. *Environmental Management* 11:265–69.

Bock, C. E. 1987. Distribution-abundance relationships of some Arizona landbirds: A matter of scale? *Ecology* 68:124–29.

Boyce, S. G. 1977. *Management of eastern hardwood forests for multiple benefits (DYNAST–MB)*. USDA Forest Service Research Paper SE–168.

Boyce, S. G. 1980. *Management of forests for optimal benefits (DYNAST–OB).* USDA Forest Service Research Paper SE–204.
Boyce, S. G. 1985. *Forestry decisions.* USDA Forest Service General Technical Report SE–35.
Boyce, S. G., and N. D. Cost. 1978. *Forest diversity, new concepts and applications.* USDA Forest Service Research Paper.
Boyle, K. A., and T. T. Fendley. 1987. *Habitat suitability index models: Bobcat.* USDI Fish and Wildlife Service Biological Report 82(10.147).
Brand, G. J., S. R. Shifley, and L. F. Ohmann. 1984. Linking wildlife and vegetation models to forecast the effects of management. In *Wildlife 2000: Modeling habitat relationships of terrestrial vertebrates*, ed. J. Verner, M. L. Morrison, and C. J. Ralph, 383–87. Madison: Univ. of Wisconsin Press.
Browder, J. A. 1978. A modeling study of water, wetlands, and wood storks. In *Wading birds*, ed. A. Sprunt, J. Ogden, and S. Winckler, 325–46. New York: National Audubon Society.
Brown, E. R., tech. ed. 1985. *Management of wildlife and fish habitats in forests of western Oregon and Washington.* Part 1: *Chapter narratives.* USDA Forest Service Publication no. R6–F&WL–192–1985.
Buechner, M. 1987. A geometric model of vertebrate dispersal: Tests and implications. *Ecology* 68:310–18.
Bunnell, F. L. 1989. *Alchemy and uncertainty: What good are models?* USDA Forest Service General Technical Report PNW–GTR–232.
Burgman, M. A., H. R. Akcakaya, and S. S. Loew. 1988. The use of extinction models for species conservation. *Biological Conservation* 43(1):9–25.
Capen, D. E., ed. 1981. *The use of multivariate statistics in studies of wildlife habitat.* USDA Forest Service General Technical Report RM–87.
Caraco, T. 1979. Ecological response of animal group size frequencies. In *Statistical distributions in ecological work*, ed. J. K. Ord, G. P. Patil, and C. Tailie, 371–86. Burtonsville, Md.: International Co-operative Publishing.
Carey, A. B. 1984. A critical look at the issue of species-habitat dependency. In *New forests for a changing world*, 346–51. Washington, D.C.: Society of American Foresters.
Caswell, H. 1976. The validation problem. In *Systems analysis and simulation in ecology.* Vol. 4, ed. B. C. Patten, 313–25. New York: Academic Press.
Clutter, J. L., J. C. Fortson, L. V. Pienaar, G. H. Brister, and R. L. Bailey. 1983. *Timber management: A quantitative approach.* New York: John Wiley and Sons.
Cobb, L., and S. Zacks. 1985. Applications of catastrophe theory for statistical modeling in the biosciences. *Journal of the American Statistical Association* 80:793–802.
Cohen, J. E. 1978. *Food webs and niche space.* Princeton: Princeton Univ. Press.
Cohen, P. R., M. L. Greenberg, D. M. Hart, and A. E. Howe. 1989. Trial by fire: Understanding the design requirements for agents in complex environments. *AI Magazine* 10(3):32–48.
Cole, B. J. 1981. Colonizing abilities, island size, and the number of species on archipelagoes. *American Naturalist* 117:629–38.
Conner, R. N. 1988. Wildlife populations: Minimally viable or ecologically functional? *Wildlife Society Bulletin* 16:80–84.
Connor, E., and D. Simberloff. 1979. The assembly of species communities: Chance or competition? *Ecology* 60:1132–40.

Cook, J. G., and L. L. Irwin. 1985. Validation and modification of a habitat suitability model for pronghorns. *Wildlife Society Bulletin* 13:440–48.

Coulson, R. N., and M. C. Saunders. 1987. Computer-assisted decision-making as applied to entomology. *Annual Review of Entomology* 32:415–37.

Covington, W. W., D. B. Wood, D. L. Young, D. P. Dykstra, and L. D. Garrett. 1988. TEAMS: A decision support system for multiresource management. *Journal of Forestry* 86(8):25–33.

Culver, D. C. 1981. On using Horn's Markov succession model. *American Naturalist* 117:572–574.

Curtis, R. O., G. W. Clendenen, and D. J. DeMars. 1981. *A new stand simulator for coast Douglas-fir: DF–SIM user's guide*. USDA Forest Service General Technical Report PNW–128.

Dale, V. H. and M. Hemstrom. 1984. *CLIMACS: A computer model of forest stand development for western Oregon and Washington*. USDA Forest Service Research Paper PNW–327.

Dane, C. W., N. C. Meador, and J. B. White. 1977. Goal programing in land-use planning. *Journal of Forestry* 75 (June):325–29.

Davis, J. R., J. R. L. Hoare, and P. M. Nanninga. 1986. Developing a fire management expert system for Kakadu National Park, Australia. *Journal of Environmental Management* 22:215–27.

Davis, L. S. 1980. Strategy for building a location-specific multi-purpose information system for wildland management. *Journal of Forestry* 78:402–8.

Davis, L. S. and L. I. DeLain. 1984. Linking wildlife-habitat analysis to forest planning with ECOSYM. In *Wildlife 2000: Modeling habitat relationships of terrestrial vertebrates*, ed. J. Verner, M. L. Morrison, and C. J. Ralph, 361–69. Madison: Univ. of Wisconsin Press.

DeGraaf, R. M., and N. L. Chadwick. 1987. Forest type, timber size class, and New England breeding birds. *Journal of Wildlife Management* 51:212–16.

DeGraaf, R. M., N. G. Tilghman, and S. H. Anderson. 1985. Foraging guilds of North American birds. *Environmental Management* 9:493–536.

Dennis, B., B. E. Brown, A. R. Stage, H. E. Burkhart, and S. Clark. 1985. Problems of modeling growth and yield of renewable resources. *American Statistician* 39:374–83.

Dennis, B., P. L. Munholland, and J. M. Scott. In press. Estimation of growth and extinction parameters for endangered species. *Ecological Monographs*.

Deutsch, K. W. 1966. On theories, taxonomies, and models as communication codes for organizing information. *Behavioral Science* 11(Jan.):1–17.

Dixon, K. R., T. A. Young, K. L. Knutson, J. R. Eby, and H. L. Allen. 1990. A comparison of three methods of estimating fractal dimension as a measure of fragmentation of old-growth habitat using a GIS. *WDW Research Notes* 1(1):1–3.

Dobson, A. P. 1988. Restoring island ecosystems: The potential of parasites to control introduced mammals. *Conservation Biology* 2:31–39.

Dueser, R. D., and J. H. Porter. 1986. Habitat use by insular small mammals: Relative effects of competition and habitat structure. *Ecology* 67:195–201.

Ebeling, W., and M. A. Jimenez-Montano. 1980. On grammars, complexity, and information measures of biological macromolecules. *Mathematical BioScience* 52:53–71.

Eberhardt, L. L. 1977. Applied systems ecology: Models, data, and statistical methods. In *New directions in the analysis of ecological systems*. Part 1, ed.

G. S. Innis, 43–55. Simulation Councils Proceedings. La Jolla, Calif.: Simulation Councils.

Ek, A. R., and R. A. Monserud. 1974. FOREST: A computer model for simulating the growth and reproduction of mixed species forest stands. School of Natural Resources Research Report A2635, Univ. of Wisconsin, Madison.

Erdelen, M. 1984. Bird communities and vegetation structure. Part 1, Correlations and comparisons of simple and diversity indices. *Oecologia* 61:277–84.

Errington, P. L. 1963. The phenomenon of predation. *American Scientist* 51:180–92.

Estes, J. E., C. Sailer, and L. R. Tinney. 1986. Applications of artificial intelligence techniques to remote sensing. *Professional Geographer* 38:133–41.

Faanes, C. A. 1984. Wooded islands in a sea of prairie. *American Birds* 38:3–6.

Fagen, R. 1988. Population effects of habitat change: A quantitative assessment. *Journal of Wildlife Management* 52:41–46.

Fahrig, L., and G. Merriam. 1985. Habitat patch connectivity and population survival. *Ecology* 66:1762–68.

Fahrig, L., and J. Paloheimo. 1988. Effect of spatial arrangement of habitat patches on local population size. *Ecology* 69:468–75.

Farmer, A. H., M. J. Armbruster, J. W. Terrell, and R. L. Schroeder. 1982. Habitat models for land use planning: Assumptions and strategies for development. *Transactions of the North American Wildlife and Natural Resources Conference* 47.

Ferson, S. 1988. Microcomputer software for stochastic demography and ecological risk analysis. *American Statistician* 42:273.

Field, D. B. 1977. Linear programming: Out of the classroom and into the woods. *Journal of Forestry* 75 (June):330–34.

Fight, R. D., J. M. Chittester, and G. W. Clendenen. 1984. *DF–SIM with economics: A financial option for the DFSIM Douglas-fir simulator*. USDA Forest Service General Technical Report PNW–175.

Flood, B. S., M. E. Sangster, R. S. Sparrow, and T. S. Baskett. 1977. *A handbook for habitat evaluation procedure*. USDI Fish and Wildlife Service Research Publication no. 132.

Folse, L. J., Jr. 1979. Analysis of community census data: A multivariate approach. In *The role of insectivorous birds in forest ecosystems*, ed. J. G. Dickson, R. N. Conner, R. R. Fleet, and J. C. Kroll, 9–22. New York: Academic Press.

Fowles, R. 1988. MICRO–EGA. *American Statistician* 42:274.

Franklin, J. F., and R. T. T. Forman. 1987. Creating landscape patterns by forest cutting: Ecological consequences and principles. *Landscape Ecology* 1: 5–18.

French, N. R. 1977. Ecosystem trophic relationships in habitat classification. In *Classification, inventory, and analysis of fish and wildlife habitat*, 443–60. USDI Fish and Wildlife Service FWS/OBS–78/76.

Garsd, A. 1984. Spurious correlation in ecological modelling. *Ecological Modelling* 23:191–201.

Garton, E. O. 1979. Implications of optimal foraging theory for insectivorous forest birds. In *The role of insectivorous birds in forest ecosystems*, ed. J. G. Dickson, R. N. Conner, R. R. Fleet, and J. C. Kroll, 107–18. New York: Academic Press.

Gass, S. I. 1977. Evaluation of complex models. *Computers and Operations Research* 4:27–35.

Gentiol, S., and G. Blake. 1981. Validation of complex ecosystem models. *Ecological Modelling* 14:21–38.

George, T. L. 1987. Greater land bird densities on island vs. mainland: Relation to nest predation level. *Ecology* 68:1393–400.

Goldbeck, A. L. 1986. *Evaluating statistical validity of research reports: A guide for managers, planners, and researchers.* USDA Forest Service General Technical Report PSW–87.

Goldberg, M., D. G. Goodenough, M. Alvo, and G. M. Karam. 1985. A hierarchical expert system for updating forestry maps with LANDSAT data. *Proceedings of the IEEE* 73:1054–63.

Goodenough, D. G., M. Goldberg, G. Plunkett, and J. Zelek. 1987. An expert system for remote sensing. *IEEE Transactions on Geoscience and Remote Sensing* GE–25(3):349–59.

Gordon, G. 1978. *System simulation.* 2d ed. Englewood Cliffs, N.J.: Prentice-Hall.

Graklandoff, G. A. 1985. Expert system technology applied to cartographic processes: Considerations and possibilities. In *Proceedings of the American Society for Photogrammetry and Remote Sensing*, 613–24. Indianapolis, Ind.

Haefner, J. W. 1974. Generative grammars that simulate ecological systems. Ph.D. diss., Oregon State Univ., Corvallis.

Hall, C. A. S., and J. W. Day. 1977. Systems and models: Terms and basic principles. In *Ecosystem modeling in theory and practice*, ed. C. A. S. Hall and J. W. Day, 6–36. New York: Wiley-Interscience.

Hansen, A. J. 1986. Fighting behavior in bald eagles: A test of game theory. *Ecology* 67:787–97.

Hariston, N. G. 1981. An experimental test of a guild: Salamander competition. *Ecology* 62:65–72.

Harvey, P. H., R. K. Colwell, J. W. Silvertown, and R. M. May. 1983. Null models in ecology. *Annual Review of Ecology and Systematics* 14:189–211.

Hassell, M. P. 1987. Detecting regulation in patchily distributed animal populations. *Journal of Animal Ecology* 56:705–13.

Hassler, C. C., S. A. Sinclair, and E. Kallio. 1986. Logistic regression: A potentially useful tool for researchers. *Forest Products Journal* 36:16–18.

Hastings, A. 1990. Spatial heterogeneity and ecological models. *Ecology* 71:426–28.

Hemstrom, M., and V. D. Adams. 1982. Modeling long-term forest succession in the Pacific Northwest. In *Forest succession and stand development research in the Northwest*, ed. J. E. Means, 14–23. Proceedings of the symposium, March 26, 1981, Forest Research Laboratory, Oregon State Univ., Corvallis.

Henderson, J. A., L. S. Davis, and E. M. Ryberg. 1978. *ECOSYM, a classification and information system for wildland management.* Dept. of Forest Resources, Utah State Univ., Logan.

Hillier, F. S., and G. J. Lieberman. 1980. *Introduction to operations research.* San Francisco: Holden-Day.

Hokans, R. H. 1984. An artificial intelligence application to timber harvest schedule implementation. *Interfaces* 14:77–84.

Holl, S. A. 1982. Evaluation of bighorn sheep habitat. *Desert Bighorn Council Transactions*, pp. 47–49.

Holling, C. S. 1984. *Adaptive environmental assessment and management*. New York: John Wiley and Sons.

Holm, E. 1988. Environmental restraints and life strategies: A habitat templet matrix. *Oecologia* 75(1):141–45.

Holmes, R. T., R. E. bonney, Jr., and S. W. Pacala. 1979. Guild structure of the Hubbard Brook bird community: A multivariate approach. *Ecology* 60: 512–20.

Holthausen, R. S. 1984. Use of vegetation projection models for management problems. In *Wildlife 2000: Modeling habitat relationships of terrestrial vertebrates*, ed. J. Verner, M. L. Morrison, and C. J. Ralph, 371–75. Madison: Univ. of Wisconsin Press.

Holthausen, R. S., and N. L. Dobbs. 1985. Computer-assisted tools for habitat capability evaluations. Paper presented at Society of American Foresters conference, July 30, Fort Collins, Colo.

Hoover, R. L., and D. L. Wills, eds. 1984. *Managing forested lands for wildlife*. Denver: Colorado Division of Wildlife in cooperation with USDA Forest Service, Rocky Mountain Region.

Horn, H. S. 1975. Markovian properties of forest succession. In *Ecology and evolution of communities*, ed. M. L. Cody and J. M. Diamond, 196–211. Cambridge Mass.: Harvard Univ. Press.

James, F. C. 1971. Ordinations of habitat relationships among breeding birds. *Wilson Bulletin* 83:215–36.

Johnson, D. H. 1985. Improved estimates from sample surveys with empirical Bayes methods. *Proceedings of the American Statistical Association*, pp. 395–400.

Johnson, D. H. 1989. An empirical Bayes approach to analyzing recurring animal surveys. *Ecology* 70:945–52.

Karr, J. R., and K. E. Freemark. 1985. Disturbance and vertebrates: An integrative perspective. In *The ecology of natural disturbance and patch dynamics*, ed. S. T. A. Pickett and P. S. White, 153–68. Orlando Fla.: Academic Press.

Keane, R. E., S. F. Arno, and J. K. Brown. 1990. Simulating cumulative fire effects in Ponderosa pine/Douglas-fir forests. *Ecology* 71:189–203.

Kimmins, J. P. 1987. *Forest ecology*. New York: Macmillan.

Kirkman, R. L., J. A. Eberly, W. R. Porath, and R. R. Titus. 1984. A process for integrating wildlife needs into forest management planning. In *Wildlife 2000: Modeling habitat relationships of terrestrial vertebrates*, ed. J. Verner, M. L. Morrison, and C. J. Ralph, 347–50. Madison: Univ. of Wisconsin Press.

Knopf, F. L., J. A. Sedgwick, and R. W. Cannon. 1988. Guild structure of a riparian avifauna relative to seasonal cattle grazing. *Journal of Wildlife Management* 52:280–90.

Kolasa, J. 1989. Ecological systems in hierarchical perspective: Breaks in community structure and other consequences. *Ecology* 70:36–47.

Kourtz, P. H. 1984. Decision-making for centralized forest fire management. *Forestry Chronicle* 60:320–27.

Kourtz, P. 1987. Expert system dispatch of forest fire control resources. *AI Applications in Natural Resource Mangement* 1(1):1–8.

Krebs, J. R., J. C. Ryan, and E. L. Charnov. 1974. Hunting by expectation or optimal foraging? A study of patch use by chickadees. *Animal Behavior* 22: 953–64.

Lack, D. 1954. The natural regulation of animal numbers. London: Oxford Univ. Press.

Lack, D. 1966. Population studies of birds. Oxford: Clarendon Press.

Lancia, R. A., S. D. Miller, D. A. Adams, and D. W. Hazel. 1982. Validating habitat quality assessment: An example. *Transactions of the North American Wildlife and Natural Resources Conference* 47:96–110.

Landres, P. B., and J. A. MacMahon. 1980. Guilds and community organization: Analysis of an oak woodland avifauna in Sonora, Mexico. *Auk* 97:351–65.

Landres, P. B., J. Verner, and J. W. Thomas. 1988. Ecological uses of vertebrate indicator species: A critique. *Conservation Biology* 2:316–28.

Law, A. M. 1981. Validation of simulation models. Part 2, Comparison of real-world and simulation output data. Univ. of Wisconsin Dept. of Industrial Engineering Technical Report 78–15, Madison.

Law, A. M., and W. D. Kelton. 1982. *Simulation modeling and analysis*. New York: McGraw-Hill.

Laymon, S. A., and R. H. Barrett. 1986. Developing and testing habitat-capability models: Pitfalls and recommendations. In *Wildlife 2000: Modeling habitat relationships of terrestrial vertebrates*, ed. J. Verner, M. L. Morrison, and C. J. Ralph, 87–91. Madison: Univ. of Wisconsin Press.

Laymon, S. A., and J. F. Reid. 1986. Effects of grid-cell size on tests of a spotted owl HSI model. In *Wildlife 2000: Modeling habitat relationships of terrestrial vertebrates*, ed. J. Verner, M. L. Morrison, and C. J. Ralph, 93–96. Madison: Univ. of Wisconsin Press.

Laymon, S. A., H. Salwasser, and R. H. Barrett. 1985. *Habitat suitability index models: Spotted owl*. USDI Fish and Wildlife Service Biological Report 82(10.113).

Leary, R. A. 1979. *A generalized forest growth projection system applied to the Lakes States region*. USDA Forest Service General Technical Report NC–49.

Lee, K. N., and J. Lawrence. 1986. Adaptive management: Learning from the Columbia River basin fish and wildlife program. *Environmental Law* 16: 431–60.

Leggett, R. W., and L. R. Williams. 1981. A reliability index for models. *Ecological Modelling* 13:303–12.

Levins, R. 1968. *Evolution in changing environments*. Princeton: Princeton Univ. Press.

Levins, R. 1975. Evolution in communities near equilibrium. In *Ecology and evolution of communities*, ed. M. L. Cody and J. M. Diamond, 16–50. Cambridge, Mass.: Harvard Univ. Press.

Lynch, J. F., and D. F. Whigham. 1981. *Configuration of forest patches necessary to maintain bird and plant communities*. Smithsonian Institution Report, Feb. 1982, 82(18):3546. Washington, D.C.

Lynch, J. F., and D. F. Whigham. 1984. Effects of forest fragmentation on breeding bird communities in Maryland, USA. *Biological Conservation* 28: 287–324.

McCracken, R. J., and R. B. Cate. 1986. Artificial intelligence, cognitive sciences, and measurement theory applied in soil classification. *Soil Society of America Journal* 50:557–61.

MacKenzie, D. I., and S. G. Sealy. 1981. Nest site selection in eastern and western kingbirds: A multivariate approach. *Condor* 83:310–21.

McKeown, D. M. 1987. The role of artificial intelligence in the integration of remotely sensed data with geographic information systems. *IEEE Transactions on Geoscience and Remote Sensing* GE–25(3):330–48.

McNay, R. S., R. E. Page, and A. Campbell. 1987. Application of expert-based decision models to promote integrated management of forests. *Transactions of the North American Wildlife and Natural Resources Conference* 52: 82–91.

Maguire, L. A., M. L. Shaffer, and A. R. Tipton. 1988. Stochastic population simulation. *Conservation Biology* 2:6–7.

Mandelbrot, B. B. 1977. *The fractal geometry of nature*. San Francisco: W. H. Freeman.

Mankin, J. B., R. V. O'Neill, H. H. Shugart, Jr., and B. W. Rust. 1977. The importance of validation in ecosystem analysis. In *New directions in the analysis of ecological systems*. Part 1, ed. G. S. Innis, 63–71. Simulation Councils Proceedings.

Mannan, R. W., M. L. Morrison, and E. C. Meslow. 1984. Use of guilds in forest bird management: A caution. *Wildlife Society Bulletin* 12:426–30.

Marcot, B. G. 1979. *California Wildlife/Habitat Relationships Program, North Coast/Cascades Zone*. 5 vols. Eureka, Calif.: USDA Forest Service, Pacific Southwest Region.

Marcot, B. G. 1980. Use of a habitat/niche model for old growth management: A preliminary discussion. In *Management of western forests and grasslands for nongame birds*, ed. R. M. DeGraaf, 390–402. USDA Forest Service General Technical Report INT–86.

Marcot, B. G. 1986a. Summary: Biometric approaches to modeling—the manager's viewpoint. In *Wildlife 2000: Modeling habitat relationships of terrestrial vertebrates*, ed. J. Verner, M. L. Morrison, and C. J. Ralph, 203–4. Madison: Univ. of Wisconsin Press.

Marcot, B. G. 1986b. Use of expert systems in wildlife-habitat modeling. In *Wildlife 2000: Modeling habitat relationships of terrestrial vertebrates*, ed. J. Verner, M. L. Morrison, and C. J. Ralph, 145–50. Madison: Univ. of Wisconsin Press.

Marcot, B. G. 1988. 1st-Class Expert Systems: 1st-Class. *AI Expert* 3:77–80.

Marcot, B. G., R. S. McNay, and R. E. Page. 1988. *Use of microcomputers for planning and managing silviculture-habitat relationships*. USDA Forest Service General Technical Report PNW–GTR–228.

Marcot, B. G., M. G. Raphael, and K. H. Berry. 1983. Monitoring wildlife habitat and validation of wildlife-habitat relationships models. *Transactions of the North American Wildlife and Natural Resources Conference* 48: 315–29.

Marks, H. M., and S. M. Woodruff. 1979. Allocations using empirical Bayes estimators. *Proceedings of the American Statistical Association*, pp. 309–13.

Martin, T. E. 1988. Habitat and area effects on forest bird assemblages: Is nest predation an influence? *Ecology* 69:74–84.

Maurer, B. A., L. B. McArthur, and R. C. Whitmore. 1981. Effects of logging on guild structure of a forest bird community in West Virginia. *American Birds* 35:11–13.

Meldahl, R. 1986. Alternative modeling methodologies for growth and yield projection systems. In *Data management issues in forestry*, 27–31. Proceedings of a computer conference, April 7–9, Atlanta, Ga. Florence, Ala.: Forest Resources Systems Institute.

Mihram, G. A. 1976. Simulation methodology. *Theory and Decision* 7:67–94.

Mills, W. L. 1987. Expert systems: Applications in the forest products industry. *Forest Products Journal* 37(9):40–44.

Milsum, J. H. 1966. *Biological control systems analysis*. New York: McGraw-Hill.

Moeur, M. 1985. *COVER: A user's guide to the CANOPY and SHRUBS extension of the stand prognosis model*. USDA Forest Service General Technical Report INT–190.

Morris, C. N. 1983. Parametric empirical Bayes inference: Theory and applications. *Journal of the American Statistical Association* 78:47–55.

Morrison, M. L., I. C. Timossi, and K. A. With. 1987. Development and testing of linear regression models predicting bird-habitat relationships. *Journal of Wildlife Management* 51:247–53.

Murray, J. D. 1988. Spatial dispersal of species. *Trends in Ecology and Evolution* 3(11):307–9.

Naylor, T. H., and J. M. Finger. 1967. Verification of computer simulation models. *Management Science* 14:B92–B106.

Neitro, W. A., V. W. Binkley, S. P. Cline, R. W. Mannan, B. G. Marcot, D. Taylor, and F. F. Wagner. 1985. Snags (wildlife trees). In *Management of wildlife and fish habitats in forests of western Oregon and Washington*. Part 1, *Chapter narratives*, ed. E. R. Brown, 129–69. USDA Forest Service, Pacific Northwest Region.

Noss, R. F., and L. D. Harris. 1986. Nodes, networks, and MUMs: Preserving diversity at all scales. *Environmental Management* 10:299–309.

O'Keefe, J. H., D. B. Danilewitz, and J. A. Bradshaw. 1987. An "expert system" approach to the assessment of the conservation status of rivers. *Biological Conservation* 40:69–84.

Oliver, W. W., and R. F. Powers. 1978. *Growth models for ponderosa pine*. Part 1, *Yield of unthinned plantations in northern California*. USDA Forest Service Research Paper PSW–133.

O'Neil, L. J., T. H. Roberts, J. S. Wakeley, and J. W. Teaford. 1988. A procedure to modify habitat suitability index models. *Wildlife Society Bulletin* 16: 33–36.

Overton, W. S. 1977. A strategy of model construction. In *Ecosystem modeling in theory and practice*, ed. C. A. S. Hall and J. W. Day, 49–73. New York: John Wiley and Sons.

Partridge, L. 1978. Habitat selection. In *Behavioural ecology: An evolutionary approach*, ed. J. R. Krebs and N. B. Davies, 351–76. Sunderland, Mass.: Sinauer Associates.

Patten, B. C., and G. T. Auble. 1981. System theory of the ecological niche. *American Naturalist* 117:893–922.

Patton, D. R. 1978. *RUN WILD: A storage and retrieval system for wildlife habitat information*. USDA Forest Service General Technical Report RM–51.

Patton, D. R. 1987. Is the use of "management indicator species" feasible? *Western Journal of Applied Forestry* 2:33–34.

Pennycuick, C. J., and N. C. Kline. 1986. Units of measurement for fractal extent, applied to the coastal distribution of bald eagle nests in the Aleutian Islands, Alaska. *Oecologia* 68:254–58.

Perry, J. N. 1988. Some models for spatial variability of animal species. *Oikos* 51(2):124–30.

Pickett, S. T. A., and P. S. White, eds. 1985. *The ecology of natural disturbance and patch dynamics*. Orlando, Fla.: Academic Press.

Pleasants, J. M. 1990. Null-model tests for competitive displacement: The fallacy of not focusing on the whole community. *Ecology* 71:1078–84.

Porter, W. F., and K. E. Church. 1987. Effects of environmental pattern on habitat preference analysis. *Journal of Wildlife Management* 51:681–85.

Powers, J. E. 1979. Planning for an optimal mix of agricultural and wildlife land use. *Journal of Wildlife Management* 43:493–502.

Pyke, G. H. 1984. Optimal foraging theory: A critical review. *Annual Review of Ecology and Systematics* 15:523–75.

Quinn, J. F., and A. Hastings. 1987. Extinction in subdivided habitats. *Conservation Biology* 1:198–209.

Rabinovich, J. E., M. J. Hernandez, and J. L. Cajal. 1985. A simulation model for the management of vicuna populations. *Ecological Modelling* 30:275–95.

Radloff, D. L., and D. R. Betters. 1978. Multivariate analysis of physical site data for wildlife classification. *Forest Science* 24:2–10.

Raedeke, K. J., and J. F. Lehmkuhl. 1984. A simulation procedure for modeling the relationships between wildlife and forest management. In *Wildlife 2000: Modeling habitat relationships of terrestrial vertebrates*, ed. J. Verner, M. L. Morrison, and C. J. Ralph, 377–81. Madison: Univ. of Wisconsin Press.

Ramm, C. W., and C. L. Miner. 1986. Growth and yield programs used on microcomputers in the North Central Region. *Northern Journal of Applied Forestry* 3:44–45.

Raphael, M. G., and R. H. Barrett. 1983. Diversity and abundance of wildlife in late successional Douglas-fir forests. Paper presented at the Society of American Foresters symposium, October 18, Portland, Oreg.

Raphael, M. G., and B. G. Marcot. 1986. Validation of a wildlife-habitat-relationships model: Vertebrates in a Douglas-fir sere. In *Wildlife 2000: Modeling habitat relationships of terrestrial vertebrates*, ed. J. Verner, M. L. Morrison, and C. J. Ralph, 129–38. Madison: Univ. of Wisconsin Press.

Rashevsky, N. 1965. The representation of organisms in terms of predicates. *Bulletin of Mathematical Biophysics* 27:477–91.

Rauscher, M. 1988. Using AI methodology to advance forest science. *AI Applications in Natural Resource Management* 2(1):58–59.

Rescigno, A. 1964. On some topological properties of the systems of compartments. *Bulletin of Mathematics and Biophysics* 26:31–38.

Rescigno, A., and G. Segre. 1965. On some metric properties of the systems of compartments. *Bulletin of Mathematics and Biophysics* 27:315–23.

Rewa, C. A., and E. D. Michael. 1984. Use of habitat evaluation procedures (HEP) in assessing guild habitat value. *Transactions of the Northeast Section of the Wildlife Society* 41:122–29.

Riechert, S. E., and P. Hammerstein. 1983. Game theory in the ecological context. *Annual Review of Ecology and Systematics* 14:377–409.

Roberts, R. H., and L. J. O'Neil. 1985. Species selection for habitat assessments. *Transactions of the North American Wildlife and Natural Resources Conference* 50:352–62.

Robinson, V. B., A. U. Frank, and M. A. Blaze. 1986. Expert systems applied to problems in geographic information systems: Introduction, review and prospects. *Computers, Environment and Urban Systems* 11(4):161–73.

Rogers, L. L., and A. W. Allen. 1987. *Habitat suitability index models: Black bear, Upper Great Lakes region*. USDI Fish and Wildlife Service Biological Report 82(10.144).

Roozen, H. 1987. Equilibrium and extinction in stochastic population dynamics. *Bulletin of Mathematical Biology* 49(6):671–96.

Rosen, R. 1967. *Optimality principles in biology*. New York: Plenum Press.

Rosenzweig, M. L. 1987. Habitat selection as a source of biological diversity. *Evolutionary Biology* 1(4):315–30.

Roughgarden, J. 1983. Competition and theory in community ecology. In *Ecology and evolutionary biology*, ed. G. W. Salt, 3–21. Chicago: Univ. of Chicago Press.

Rudis, V. A., and A. R. Ek. 1981. Optimization of forest island spatial patterns: Methodology for analysis of landscape pattern. In *Forest island dynamics in man-dominated landscapes*, ed. R. L. Burgess and D. M. Sharpe, 241–56. New York: Springer-Verlag.

Ruggiero, L. F., R. S. Holthausen, B. G. Marcot, K. B. Aubry, J. W. Thomas, and E. C. Meslow. 1988. Ecological dependency: The concept and its implications for research and management. *Transactions of the North American Wildlife and Natural Resources Conference* 53:115–26.

Salwasser, H., and F. B. Samson. 1985. Cumulative effects analysis: An advance in wildlife planning and management. *Transactions of the North American Wildlife and Natural Resources Conference* 50:313–21.

Salwasser, H., and J. C. Tappeiner II. 1981. An ecosystem approach to integrated timber and wildlife habitat management. *Transactions of the North American Wildlife and Natural Resources Conference* 46:473–87.

Savidge, J. A. 1987. Extinction of an island forest avifauna by an introduced snake. *Ecology* 68:660–68.

Schamberger, M., A. H. Farmer, and J. W. Terrell. 1982. *Habitat suitability index models: Introduction*. USDI Fish and Wildlife Service FWS/OBS–82/10.

Schellenberger, R. E. 1974. Criteria for assessing model validity for managerial purposes. *Decision Science* 5:644–53.

Schindler, D. W. 1987. Detecting ecosystem responses to anthropogenic stress. *Canadian Journal of Fisheries and Aquatic Science* 44:6–25.

Schmoldt, D. L., and G. L. Martin. 1986. Expert systems in forestry: Utilizing information and expertise for decision making. *Computers and Electronics in Agriculture* 1:233–50.

Schroeder, R. L. 1983. *Habitat suitability index models: Yellow warbler*. USDI Fish and Wildlife Service FWS/OBS–82/10.27.

Schroeder, R. L. 1987. *Community models for wildlife impact assessment: A review of concepts and approaches*. National Ecology Center, USDI Fish and Wildlife Service Biological Report 87(2).

Schuler, A. T., H. H. Webster, and J. C. Meadows. 1977. Goal programming in forest management. *Journal of Forestry* 75 (June):320–24.

Scott, M. J., B. Csuti, J. D. Jacobi, and J. E. Estes. 1987. Species richness: A geographic approach to protecting future biological diversity. *BioScience* 37(11):782–88.

Severinghaus, W. D. 1981. Guild theory development as a mechanism for assessing environmental impact. *Environmental Management* 5:187–90.

Shannon, R. E. 1975. Systems simulation: The art and science. Englewood Cliffs, N.J.: Prentice-Hall.

Shettleworth, S. J., J. R. Krebs, D. W. Stephens, and J. Gibbon. 1988. Tracking a fluctuating environment: A study of sampling. *Animal Behavior* 36(1): 87–105.

Short, H. L. 1982. *Technique for structuring wildlife guilds to evaluate impacts on wildlife communities*. USDI Fish and Wildlife Service Special Scientific Report, Wildlife no. 244.

Short, H. L. 1983. *Wildlife guilds in Arizona desert habitats*. USDI Bureau of Land Management Technical Note no. 362. Fort Collins, Colo.

Shugart, H. H., Jr. 1984. *A theory of forest dynamics*. New York: Springer-Verlag.

Shugart, H. H., Jr., and S. W. Seagle. 1985. Modeling forest landscapes and the role of disturbance in ecosystems and communities. In *The ecology of natural disturbance and patch dynamics*, ed. S. T. A. Pickett and P. S. White, 353–68. Orlando, Fla.: Academic Press.

Simberloff, D., and J. Cox. 1987. Consequences and costs of conservation corridors. *Conservation Biology* 1:63–71.

Smith, R. M. 1984. Habitat-simulation models: Integrating habitat-classification and forest-simulation models. In *Wildlife 2000: Modeling habitat relationships of terrestrial vertebrates*, ed. J. Verner, M. L. Morrison, and C. J. Ralph, 389–93. Madison: Univ. of Wisconsin Press.

Smith, T. M., H. H. Shugart, Jr., and D. C. West. 1981. Use of forest simulation models to integrate timber harvest and nongame bird management. *Transactions of the North American Wildlife and Natural Resources Conference* 46:501–10.

Solis, D. M., and R. J. Gutierrez. 1982. *Spotted owl habitat use on the Six Rivers National Forest, Humboldt County, California*. Cooperative Agreement, Final Report, Six Rivers National Forest, USDA Forest Service, Eureka, Calif.

Sousa, P. J. 1987. *Habitat suitability index models: Hairy woodpecker*. USDI Fish and Wildlife Service Biological Report 82(10.146).

Stamps, J. A., M. Buechner, and V. V. Krishnan. 1987. The effects of edge permeability and habitat geometry on emigration from patches of habitat. *American Naturalist* 129:533–52.

States, J. B., P. T. Haug, T. G. Shoemaker, L. W. Reed, and E. B. Reed. 1978. *A systems approach to ecological baseline studies*. USDI Fish and Wildlife Service FWS/OBS–78/21.

Stock, M. 1987. AI and expert systems: An overview. *AI Applications in Natural Resource Management* 1(1):9–17.

Stone, N. D., R. N. Coulson, R. E. Frisbie, and D. K. Li. 1986. Expert systems in entomology: Three approaches to problem solving. *Bulletin of the Entomological Society of America* 32:161–66.

Strebel, D. E. 1985. Environmental fluctuations and extinction: Single species. *Theoretical Population Biology* 27:1–26.

Strong, D., L. Szyska, and D. Simberloff. 1979. Tests of community-wide character displacement against null hypotheses. *Evolution* 33:897–913.

Sweeney, J. M. 1984. Refinement of DYNAST's forest structure simulation. In *Wildlife 2000: Modeling habitat relationships of terrestrial vertebrates*, ed. J. Verner, M. L. Morrison, and C. J. Ralph, 357–60. Madison: Univ. of Wisconsin Press.

Swift, B. L., J. S. Larson, and R. M. DeGraaf. 1984. Relationship of breeding bird density and diversity to habitat variables in forested wetlands. *Wilson Bulletin* 96:48–59.

Tappeiner, J. C., J. C. Gourley, and W. H. Emmingham. 1985. *A user's guide for on-site determinations of stand density and growth with a programmable calculator*. College of Forestry, Forest Research Lab, Special Publication no. 11, Oregon State Univ., Corvallis.

Taylor, A. D. 1990. Metapopulations, dispersal, and predator-prey dynamics: An overview. *Ecology* 71:429–33.

Terrell, J. W., ed. 1984. *Proceedings of a workshop on fish habitat suitability index models*. USDI Fish and Wildlife Service Biological Report 85(6).

Thieme, R. H. 1987. Forest road planning: A knowledge-based approach. Ph.D. diss., Purdue Univ., West Lafayette, Ind.

Thieme, R. H., D. D. Jones, and H. G. Gibson. 1986. *A knowledge-based structure for forest road layout*. American Society of Agricultural Engineers Paper no. 86–5042. St. Joseph, Mich.

Thieme, R. H., D. D. Jones, H. G. Gibson, J. D. Fricker, and T. W. Reisinger. 1987. Knowledge-based forest road planning. *AI Applications in Natural Resource Management* 1:1.

Thomas, J. W., ed. 1979. *Wildlife habitats in managed forests: The Blue Mountains of Oregon and Washington*. USDA Forest Service Agricultural Handbook no. 553.

Tilman, D. 1982. *Resource competition and community structure*. Princeton: Princeton Univ. Press.

Toft, C. A., and P. J. Shea. 1984. Detecting community-wide patterns: Estimating power strengthens statistical inference. In *Ecology and evolutionary biology*, ed. G. W. Salt, 38–45. Chicago: Univ. of Chicago Press.

Toth, E. F., D. M. Solis, and B. G. Marcot. 1986. A management strategy for habitat diversity: Using models of wildlife-habitat relationships. In *Wildlife 2000: Modeling habitat relationships of terrestrial vertebrates*, ed. J. Verner, M. L. Morrison, and C. J. Ralph, 139–44. Madison: Univ. of Wisconsin Press.

Townsend, C. R., and R. N. Hughes. 1981. Maximizing net energy returns from foraging. In *Physiological ecology: An evolutionary approach to resource use*, ed. C. R. Townsend and P. Calow, 86–108. Sunderland, Mass.: Sinauer Associates.

Udvardy, M. D. F. 1984. A biogeographical classification system for terrestrial environments. In *World Congress on National Parks: National Parks Conservation and Development*, 34–38. Washington, D.C.: Smithsonian Institution Press.

U.S. Fish and Wildlife Service. 1980. *Habitat evaluation procedures (HEP)*. Ecological Services Manual no. 102. Washington, D.C.: GPO.

VanHorn, R. 1969. Validation. In *The design of computer simulation experiments*, ed. T. H. Naylor, 232–51. Durham, N.C.: Duke Univ. Press.
VanHorn, R. L. 1971. Validation of simulation results. *Management Science* 17: 247–58.
Verner, J. 1983. An integrated system for monitoring wildlife on the Sierra National Forest. *Transactions of the North American Wildlife and Natural Resources Conference* 48:355–66.
Verner, J., and A. S. Boss. 1980. *California wildlife and their habitats: Western Sierra Nevada*. USDA Forest Service General Technical Report PSW–37.
Vogel, D. 1989. An integrated GIS/expert system. *ARC News* (Fall):14–15.
Wakeley, J. S., and L. J. O'Neil. 1988. *Techniques to increase efficiency and reduce effort in applications of the habitat evaluation procedures (HEP)*. Environmental Impact Research Program Technical Report EL–88–13. Vicksburg, Miss.: Dept. of the Army, U.S. Army Corps of Engineers.
Walters, C. 1986. *Adaptive management of renewable resources*. New York: Macmillan.
Walters, C., and R. Hilborn. 1978. Ecological optimization and adaptive management. *Annual Review of Ecology and Systematics* 9:157–88.
Walters, C. J. 1971. Systems ecology: The systems approach and mathematical models in ecology. In *Fundamentals of ecology*, 3d ed., ed. E. P. Odum, 276–92. Philadelphia: Saunders.
Walters, C. J., and J. E. Gross. 1972. Development of big game management plans through simulation modeling. *Journal of Wildlife Management* 36:128–34.
Watt, K. E. F. 1966. *Systems analysis in ecology*. New York: Academic Press.
Weaver, J., R. Escano, D. Mattson, T. Puchlerz, and D. Despain. 1985. A cumulative effects model for grizzly bear management in the Yellowstone ecosystem. In *Proceedings: Grizzly bear habitat symposium*, comp. G. P. Contreras and K. E. Evans, 234–36. USDA Forest Service General Technical Report INT–207.
Weiger, R. G. 1975. Simulation models of ecosystems. *Annual Review of Ecology and Systematics* 6:311–38.
White, W. B., and B. W. Morse. 1987. ASPENEX: An expert system interface to a geographic information system for aspen management. *AI Applications in Natural Resource Management* 1(2):49–53.
Wilcove, D. S. 1987. From fragmentation to extinction. *Natural Areas Journal* 7: 23–29.
Williams, B. K. 1985. Optimal management strategies in variable environments: Stochastic optimal control methods. *Journal of Environmental Management* 21:95–115.
Williams, G. L., D. R. Russell, and W. K. Seitz. 1977. Pattern recognition as a tool in the ecological analysis of habitat. In *Classification, inventory, and analysis of fish and wildlife habitat*, 521–31. USDI Fish and Wildlife Service FWS/OBS–78/76.
Williamson, J. F. 1983. Woodplan: Microcomputer programs for forest management. *Proceedings of a National Workshop on Computer Uses in Fish and Wildlife Programs*, pp. 128–30. Workshop held December, Virginia Polytechnic Institute and State Univ., Blacksburg.
Wisdom, M. J., L. R. Bright, C. G. Carey, W. W. Hines, R. J. Pedersen, D. A. Smithey, J. W. Thomas, and G. W. Witmer. 1986. *A model to evaluate*

elk habitat in western Oregon. USDA Forest Service, Pacific Northwest Region, Publication no. R6–F&WL–216–1986.

Wu, L. S., and D. B. Botkin. 1980. Of elephants and men: A discrete, stochastic model for long-lived species with complex life histories. *American Naturalist* 116:831–49.

Wykoff, W. R., N. L. Crookston, and A. R. Stage. 1982. *User's guide to the stand prognosis model.* USDA Forest Service General Technical Report INT–133.

Yahner, R. H., and D. P. Scott. 1988. Effects of forest fragmentation on depredation of artificial nests. *Journal of Wildlife Management* 52:158–61.

Yost, R., S. Itoga, Z. C. Li, and P. Kilham. 1986. Soil acidity management with expert systems. Paper presented at the IBSRAM Asian Regions Seminar on Soil Acidity, Land Clearing, and Vertisols, November 1986, Khon Kaen Univ., Thailand.

7 Multivariate Assessment of Wildlife Habitat

Introduction

Multivariate analysis is a branch of statistics used to analyze multiple measurements that have been made on one or more samples of individuals.

Multivariate analysis is distinguished from other classes of statistical procedures in that multiple variables are considered in combination as a system of measurement. Because the variates are typically dependent among themselves, we cannot separate them and examine each individually (Cooley and Lohnes 1971:3). Multivariate techniques have become one of the most important analytical tools available to scientists in all disciplines, and they are especially important to wildlife biologists concerned with the evaluation of wildlife-habitat relationships.

In this chapter we first examine the rationale for using multivariate statistical techniques, relating these methods of analysis to our conceptualization of wildlife-habitat studies. We then review the all-important assumptions associated with these methods. We follow with a classification and discussion of many current techniques and their applications to wildlife studies. Although most multivariate techniques involve parametric linear models, nonlinear and nonparametric techniques are gaining in popularity; we discuss all of these broad classes. A formal course in multivariate analysis is not a prerequisite for understanding this chapter; many good statistics texts are available, and they should be consulted for a general statistical overview of methods and definitions (see, Morrison 1967; Cooley and Lohnes 1971; Pimentel 1979; Marascuilo and Levin 1983; Afifi and Clark 1984; Dillon and Goldstein 1984).

Conceptual Background

Multivariate statistical techniques were not originally designed for analysis of wildlife habitat and behavioral data. Indeed, multivariate techniques have been in use since the late 1880s, with a progression of new, more sophisticated methods following throughout the 1900s (see Cooley and Lohnes 1971:4). Only since the 1960s have researchers placed emphasis on the quantitative analyses of wildlife habitat. The appropriateness of multivariate analyses for examining wildlife habitat soon became evident. Multivariate techniques apply well to the analysis of wildlife-habitat data.

The application of multivariate analyses to wildlife data is the product of a synthesis beginning in the early 1970s that united several lines of ecological research. As outlined by Shugart (1981), this synthesis involved linking two analytical tools with three ecological concepts (niche theory, microhabitat, individual responses; fig. 7.1). The availability of high-speed computers along with prewritten multivariate computer programs (that is, "canned" statistical computer packages) allowed easy access to this general field of analysis. Hutchinson's reformulation of the niche concept (1957) in terms of an "*n*-dimensional hypervolume," as discussed in Chapter 2, urged ecologists to alter their view of wildlife habitat and the ways in

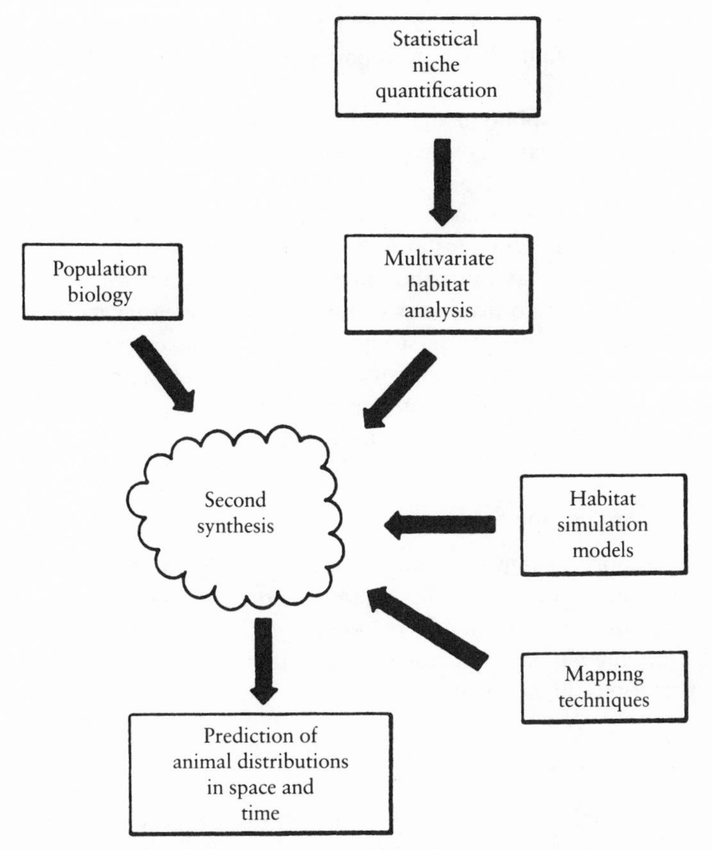

Figure 7.1. Schematic diagram of the scientific research elements that combined in a synthesis to produce multivariate habitat analysis. (From Shugart 1981, fig. 1.)

which they analyzed data. The realization that small spatial-scale studies could reveal much about animal interactions, as well as an emphasis on the importance of individual organism responses in determining species distributions, also led to multivariate habitat analysis.

Green (1971) was one of the first ecologists to formally apply multivariate analyses to Hutchinson's concept. The *n*-dimensional concept of the niche and the *n*-dimensional sample space of multivariate analysis are analogous in many ways, and this similarity led to an obvious application of multivariate methods to ecological data (Shugart 1981).

Shugart (1981) also noted that, by the mid-1960s, increased interest in habitat selection as a behavioral phenomena tended to produce data sets with multiple variables recorded for each observation (recall our discussion of behavioral data in Chapter 5). Such data were also well suited for examination by multivariate techniques, so the interest in these methods expanded further.

Species often respond to an environmental variable or gradient in a nonlinear fashion. The form of this function can take many shapes, from symmetric to asymmetric, from unimodal to bimodal (Johnson 1981). Nonlinear responses arise because animals simply cannot respond to all levels of a variable in the same fashion. For example, soil moisture determines, in part, the density and distribution of amphibians; but we would expect amphibian density to decline as conditions become either too dry or too wet. Similarly, because of interspecific interactions, an ever-increasing snag density does not result in an ever-increasing density of cavity-nesting birds. Thus, we include in this discussion the important area of nonlinear analysis and its ramifications for wildlife management.

Assumptions

Four common assumptions are associated with parametric multivariate analyses: multivariate normality, equality of the variance-covariance matrices (group dispersions), linearity, and independence of error terms (residuals). Violating any of these assumptions will bias or taint analysis results, interpretations, and conclusions.

Multivariate Normality

In many biological situations the values for one or more variables will be multimodal, skewed, or otherwise not normal. Researchers use tests of skewness and kurtosis, tests of equality of variances, and related tests to assess how well data conform to the assumptions in univariate analysis. The assumption of multivariate normality is more than simply an assumption that each variable is itself normally distributed, as in the univariate sense. Rather, it also must be true that the projections of the data points on any line in the multivariate space are normally distributed (Morrison 1967:90). Unfortunately, tests of this assumption are cumbersome and yield only approximations. Variable-by-variable examination of normality, plus linear combinations of the variables, however, will certainly identify variables that depart greatly from normality. Standard univariate transformations (such as log and square-root transformations) can be applied as appropriate.

Equality of Group Dispersions

Only in the study of single statistical populations is multivariate normality independent of equality of dispersions. In biology, as means increase, often so do variances. In ecology, inequality of dispersions of biological responses along environmental gradients has led to many prefixes, such as *steno-*, *dys-*, *oligo-*, and *eury-* (Pimentel 1979:176–77). Pimentel noted that, as biologists, we seem too concerned with comparing means (1979:177). If populations have unequal dispersions, then the populations are different even if their central tendencies are equal. Thus, trying to force normality to meet the formal assumption of equality of dispersions is biologically unsound. In fact, the magnitude of this variation (dispersion) can be used as a multivariate measure of niche breadth (see Carnes and Slade 1982). It is, however, possible to compare populations having unequal dispersions. The test for homoscedasticity—where variance of values of a variable does not change over the value's range—is extremely powerful: even minor differences between group dispersions likely will be discovered, and large samples typically lead to a rejection of the hypothesis of equality. Subsequent multivariate analyses and interpretations thus are questionable. Fortunately, research indicates that tests of equality of group centroids are rather insensitive to moderate departures from multivariate normality and homoscedasticity. If sample sizes are large and equal between groups, then the inequality of dispersions has no real effect on Type I and Type II errors; valid results and appropriate interpretations should be produced (Pimentel 1979:177). Application of this rationale, however, requires careful planning so that equal samples can be collected along intervals of the variable's values and, further, that thorough evaluation of sample sizes and validation of results are obtained. Unfortunately, researchers have taken neither of these steps in most wildlife-habitat studies.

Linearity

Linearity is important in multivariate analysis in two main ways. First, most models are based on a linear model (relationship is best explained by a straight line). Second, the correlation coefficient, which forms the basis of most multivariate calculations, is sensitive only to the linear component of the relationship between two variables. The relationship between a dependent and an independent variable can, however, be decomposed into several independent trends, including linear, quadratic, cubic, and those of higher order. Thus, a relationship that is strongly quadratic may appear only weakly linear. If only the linear components appear in the correlation matrix, then the true nature and extent of the relationship between variables may not be clear. Fortunately, many relationships can be approximated using linear models even though nonlinear components may exist. The data transformations noted above may help to linearize a nonlinear relationship. Nonlinearity changes the probabilities in tests of significance; for example, one is likely to fail to reject null hypotheses of equality of group centroids but to reject null hypotheses of equality of group dispersions (Pimentel 1979:178–79). Data sets with wide ranges of observations for one or more variables are likely to be strongly nonlinear (such as when one is examining habitat use across seral stages or using multiple age classes). Pimentel outlined several steps for examining and identifying this possibility.

Independence of Residuals

Independence and random sampling are cornerstones of all biological investigations, whether they are later analyzed by uni- or multivariate techniques. In studies of biological populations in which the population of interest is hard to define and sampling procedures (such as trapping) may affect the probability of an individual being included in the sample, truly random sampling is difficult to achieve. Researchers must specifically guard against introducing systematic bias into any sample. Behavioral responses may vary among sexes, age classes, locations, and the like, rendering the gathering of a truly random sample highly unlikely.

Because these responses may differentially influence the locations of animals at the time of observation, most observations are not likely to be true random samples. Thus, we must tightly restrict the definition of the population sampled. For example, populations designated by sex, age, time, and location should provide reliable interpretations. When these factors are ignored, unequal representation of the factors between samples might imply differences that do not exist between populations (Pimentel 1979:176). Unfortunately, few published studies specifically or adequately define their statistical, sampled population. Failure to define the population has clear and adverse management implications: users will be unable to appropriately apply results to the proper time and place, thus overextending results beyond the proper target population (Tacha, Warde, and Burnham 1982).

Other Considerations

The scale of data collected plays a role in determining which multivariate procedures to use. Most multivariate techniques perform best using at least interval-scale data. Nominal- and ordinal-scale data are commonly collected in wildlife research, especially that derived from behavioral observations. Nominal- and ordinal-scale multivariate procedures (that is, distribution-free or nonparametric statistical techniques), however, are only beginning to gain popularity. It is a common misconception, though, that only nonparametric techniques can be applied to nominal or ordinal data.

The validity of statistical conclusions depends only on whether the data to which they are applied meet the distributional assumptions (that is, multivariate normality and equal dispersions) used to derive them, not on the scaling procedures used to obtain the data. That is, the tests are "blind" to the numbers "fed" to them (Harris 1985:326). This does not mean, however, that the researcher can simply ignore the type of data used. The researcher must take this factor into account when considering the kinds of theoretical statements and generalizations made on the basis of significance tests (Harris 1985:328). The type and magnitude of the violation of assumptions can influence the significance associated with a test. Harris gave this basic principle: Consider the range of possible data transformations, linear or otherwise, that could be applied to the data without changing important properties or losing information contained in the data. Then restrict generalizations and theoretical statements to those whose truth or falsehood would not be affected by the most extreme of the permissible transformations. Nonparametric alternatives to parametric tests, if available, should be used when the data show evidence of gross violation of multivariate normality or equality of group dispersions (Harris 1985:328).

Exploratory versus Confirmatory Analysis

Scientists often collect wildlife data in observational studies not specifically designed to test a statistical hypothesis. Here, the researcher does not have sufficient information to state and test *a priori* a null hypothesis. Analysis of such data is often called exploratory analysis. In contrast, if a researcher has some prior information regarding the theoretical structure of the data—such as that derived from an earlier exploratory study—then "confirmatory" studies and analyses are performed. If data failed to meet the assumptions associated with the particular analytical method(s) used, however, then results of these studies are often stated in terms of "exploratory" or "descriptive" analyses. That is, because formal tests of null hypotheses require that all assumptions are met, one cannot technically test these hypotheses under violations of assumptions. Studies that fail to satisfy these multivariate assumptions usually conclude that results should be viewed as exploratory or descriptive and/or that because multivariate tests are robust to violations of assumptions, formal hypothesis testing is appropriate. We agree with the latter point regarding the robustness of the tests but add three cautions: authors must clearly state the severity of the violations of assumptions, provide details on the specific form of these violations, and interpret how these violations could bias results, interpretations, and conclusions. For example, analyzing a clearly curvilinear relationship by a linear technique might mask important ecological relationships and result in inappropriate management recommendations over at least part of the range of the variables' values.

Classification of Multivariate Techniques

We can divide multivariate techniques into two general categories: one group, data reduction, or ordination procedures, also called dependence models; and two or more group, classification procedures, also called interdependence models. Both categories, while containing primarily parametric statistical procedures, also include nonparametric methods. Multivariate techniques also can be subdivided into linear and nonlinear techniques.

There is much confusion about what constitutes a multivariate technique. If we define multivariate techniques as those applied when two or more variables are used either as independent or dependent variables, then higher-order ANOVA and possible one-way ANOVA must be included. ANOVA, involving only one control variable, however, has traditionally not been included in treatments of multivariate statistics. If we define multivariate techniques as those applied to multiple dependent measures, then multiple regression is excluded, and this again runs counter to standard practice (see Harris 1985:10).

Examination of table 7.1, a flow chart of many of the most common statistical methods, quickly reveals the interrelated nature of this broad class of methods. Methods are broadly classified by the number of dependent and independent variables of interest and the goal of the researcher in analyzing the data set. That is, does the researcher wish to examine the structure, or interdependence, of variables (PCA); the relationship (corelation or correlation) among variables (multiple *R*); separating groups (MANOVA); or developing predictive equations (DA)? These are not mutually exclusive techniques; several are often used to analyze a single data set. Dillon and Goldstein (1984, fig. 1.5–1) and Harris (1985, table 1.1) gave similar classifications of multivariate techniques. Below we outline some of the specific methods found within these categories and provide examples of multivariate analyses applied to studies of wildlife-habitat relationships. We follow the general structure and discussion given by Neff and Marcus (1980). Their document, written for specific application to

Table 7.1. Classification of common statistical techniques

Major research question	Number of dependent variables	Number of independent variables	Analytical method[a]
Degree of relationship among variables	One	One	Bivariate r
		Multiple	Multiple R
	Multiple	Multiple	Canonical R
Significance of group differences	One	One	1-way ANOVA; t-test
		Multiple	Factorial ANOVA
	Multiple	One	1-way MANOVA; T^2
		Multiple	Factorial MANOVA
Prediction of group membership	One	Multiple	1-way DA; logistic regression
	Multiple	Multiple	Factorial DA
Structure	Multiple	Multiple	PCA; FA; metric and nonmetric multidimensional scaling; correspondence analysis

Source: *Using Multivariate Statistics* by Barbara G. Tabachnick and Linda S. Fidell. Copyright © 1983 by Barbara G. Tabachnick and Linda S. Fidell. Adapted with the permission of Harper Collins Publishers Inc.

[a]Analysis of variance (ANOVA); multivariate analysis of variance (MANOVA); Hotelling's T^2; discriminant analysis (DA); principal components analysis (PCA); factor analysis (FA).

systematics, covers major multivariate methods in a well-organized and lucid manner. Unfortunately, it is not widely available.

A survey of wildlife-habitat literature reveals that most biometric approaches have been restricted to principal components analysis (PCA), multiple regression (MR), and discriminant analysis (DA). Further, most of this research has been based on observational studies and has used multivariate statistics in an exploratory sense (Noon 1986). We will concentrate our discussion on describing how PCA, MR, and DA are used and interpreted. We will, however, include examples of other multivariate analyses that have application to wildlife-habitat research. Two published proceedings of conferences that dealt with wildlife-habitat data should be consulted for many specific examples: the works of Capen (1981) and Verner, Morrison, and Ralph (1986) (see also Morrison et al. 1990).

Data Structure: Ordination and Clustering

Methods within this broad category of analyses seek to reduce a complex data set to a few numbers of dimensions (axes) that are internally correlated but unique (not correlated) with regard to other derived dimensions (that is, to modify the structure of the data set). Methods of data reduction and ordination fall into this category. A separate technique computationally, usually called cluster analysis, is often discussed along with ordination (see Pimentel 1979, chap. 8). These two methods are similar in that no groups are *a priori* assumed to exist. Technically, this applies to divisive clustering techniques, which begin by assuming that all cases are statistically indistinguishable and proceed to identify the smallest number of groups for which this is true at various levels of similarity. This does not apply to agglomerative clustering techniques, which assume that each case initially belongs to its own unique group and proceed to combine cases as this proves false at various levels of similarity. The initial lack of designation of groups contrasts with true classification techniques, in which groups are known prior to analysis and individuals are assigned to the group of closest resemblance; this includes methods listed in table 7.1, such as DA. Ordination techniques can, however, be used to identify groups (clusters) in an ad hoc sense. Thus, we will discuss both ordination and clustering in this section. We will also provide some detail on the methods used to compute PCA so readers will better understand the statistical concepts associated with interpreting the multivariate methods we discuss throughout this chapter.

Ordination involves placing observations into a theoretically continuous sequence that reflects some fundamental property of the observations. The interpretation of any ordination should depend on the particular study. Data tend to follow, however, three general types of patterns: serial, nodal, and polynomial. On serial axes, data are plotted from one extreme to another without any obvious breaks. Such distributions indicate an underlying gradient in the data. Nodal axes indicate that data are grouped into two or more "nodes" or clusters, and these nodes may be poorly or clearly defined. On polynomial axes, data follow a curved pattern and may or may not be divided into nodes. These curves result from the second axis not being clearly independent from the first, reflecting both a nonlinear and a nonnormal condition (Pimentel 1979:136–37). Polynomial patterns of observations must be reorganized if the appropriate analytical procedure is to be used; this fact will become increasingly evident as we move through this chapter.

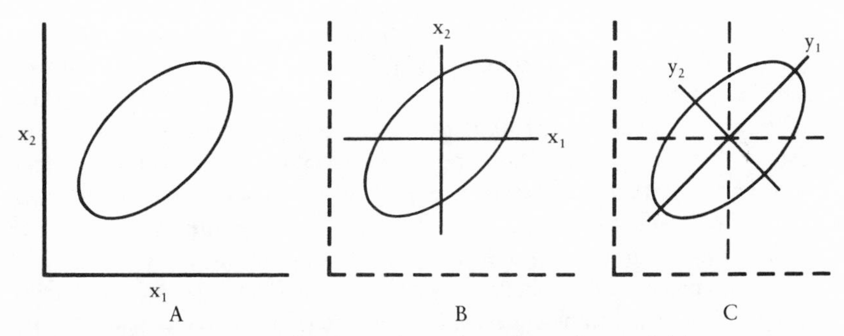

Figure 7.2. Bivariate example of transformation of raw data variable axes to component axes. *A*, raw data axes; *B*, change of original Cartesian coordinate axes to deviation axes (the origin of the x_1 and x_2 axes is at the means of x_1 and x_2); *C*, transformation from deviation axes to component score axis—there is zero correlation between y_1 and y_2. (From R. A. Pimentel, *Morphometrics*, Copyright © 1979 by Kendall/Hunt Publishing Company. Reprinted by permission of Kendall/Hunt Publishing Company.)

Principal Components Analysis

A multivariate data set can be visualized as a cloud of points in an n-dimensional variable space (for example, each point is an observation, such as species density, in n different study plots), with each axis representing a variable; in general, it comprises the various x_1, x_2 coordinates in a two-dimensional plot (fig. 7.2). PCA is a method that identifies new sets of orthogonal (mutually perpendicular and thus not correlated) axes in the direction of greatest variance among observations. The first axis is the line in a direction through the observations such that the projections of the observations onto the axis have maximum variance. In the univariate sense this is related to least squares regression. The second axis is in the direction of greatest variance perpendicular to the first axis; that is, that does not duplicate the variance "explained" by the first axis. These are the principal component axes; figure 7.2 shows the bivariate case where the x's and y's are the new axes. It becomes increasingly difficult to graphically depict PCA beyond three axes.

One usually begins PCA by mean-centering the raw data set, A; this produces a new data matrix, B. From B the variance-covariance matrix, S, is determined. The variance-covariance, or covariance, matrix contains p columns and p rows that correspond to p variables; in this $p \times p$ matrix, the off-diagonal elements are covariances, and diagonal elements are variances. This matrix is also called the dispersion matrix, since it measures the absolute variation within and between variables, and is the computational backbone of many linear multivariate methods (such as MANOVA and DA). The so-called eigenvectors of S are vectors in the directions of the principal axes; that is, they are orthogonal axes in the directions of the maximum variance among individuals in the B matrix. Eigenvectors are best combinations of correlated predictor variables that account for most of the variation in the response variable. The first principal component axis (y in fig. 7.2), often designated PCI, is a linear combination of the original observations weighted by the elements of the first eigenvector.

One can consider the collection of eigenvectors for a data set as a matrix, E, with as many columns or vectors as the rank of the matrices B and S. The matrix E as a whole is sometimes called the weighting or coefficient matrix, because it contains all of the weights or coefficients for each original or mean-centered variable used to determine the position of each observation on each of the new axes. The entire set of scores, Z, is then the entire set of coordinates of the original observations in this new coordinate system defined by the principal component axes, such that $Z = B \times E$. The amount of the total variance among the observations attributable to or explained by each component is given by the eigenvalue associated with the eigenvector corresponding to that component. The sum of all eigenvalues is equal to the sum of all the variances of the original variables. The eigenvalues are then ordered so that the largest eigenvalue is associated with the first eigenvector, since the first component axis is in the direction of the maximum variance among observations.

There are two major ways to perform PCA, one using a correlation matrix and the other using the variance-covariance matrix. Results obtained from these methods cannot be easily transformed from one to the other (unless data are standardized). Thus, results of PCA are influenced by the computational method used. We will not detail these specific methods here, but researchers should carefully explain their specific computational methods in all research publications; this applies to all statistical techniques. Further, researchers must be careful to understand the methods used by other researchers when comparing results among studies; perceived differences in results might be due to computational, rather than biological, differences.

Neff and Marcus (1980) noted that much confusion exists in the terminology used to describe specific computational steps in multivariate statistics. Thus, it is worthwhile to contrast several of the terms frequently used—and intermixed—in the biological literature. A principal component of a data matrix is a new variable produced from the linear combination of the original variables using the elements of the corresponding eigenvector as the coefficients (weights). The elements of the eigenvector are not eigenvalues. Each eigenvector has a single eigenvalue, equal to the variance of the observations applied to each corresponding principal component axis. Eigenvectors are also called latent vectors or roots; eigenvalues are also called characteristic vectors or roots.

The contribution of a variable to each principal component is given by the magnitude of the eigenvector coefficients, with the algebraic sign indicating the direction of the effect. Computing the component loadings aids the interpretation of principal components. Loadings give the ordinary product-moment correlation of each variable and the respective component (see Dillon and Goldstein 1984:30–31). Because PCA is frequently used to generate a reduced set of variables that account for most of the variability in the original data set and that can be used for subsequent analysis, researchers must decide how many components to interpret. Unfortunately, no universally accepted method for doing so exists. A standard, but not required, procedure is to interpret only those components with associated eigenvalues greater than 1.0. Other workers have used all components that were above some arbitrarily large proportion of the variation and/or those that were obviously high relative to other components (for example, PCI might explain 32 percent; PCII, 18 percent; but PCIII, only 4 percent). Dillon and Goldstein (1984, chap. 2) outlined several methods for interpreting PCA results and for determining the number of components to retain; we also outline some ecological applications below. Typically, however, most of the variation is accounted for by the first three PC axes, and often by the first two. Interpreting the subsequent components biologically becomes increasingly difficult, as they often represent noise in the data. Thus, let understandability and common sense guide how many PCs to explain.

Collins (1983) examined geographic variation in habitat use of black-throated green warblers. His objectives were to determine if habitat structure differed at several points in the species range and, if so, which structural variables changed and which were similar between sites. He used PCA to analyze habitat structure for fifty samples at the five sites he analyzed. He extracted four components with eigenvalues greater than 1.0 from the correlation matrix of habitat variables (table 7.2). Principal component I (PCI; 28.6 percent of the variance) was interpreted as a forest-height component separating habitats with large trees (T6, T7), tall canopies (CH), and a low amount of shrub cover (SC) from areas characterized by smaller trees (T1–T3). Note that both magnitude and direction (+ or −) were used to infer this relationship. PCII (13.7 percent) was a deciduous to coniferous component in which deciduous samples had greater tree species richness (SPT) and canopy cover (CC), while coniferous stands had fewer tree species (CO) and greater ground cover (GC). PCIII (12.0 percent) had large negative loadings for medium-sized trees and the percentage of coniferous vegetation (T4–T5, CO). PCIV (9.9 percent) had high positive

Table 7.2. Summary of the first four principal components

Variable	Mnemonic	Principal component			
		I	II	III	IV
Eigenvalue		3.72	1.79	1.56	1.29
Percentage of variance		28.6	13.7	12.0	9.9
Percentage ground cover	GC	0.111	0.577	0.261	−0.051
Percentage shrub cover	SC	−0.532	−0.314	0.160	−0.294
Percentage canopy cover	CC	−0.072	−0.496	−0.329	0.551
Percentage conifer	CO	0.238	0.709	−0.423	0.258
Canopy height	CH	−0.739	0.466	−0.281	0.111
Number of tree species	SPT	0.056	−0.510	−0.016	0.508
Trees 7.5–15 cm dbh	T1	0.677	0.019	0.264	0.299
Trees 15.1–23 cm dbh	T2	0.876	0.205	−0.154	0.160
Trees 23.1–30 cm dbh	T3	0.818	0.059	−0.368	−0.032
Trees 30.1–38 cm dbh	T4	0.267	−0.196	−0.549	−0.601
Trees 38.1–53 cm dbh	T5	−0.397	−0.076	−0.711	0.002
Trees 53.1–68 cm dbh	T6	−0.657	0.208	−0.167	0.270
Trees > 68 cm dbh	T7	−0.511	0.169	0.166	0.175

Source: Collins 1983, table 2.
Note: Values are correlations with original variables.

loadings for CC and SPT and a negative loading for T4, suggesting that more tree species occurred as the canopy closed.

Collins plotted the distribution of sites in the space defined by the first two PCs (fig. 7.3). The plot showed that warbler habitat varied considerably among sampling sites. Note that this two-dimensional ordination helped to separate sites that would overlap broadly if plotted on only one axis; a three-dimensional plot would even more fully depict the separation identified by the first three PCs.

Maurer, McArthur, and Whitmore (1981) collected data on eight habitat variables for thirty-four species of birds and subjected the resulting data set to PCA. Results (table 7.3) showed that the first three PCs accounted for most of the variation possible between the bird species sampled and their associations with the habitat variables measured. This is not to say that, because 96.4 percent of the cumulative variance was explained by PCI–PCIII, only 3.6 percent of the variation in bird habitat use remained unresolved, that is, that the model was almost perfect. Rather, this cumulative variation only indicates that nearly all of the data in this *particular* data set can be extracted in the first three components. As noted previously, it is, of course, desirable to explain most of the variance in the first few

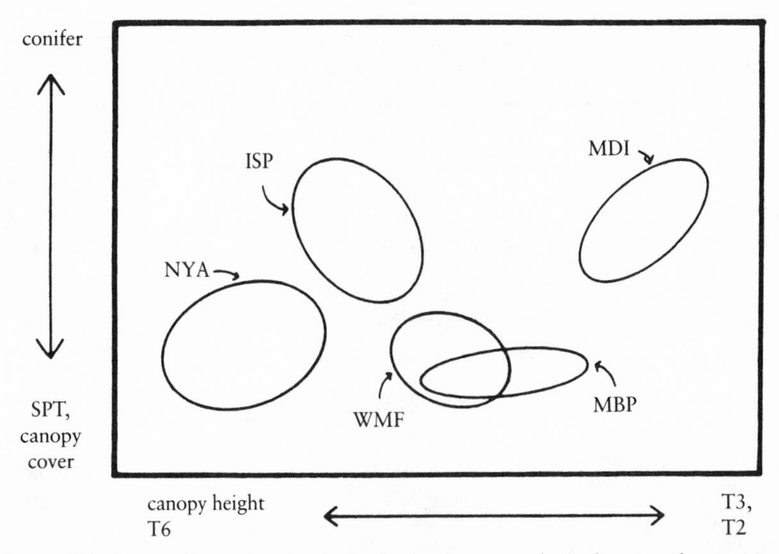

Figure 7.3. A two-dimensional principal component analysis showing the position of the five study sites within habitat structure space. The 95 percent concentration ellipses indicate the variability of within-site habitat structure. The horizontal axis is a gradient from tall, large trees and low shrub cover to smaller trees. The vertical axis is a deciduous to coniferous forest gradient. See table 7.2 for variable mnemonics. (From Collins 1983, fig. 1.)

components (dimensions). Methods to determine the reliability of models are discussed later in this chapter.

Unlike Collins (1983), Maurer, McArthur, and Whitmore used statistical significance to identify important correlations. Note, however, the wide disparity between correlations on PCI: from 0.98 for canopy layers ($P < 0.01$) to -0.38 ($P < 0.05$) for slash. For PCIII, note that while -0.33 was significant, 0.32 was not. Further, PCIII accounted for only 5.5 percent of the overall variation in the data set, and litter was both positively and significantly correlated for PCI and PCIII. Thus, we see that statistical significance, although certainly an important element to consider in evaluating results of PCA, does not necessarily produce clear patterns.

An assumption implicit in PCA and other ordination techniques is that some important biological phenomenon is present in the data set and will be represented in the first few components. Biologists are very good at explaining such phenomena even when no supporting evidence exists (for example, when the study is not replicated in time and space). Authors should be willing to state, and editors should be willing to accept, explanations that acknowledge the absence of any apparent pattern. A study by Karr and

Table 7.3. Results of principal components analysis using weighted averages of eight habitat variables for thirty-four bird species

Component	1	2	3
Variation explained	67.46%	23.47%	5.50%
Cumulative variation	67.46%	90.93%	96.43%
Variable	Correlations with original variables		
Litter	0.85**	0.23	0.42**
Slash	−0.38*	0.86**	0.32
Herb. veg.	−0.93**	−0.29	0.07
$\overline{X}$ ht. herb. veg.	−0.96**	−0.01	−0.04
Canopy layers	0.98**	−0.14	0.01
Max. canopy ht.	0.90**	−0.41*	0.08
Canopy cover	0.95**	0.11	−0.20
Trees < 12.7 cm	0.27	0.90**	−0.33*

Source: Maurer, McArthur, and Whitmore 1981, table 3.
*$P < 0.05$.
**$P < 0.01$.

Martin (1981) illustrates this point. They analyzed "real" biological data and random number matrices using PCA and found that percent variation accounted for by both data sets were similar, especially for the second and higher axes; component loadings for real and random data were often similar; and matrix size was an important determinant of the amount of variation extracted by PCA (and related sample-size effects) (table 7.4). They noted that similarity in results for real and random data did not necessarily mean that no biologically important relationships existed or could be interpreted by the PCA approach. The work of Karr and Martin identifies, however, a serious problem with exploratory studies, especially those that never progress beyond the initial stage: results should not be used simply for post facto interpretations but should help guide subsequent confirmatory research. Evaluation of biological data along with random data sets adds credence to exploratory or descriptive research. Rexstad and his colleagues (1988) conducted a similar evaluation of multivariate methods. In summary, multivariate methods such as PCA are designed to impart order to a set of data. The methods succeed regardless if the results are biologically interpretable or based on meaningless, random numbers. It is up to us to apply biological knowledge and common sense to the results.

Table 7.4. Percent of variation accounted for by the first three principal components for real and random number (in parentheses) matrices of the same dimensionality

	Percent of variation by component			
Matrix Size	I	II	III	Cumulative
6 × 56	33 (23)	22 (20)	18 (18)	73 (61)
7 × 5	45 (44)	23 (28)	11 (18)	79 (91)
8 × 21	46 (27)	17 (20)	10 (17)	73 (64)
10 × 24	57 (25)	16 (14)	12 (14)	85 (53)
10 × 46	65 (18)	12 (16)	12 (16)	85 (50)
15 × 15	—[a] (22)	—[a] (16)	11 (13)	58 (51)

Source: Karr and Martin 1981, table 2. The real data examples are from published studies using principal components analysis.
[a]Not available in original publication.

General Interpretation

As we have seen, it is often difficult to determine which variables are involved with what component and to what degree. By comparing component loadings, we can determine those variables most related to a component. Dillon and Goldstein (1984:69) observed that it is easiest to adopt heuristics for the purpose of interpreting the pattern. The procedure given by Dillon and Goldstein can simplify interpretation considerably. Although originally written specifically for factor analysis, these guides apply equally well to PCA and DA. The results in tables 7.2 and 7.3 can serve as examples.

1. Starting with the first variable and first component (or factor) and moving horizontally from left to right across the components, circle that variable loading with the largest absolute value. (For example, in table 7.3 the "litter" variable has the highest loading with PCI.) Then consider the second variable, and again moving from left to right horizontally, identify the highest absolute loading for that variable on any component and circle it. Continue this procedure for each variable.
2. After evaluating all of the variables, examine each loading for "significance." This assessment can be made on the basis of the statistical significance (see table 7.3) of the correlation coefficient (loading) or on the basis of the practical significance (table 7.2). For statistical significance, in most instances with sample sizes less than 100, the smallest loading would have to be greater than 0.30 in order to be considered significant. Practical significance involves some reasonable

or practical rule for the minimum amount of a variable's variance that must be accounted for by a component (or the number of variables included in a PC). This is especially important in ecological studies in which biological significance may require a much higher loading than indicated by mere statistical significance (the latter of which is largely based on sample size). Significant loadings should be underlined.

3. Examine this pattern matrix to identify the variables that have not been underlined and therefore do not "load" on any component (possibly "percentage shrub cover" in table 7.2 and "litter" in table 7.3). The researcher then must decide whether to rest the analysis on only those variables with significant loadings or to evaluate critically each variable with regard to the research objective and biological knowledge.
4. Based on the results of the previous steps, attempt to assign some biological meaning to the pattern of component loadings. Variables with higher loadings have greater influence; variables with negative loadings have inverse influence. Assign a name that reflects, to the extent possible, the combined meaning of the variables that load on each such component. In practice, having many variables with moderate-sized loadings complicates this step.

Other Methods

Factor Analysis

Factor analysis (FA) is also a data reduction technique. In contrast with PCA, FA centers on only that part of the total variation that a particular variable shares with the other variables that constitute the data set. In PCA, the goal was the construction of linear combinations of the original variables that accounted for a large part of the total variation in the data set. In contrast, FA finds a new set of variables, fewer in number than the original set, which express common elements among the original variables. Thus, FA refers to techniques that distinguish different types of variation. There is much confusion in the literature regarding PCA and FA. Part of this confusion involves terminology. Because the two methods are closely related, many statistical packages include PCA along with FA procedures.

The many decisions a user must make regarding specific computational steps complicate practical application of factor analysis; each step must be understood if biologically meaningful results are to be obtained. First, there are several different methods of extracting factors; the principal factor method and the maximum likelihood method are the most common (Dillon and Goldstein 1984, chap. 3). Second, factors can be rotated in various ways, from orthogonal to oblique. The goal of rotation is to reduce the number of variables with high loadings on each factor, thus obtaining a relatively simple structure. Readers should consult Dillon and Goldstein (1984:87–95) for an excellent graphical description of factor rotation. Unfortunately, any simple structure probably would not reflect the structure of the total phenomenon. Thus, factor rotation is likely only useful in considering very large variable sets.

Principal Coordinates Analysis

Principal components analysis (PCA) appears to perform best when each variable is continuous. Of course, continuous variables are not always part of biological studies. In many cases, discrete or categorical variables are the only measures possible. Both qualitative data and missing data are frequently encountered in numerical taxonomy. An ordination technique known as principal coordinates analysis (P-co-A or PCORD) is more appropriate than PCA in these situations. Like PCA, PCORD creates a new set of orthogonal axes. Unlike PCA, however, PCORD derives coordinate scores that are based solely on a matrix of individuals (OTUs, operational taxonomic units, of numerical taxonomy). This association matrix is often calculated from a matrix of interindividual (OTU) distances, distance squared, or a variety of similarity matrices (Sneath and Sokal 1973:248–49). In contrast with PCA, PCORD can tolerate moderate amounts of missing data.

PCORD has found its widest application in numerical taxonomy, but it can be applied to more habitat-and behavior-oriented studies in which similarity coefficients are used. Further, PCORD can be used to display between-group differences using Mahalanobis D^2 as the association index (Neff and Marcus 1980:77–78).

Correspondence Analysis

Correspondence analysis (CA) and reciprocal averaging (RA) are alternative names for the same technique (although some have argued that a slight difference can exist; see Digby and Kempton 1987:70). CA is a type of PCA that uses a specially scaled data matrix derived from tables of counts (contingency tables), such as the presence-absence of species or the frequency of behaviors used by a foraging animal at several different sites. The outcome of this method is the calculation of loadings for both the observations (individuals) and the variables on equivalent scales, which are then plotted together (Neff and Marcus 1980:81; Pielou 1984:176).

Data are converted to proportions or estimates of probabilities and then scaled by a form of simultaneous row and column standardization. The scaled matrix T is used to form association matrices $T'T$ and TT', which are then analyzed by standard eigenvalue and eigenvector routines following PCA procedures. The eigenvectors of T′T and TT′ are scaled to have the same units, to permit plotting the loadings on one graph (Neff and Marcus 1980:81). Details of these analytical procedures are given by Pielou (1984:176–88), Digby and Kempton (1987:70–75), and other authors of texts on multivariate analysis who emphasize ordination of ecological data. Hill's article "Correspondence Analysis" (1974) is an especially important paper to review.

Researchers have shown the efficiency of CA with heterogeneous, nonlinear data. Unfortunately, CA has the disadvantage that higher axes, while linearly independent, show higher-order correlations. Further, the ends of ordination axes derived from CA are compressed relative to middle points (Kenkel and Orloci 1986). A modification of CA, termed detrended correspondence analysis (DCA), was designed to further reduce the curvilinear nature of many ordinations (Hill and Gauch 1980). Controversy exists, however, regarding any perceived improvement that DCA offers relative to other ordination techniques (Wartenberg, Ferson, and Rohlf 1987; Whittaker 1987). Ben-Shahar and Skinner (1988) gave a detailed example of correspondence analysis and its interpretation.

Nonmetric Multidimensional Scaling

NMDS is a nonparametric method for ordination that involves a nonlinear model. Dissimilarity measures are transformed to an ascending numeric rank order (*nonmetric* refers to this nonparametric rank ordering). In addition, the (typically Euclidean) distance between all pairs of samples is obtained; these compose an additional set of dissimilarity coefficients. In general, the rank order measures and distances will not agree. The goal of NMDS is to obtain the correct ordination from this disagreement. The rankings and distances become part of an iterative procedure that extracts the best fit of the two sets of measures (Pimentel 1979:160–61). NMDS is especially appropriate when one has more faith in the ranking of data than in the actual metric values.

The major strength of NMDS is its nonlinear, nonparametric basis. A weakness is that the number of dimensions of the structure should be known. Kruskal and Wish (1978:34) provided rules for determining the number of dimensions that can be expected to give stable results at various sample sizes (see also Neff and Marcus 1980:88–90). A two- or three-dimensional solution is preferred because solutions of higher dimensions become increasingly difficult to interpret.

Comparison of Ordination Techniques

Studies have found that PCA often exhibits serious distortion of the resulting PCA axes. This is called involution of gradients, which is attributable to the use of a linear model to summarize trends related to nonlinear and nonmonotonic species responses. Studies comparing PCA, NMDS, CA, and/or DCA generally report that NMDS performs better than the other techniques, apparently owing to its nonlinear nature. This result is by no means universal, however. The advantage conferred by NMDS depends on the specific data set being considered and its departure from the assumptions associated with linear-based analyses (see Kenkel and Orloci 1986 for a review). Also, Pielou (1984:197–99) presented an example showing the relative utility of CA and DCA.

Miles (1990) compared results of PCA, FA, and CA when each method was applied to five sample data sets of bird foraging behavior. Of the three methods, PCA and CA showed a high level of consistency in the magnitude and sign of the coefficients from the first three eigenvectors; the concordance of results with FA was low, however. Further, the use of rotated axes for interpreting the foraging data was not recommended. Miles concluded that CA was the preferred method for analyzing foraging data (he did not test NMDS). Several papers in *Quantitative Ethology* (Colgan 1978) deal with the application to behavioral studies of multivariate analyses, including many ordination and clustering techniques.

It should be evident that much difference of opinion exists regarding the "best" ordination technique. Simply put, the perfect method does not exist. PCA has received the most use in the literature and has the advantage of being readily available in most canned statistical packages. The greatest problem involves not so much the specific method used, but rather the absence of any follow-up confirmatory or validation investigations. Further, problems of sample size, post facto interpretation of ordinations, and noncritical examination of model assumptions could all bias results of ordination analyses. Even if multiple ordination techniques are used to "confirm" study results, this does not justify inadequate sample sizes or biases sampling methods.

Cluster Analysis

Cluster analyses are methods for analyzing a single data set to find groups of cases that display various degrees of similarity. A distinction is usually made between algorithms that are "divisive" and partition multivariate space into regions and those that are "agglomerative" and form larger and larger groups in some type of hierarchy. Dimension-reducing methods such

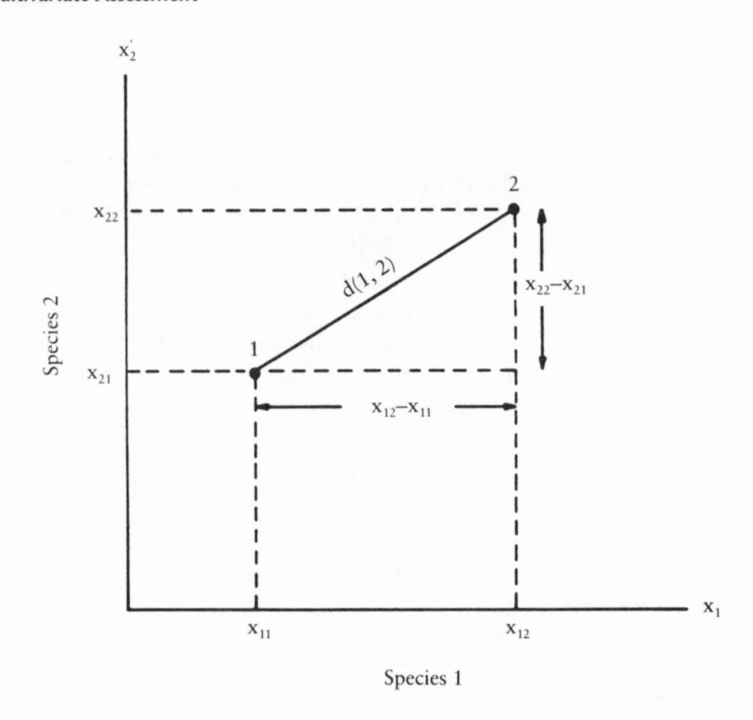

Figure 7.4. The distance between points 1 and 2, with coordinates (x_{11}, x_{21}) and (x_{12}, x_{22}), respectively, is $d(1, 2)$. (From E. C. Pielou, *The Interpretation of Ecological Data*, Copyright © 1984 by John Wiley & Sons, Inc. Reprinted by permission of John Wiley & Sons, Inc.)

as PCA and NMDS are not clustering techniques, because they do not identify discrete groups of cases. The assigning of cases to groups in ordination techniques such as PCA is an ad hoc "clustering" procedure (Neff and Marcus 1980:194–95). Cluster analysis can be viewed as either an ordination or a classification technique, depending upon its application. Pielou (1984, chap. 2) gave a detailed explanation of cluster analysis.

Wildlife-habitat studies have often incorporated cluster analysis as an initial phase of a more rigorous multivariate analysis. Researchers have used cluster analysis to quantitatively define taxonomic assemblages of animals (often inappropriately labeled "guilds" in the ecological literature; see Chapter 4). Cluster analysis can also help identify groups of species or sites as a complement to PCA or other ordination techniques.

Cluster analysis is often based on Euclidean distance, which is used as a measure of the similarity or dissimilarity between two data points (cases). Euclidean distance, graphically depicted for the two-dimensional case in figure 7.4, is simply an extension to a space of n dimensions of Pythagoras's distance theorem (Pielou 1984:14–15).

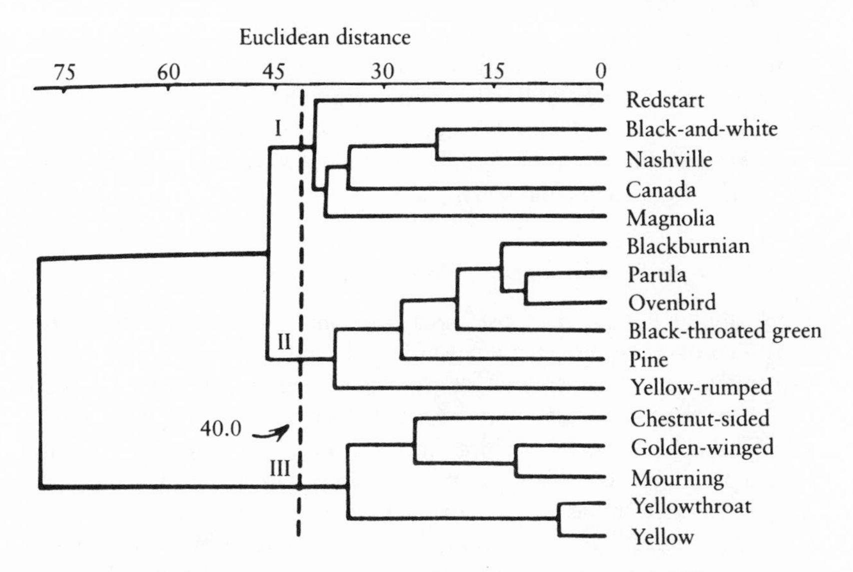

Figure 7.5. A cluster analysis of average habitat variables for sixteen species of warblers. The species fall into three habitat types: Group I contains shrub and forest edge species; Group II has mature forest species; and Group III contains open country birds. (From Collins, James, and Risser 1982, fig. 1.)

Applying cluster analysis to habitat data, Collins, James, and Risser (1982) analyzed the habitats of sixteen species of warblers using a habitat matrix of Euclidean distances (fig. 7.5). The researchers selected the Euclidean distance 40 on the dendogram as the threshold level at which relatively homogeneous groups could be identified. They defined three groups of birds based on general habitat use (shrub and forest edge, mature forest, and open country). Much variation in specific habitat use was evident, however, within each group. Further, the location (distance) at which clusters should be broken (represented by the dashed vertical line in fig. 7.5) is a user-specific decision.

Thus, the experiences of the users may cause much variation in the results of cluster analysis. That is, how well does the researcher understand the ecologies of the animals being considered, and how well does the researcher understand clustering techniques? Many different methods of cluster analysis exist, and each can result in a different representation (dendogram) of results (see Pielou 1984, fig. 2.17).

Potential users should review the detailed description of cluster analysis given by Dillon and Goldstein (1984, chap. 5) and Pielou (1984, chap. 2); Pielou specifically discusses ecological applications of cluster analysis. Both books provide specific guides for identifying and interpreting clusters, as well as numerous cautions regarding the usefulness of results. Cluster analysis, like all of the ordination techniques mentioned in this section, should only be the first step in an analysis. Consequently, it should lead to further investigation of the data and not simply to casual acceptance of the clusters obtained (Dillon and Goldstein 1984:208). Cluster analysis is, however, a useful means of initially viewing relationships among species, sites, and other groups of interest.

Assessing Relationships: Multiple Regression Analysis

Regression analysis is the most widely used method of data analysis. Regression provides three general types of results. First, regression can predict or estimate one response variable from one or more predictor variables. The estimated variable is called a predicted, criterion, or dependent variable. The one or more variables that estimate a dependent variable are termed predictors, covariates, or independent variables. Second, regression analysis can determine the best formula for predicting some relationship. Third, the success (precision) of a regression analysis can be ascertained, usually through use of correlation coefficients (Pimentel 1979:33).

Although popular, regression analysis is fraught with numerous problems that, if not adequately evaluated, can dramatically bias biological conclusions (see Pimentel 1979, chap. 3, for an especially sobering view of regression analysis). This caveat applies to all statistical procedures. Here again, the multivariate extension of simple linear regression magnifies these problems. In this section we highlight many of the most severe problems associated with multiple regression. We will concentrate on uses of multiple regression in model-building. We assume that readers are familiar with simple linear regression; most have likely at least briefly overviewed multiple regression (MR) in lower-division statistics courses. We draw from the excellent texts by Wesolowsky (1976) and Draper and Smith (1981) that are devoted to applied aspects of regression analysis.

Before beginning a statistical analysis, one should carefully plan each step, specify the objectives of the work, and provide checkpoints as the analyses progress. Because MR is easily misused and misunderstood, an organized plan for solving problems associated with this method is necessary. Draper and Smith (1981, chap. 8) provided a flow diagram of the steps necessary to ensure proper development of predictive models, a frequent goal of biologists using MR. This diagram (fig. 7.6) identifies three stages in such a plan: planning the analysis, developing the models, and verifying (validating) the initial model output (results). This pattern can be followed, of course, for any statistical analysis. We will emphasize model development and validation in this section. We covered variable selection in Chapter 4 and will address evaluation of sample sizes later in this chapter.

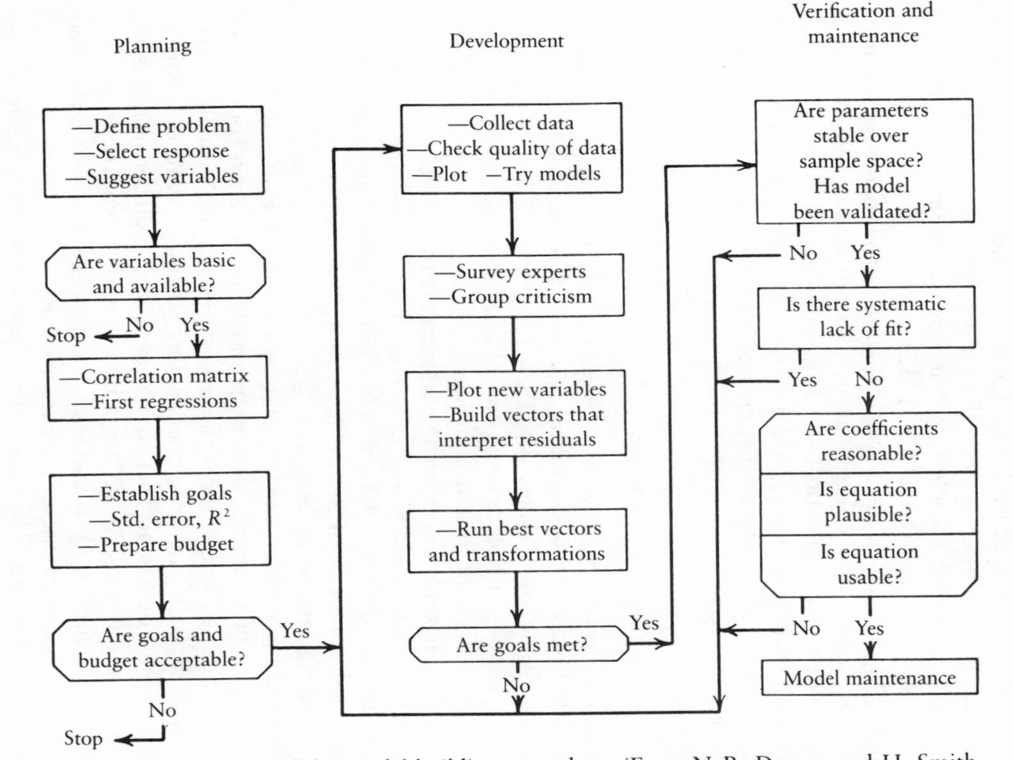

Figure 7.6. Summary of the model-building procedure. (From N. R. Draper and H. Smith, *Applied Regression Analysis*, Second Edition, Copyright © 1981 by John Wiley & Sons, Inc. Reprinted by permission of John Wiley & Sons, Inc.)

Conceptual Framework: Multiple Regression and Causation

In all statistical analyses results should at least reflect biological reality. Biologists hope to identify causal links between results of statistical analyses and a biological phenomenon. It is important to remember, however, that no statistical procedure, by itself, can prove cause and effect. Cause-and-effect relationships are derived from interpreting manipulative experiments as higher-level theory, and theory comes from outside statistics (Dillon and Goldstein 1984:213, 431–32).

Pimentel (1979, chap. 3) severely criticized biologists for gross misuse of multiple regression. Predicting a variable or demonstrating an association is an easy task with MR, but it is trivial when compared with most biological and management goals. This caution is especially appropriate here, as MR and many of the other methods described in this chapter have prediction as a goal.

Basic Statistical Concepts

Although we assume readers have a basic knowledge of linear regression, we wish to briefly review several concepts central to the understanding of regression. We follow Wesolowsky (1976:12–15).

The quality of statistics as estimators of population parameters is evaluated by the criteria of bias, mean squared error (MSE), consistency, and maximum likelihood; we outline the first three of these here. The expected value, E, of a parameter θ whose value is estimated as $\hat{\theta}$ is said to be an unbiased estimator if $E(\hat{\theta}) = \theta$. Lack of bias is a desirable, but not necessarily indispensable, property of an estimator. We seek to minimize bias through careful and rigorous study design.

The mean squared error (MSE) of an estimator $\hat{\theta}$ is $E[(\hat{\theta} - \theta)^2]$, or the expected value of its squared deviation from the parameter θ. For unbiased estimators the MSE equals the variance of the estimator. An unbiased estimator with a smaller variance than another unbiased estimator is said to be relatively more efficient. There is, however, a trade-off between bias and efficiency: the estimator $\hat{\theta}_2$, although biased, is likely more desirable than the unbiased estimator $\hat{\theta}_1$, because it has a smaller MSE and is more likely to produce close estimates for the parameter (θ).

An estimator is said to be consistent if values of $\hat{\theta}$ can be made to converge to θ by increasing sample size n. An unbiased estimator may or may not be consistent. Consistency is a desirable property because it increases the likelihood that larger samples will give better estimates. Thus, the general feeling, often encountered in ecological studies, that an increased sample size will swamp problems with our estimators is only appropriate under high consistency.

In MR we wish to establish a relationship between the dependent variable and the independent variables; the general form of the linear multiple regression model is

$$Y = B_0 + B_1X_1 + B_2X_2 + ... + B_qX_q + e$$

where the B's are constants called regression parameters, regression coefficients, or partial regression coefficients, and e is a random variable called the error term. This error term is necessary because Y_R, denoting the sample regression model, is unlikely to perfectly depict the relationship among variables (data points). In such a linear model all points would have to fall on a straight line for e to equal 0. Usually termed a residual, e is the difference between what is actually observed and what is predicted by the regression equation. A linear relationship is the popular choice for most workers because it is easy to work with and it sometimes provides a first-order approximation for a nonlinear relationship.

Regression models allow us to identify how much of the observed variation in the dependent variable(s) is explainable by the independent variables and how much is not (that is, the error term e). A measure of the relative importance of each of these sources of variation is termed the coefficient of multiple determination, R^2. R^2 is the ratio of the sum of squares due to regression (explained by regression) to the sum of squares about the $\overline{Y}$ (mean). R^2 ranges from 0 for no linear relationship to 1 for a perfectly linear relationship. The value of R^2 is thus a measure of the explanatory value of the linear relationship (Wesolowsky 1976:43).

The coefficient of multiple determination is often incorrectly regarded as a one-number summary of the quality of the regression model. Because R^2 is a statistic, however, chance influences the particular value it takes. When sample sizes are small in relation to the number of parameters being fitted, it is possible to get a large value of R^2 even when no linear relationship exists. Wesolowsky (1976:61) gave the following example of this relationship: a researcher obtains an R^2 of 0.85 after fitting an eighteen-parameter linear relationship with twenty data points. How impressive is this R^2? Wesolowsky showed that the chance of achieving this R^2 under these conditions was 0.75. Conversely, a low value of R^2 does not necessarily indicate a "bad" relationship but may simply show insufficient variation in the Y values. In figure 7.7, the line fitted through points 3–6 has an R^2 value of 0.28, whereas for points 1–2 and 7–8, R^2 equals 0.83. If the values of the independent variables have little variation, thus causing the variance in

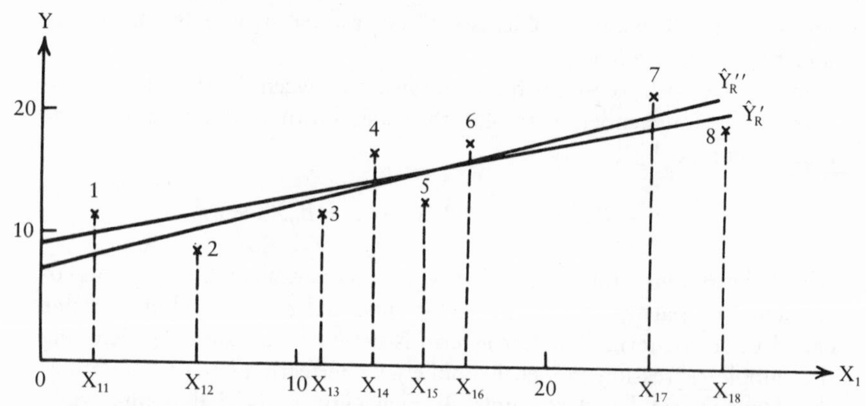

Figure 7.7. The effect of spread in the independent variable. (From Wesolowsky 1976, fig. 3.3.)

Y_R values to be small in relation to the variance in the error term, we could obtain a low R^2. We learn here that researchers must be sure to include in the study a wide enough range of values of the independent variable(s), such as snag densities, stream flow rates, or successional stages.

The significance value associated with a particular R^2 has also caused much confusion. With large sample sizes, even very small values of R^2 may be statistically significant ($P < 0.05$). This means that even though the fitted linear relationship has very little absolute explanatory value (R^2), we may still be able to reject the null hypothesis that all the slopes are equal to 0. It is not correct to say that a low R^2 means that a regression has no value. One or more of the individual coefficients may be significant, and the corresponding parameters rather than the overall regression model may be of primary interest in the research. If, however, a large number of parameters is being estimated, some of them will appear to be significant because of chance alone (Wesolowsky 1976:61–62).

It should be clear, then, that interpretation of regression equations resulting from analysis of wildlife-habitat data is not the straightforward process that it initially appears to be. The statistical significance of a model may result from inadequate sample size, an overabundance of parameters, chance, and/or a true biological relationship. When interpreting regression models, one should consider all of these factors, in addition to the violations of model assumptions discussed below.

Variable Screening

Multicollinearity

Earlier in this chapter we discussed assumptions associated with linear multivariate models; they fully apply to MR. One of the primary causes of misinterpretation and misuse of regression is multicollinearity. Multicollinearity exists when any independent variable is correlated with another independent variable or with a linear combination of other independent variables (Wesolowsky 1976:49). Multicollinearity, common and even inevitable in most disciplines, is especially evident in wildlife-habitat analyses. This is the case because we use MR as an exploratory tool to help identify important variables from a larger set of "possibly important" variables. Ecologists and wildlife biologists often measure as many variables as time and money allow, in the hope that later statistical analyses will identify some variable or combination of variables that help explain or predict some phenomenon. This, however, is actually an inappropriate application of MR. MR is not a data reduction technique per se. In application, problems that adversely influence interpretation of regression models render the use of MR as a data reduction technique highly suspect.

Correlation among independent variables causes three main problems: standard errors of regression coefficients are increased; as independent variables become increasingly correlated (near a linear relationship), computational difficulties arise (that is, precision associated with inverting a matrix with near-zero values); and the omission of variables may result in biased estimators for the regression parameters of the remaining variables if the missing variables are correlated with those remaining. This latter point negates one of the most commonly used methods of reducing multicollinearity and simplifying data sets used by ecologists and wildlife biologists. Wesolowsky (1976:49–56) detailed each of these three concerns; we will highlight the omission of missing variables.

When multicollinearity is present, all the sample regression coefficients are unbiased. The expected value of a coefficient is equal to the corresponding parameter values. The pairwise correlations aspect of multicollinearity, however, often causes biased parameter estimators. That is, if two independent variables are correlated with one another and one of these variables is omitted from the regression analysis, then the sample regression coefficient for the included variable may be biased. This occurs because the incremental explanatory power of the missing variable is removed, causing a larger sum of squared residuals, which, therefore, increases the variance of the regression coefficients. Further, it is possible to have high multicollinearity without having high correlations between any pairs of independent variables; this occurs, for example, when total canopy cover is included as a variable along with its constituent parts (see Wesolowsky 1976:50–51).

If we drop an uncorrelated independent variable, then the remaining regression coefficients will be unbiased estimators if the omitted variable had no effect on the regression model (that is, where the slope associated with this variable was zero). In ecological studies, we can seldom know if all relevant variables were included in the model. Therefore, we can rarely assume that the regression coefficients obtained are unbiased estimators for the regression parameters (Wesolowsky 1976:51). We thus find ourselves in a quandary. Ecologists usually try to include many possible measurements in a model because of the fear of missing something important; this helps ensure that at least many relevant variables are considered. The quandary arises when one attempts to use MR as a data reduction technique and discovers that the extra, often intercorrelated, variables add bias to the model.

A common procedure in wildlife studies to address the problem of multicollinearity is to identify pairs of correlated independent variables and then drop one member of the correlated pair. This is a good technique, but two problems might arise from its use. First, the regression coefficients in the resulting model will likely be biased. Second, no specific rules govern the value of r, the correlation coefficient, at which a member of a pair should be identified for deletion. Values used in the literature range from 0.5 to 0.8; 0.7 seems to be a commonly used level. The significance associated with an r-value is not especially useful, given that the r necessary to show significance ($P < 0.05$) drops as sample size rises. Thus, one is left to judge both the biological and statistical significance of the r chosen as a cutoff. Further, it is difficult to decide which member of a correlated pair to remove. Rarely do variables clearly measure precisely the same phenomenon: for example, height in coniferous trees and their diameter at breast height (dbh) are usually correlated at $r = 0.9$ or higher. Because dbh is relatively easy to measure, it is the obvious choice for inclusion in the model. In most applications, however, we initially include variables because we are unsure of their absolute and/or comparative importance in explaining a phenomenon; r-values for many pairs of variables often range between 0.4 and 0.6, and a third variable often correlates with one of a pair at a similar level. Thus, removing variables in this manner renders MR a data reduction technique.

Morrison, Timossi, and With (1987) encountered problems of multicollinearity when attempting to develop multiple regression models that predicted the relationship between numbers of birds and habitat features in a mature coniferous forest in the Sierra Nevada of California. They removed the member of a highly intercorrelated pair of variables, with an r threshold set at 0.8. The variable in a retained intercorrelated pair "was the one judged to be the most biologically meaningful and easiest to measure"

Table 7.5. Coefficient of multiple determination and prediction error for some models of bird-vegetation relationships, Sierra Nevada, California

Species	R	Adjusted R^2	Prediction error[a]
Red-breasted sapsucker	0.44	0.15	0.003
Hairy woodpecker	0.28	0.06	0.012
White-headed woodpecker	0.42	0.13	0.019
Pileated woodpecker	0.34	0.09	0.012
Hammond's flycatcher	0.41	0.15	0.090
Dusky flycatcher	0.53	0.24	0.042
Mountain chickadee	0.39	0.12	0.024
Chestnut-backed chickadee	0.17	0.02	0.084
Red-breasted nuthatch	0.27	0.06	0.078
Brown creeper	0.36	0.10	0.104

Source: Adapted from Morrison, Timossi, and With 1987, table 3 © 1987 The Wildlife Society.

[a]Values are SEs of predicted bird abundances (SSE), based on nontransformed abundance values.

(Morrison, Timossi, and With 1987:249). They noted that if the variables retained are highly intercorrelated, then use of the final regression model for prediction for values of variables that do not follow the past pattern of multicollinearity is highly suspect (see Neter and Wasserman 1974:388). Changes in this pattern of multicollinearity will arise as environmental conditions change; food abundance, rainfall, vegetation quality, predator abundance, and a host of other factors will alter the conditions under which a model was initially developed. This concept relates to our previous discussion of bias and consistency. That is, although multicollinearity does not in itself necessarily bias a model (sample regression coefficients), it may prevent extension of model results to other times and places, rendering the model ineffective for predictive purposes.

Returning to the study by Morrison, Timossi, and With, we see that they reduced their initial set of fifty-six variables to fourteen variables "with low intercorrelations" (1987:249). Overall, models resulting from their study were poor predictors of bird abundance, with most values of R^2 below 0.20 (table 7.5). Prediction error was low, however, indicating that the models they developed had low variance. Their results indicate that they sampled from a narrow range of independent variables. Because they sampled only in mature forests, there was little slope (variation) in the bird abundance–habitat features relationship (recall fig. 7.7). Also, by using MR as a data reduction technique, they lost a good deal of the explanatory potential in the variables they omitted prior to analysis. Thus, their conclusion that the habitat data used as independent variables were inadequate to track bird abundance was misleading.

Several procedures exist for detecting multicollinearity in a data set (see Neff and Marcus 1980:121–23 and all other texts cited in this section). The most widely used means of detection is termed the variance inflation factor (VIF), or tolerance. The VIF of each X_i is calculated as $1/(1 - R_i^2)$, where R_i^2 is the coefficient of multiple determination of X_i regressed in the absence of the explanatory variables. Although a high VIF does provide some overall indication of multicollinearity, it cannot distinguish among several coexisting near-dependences, and unfortunately it is itself numerically unstable when multicollinearity exists (Dillon and Goldstein 1984:272–73). Most canned statistical packages calculate VIF for each variable, usually under the term *tolerance* (for SPSS, see Norusis 1985:40–41). Thus, tolerance can serve as one indication of overall severe multicollinearity. Dillon and Goldstein (1984:273) provide several other observations that should be considered in combination with the VIF:

1. Certain regression coefficients (of variables significantly correlated with one already in the model) have signs opposite to what you would expect (for example, numbers of canopy-dwelling birds "should not" decrease with increasing canopy cover).
2. Several of the simple correlations between predictor and response variables are high, but the corresponding parameter estimates for these variables are not statistically significant.
3. The R^2-value of the model is high, but the partial correlation coefficients are low.

None of these conditions, however, is either necessary or sufficient for multicollinearity. Other procedures for detecting multicollinearity are available (Dillon and Goldstein 1984:272–80) but are beyond the scope of our discussion. None of the procedures, however, provides a single best means of detection, and none solves the problem of multicollinearity.

Some statisticians suggest leaving most if not all of the predictor (explanatory) variables in a regression model, removing only the most obvious cases of intercorrelation (for example, the case of two variables, dbh and tree height, measuring essentially the same feature of pertinence to birds). This procedure will preclude the use of MR as a data reduction technique and increase the likelihood of including many biologically relevant variables. MR has its greatest utility as a predictive tool. That is, one can use preliminary studies or data from the literature to form specific hypotheses regarding the response of animals to some set of independent habitat variables; this can best be set in an experimental context. Then, only those variables previously shown to influence the relationship under study can be used for further model-building and field study.

In using standard regression analysis, we must face the problems of multicollinearity by carefully examining model components and results. Further, validating model results will help determine the usefulness of a model in the presence of multicollinearity. Several modifications of standard MR are available, however, that at least in part reduce the adverse influences of multicollinearity. These methods include ridge regression and principal components regression (Draper and Smith 1981:313–25, 327–36). Draper and Smith detailed the uses and problems associated with ridge regression, a promising method that has not received popular application in wildlife studies. Principal components regression has received some limited attention in wildlife studies and, because of its clear relation to our previous discussion of PCA, will be discussed here.

Principal Components Regression

As the name implies, principal components regression (PCR) involves the regression of the dependent variable on the principal components of the independent variables. PCR has a major advantage over MR. With PCR, all of the information in the data set is retained (recall that no intercorrelated variables are removed under PCA), ensuring the retention of potentially important sources of information. A major disadvantage, however, is the inclusion of possibly irrelevant information in the final PCR models, which may in turn result in a poorer fit to the data than one would obtain with a standard MR (Maurer 1986).

Maurer (1986) provided a good example of PCR. He used both MR and PCR to evaluate the reliability of predictive models relating bird densities to measures of habitat. He found a trade-off between the techniques: standard MR sacrificed generality for increased precision, whereas PCR models were less sensitive to specific data sets but fit the data with less precision. Maurer concluded that there was no reason to select one procedure over the other on biological grounds. We think, however, that PCR is preferable to standard MR if the researcher is still in the exploratory stage of model development and hindcasting (see Chapter 6). The major statistical packages provide either direct or indirect means of computing PCR (Draper and Smith 1981:329–32 gave an example using BMDP).

Serial Correlation

Serial correlation or autocorrelation occurs most often with observations taken over time. Serial correlation may result from unrecognized nonlinearity in the regression relationship, from missing variables, or from sampling at intervals of time that correspond to cyclic changes in a biological variable, such as daily behavioral patterns. If one does not compensate for serial correlation, then the standard errors of the sample regression coefficients may be underestimated and the R^2-value could also be higher than it should be. Although the parameter estimators are still unbiased (as under multicollinearity), they will no longer be the minimum variance estimators; better estimators can be found (Wesolowsky 1976:136–37).

Ecological data is especially prone to serial correlation. As we observed earlier, biologists often lump data over broad time periods (such as "seasons") in an attempt to obtain adequate overall sample sizes. Such a process is questionable not only on biological grounds but also on statistical grounds, because of the problems of serial correlation. Wesolowsky (1976:136–47), Draper and Smith (1981:153–69), and others outlined methods for reducing the influence of serial correlation. A method called the Durbin-Watson test, available in most statistical packages, is used to identify certain types of serial correlation. When the Durbin-Watson test indicates serial correlation, one should examine the data for nonlinearity and consider the possibility of missing variables. If the problem cannot be eliminated in this way, one can apply various approximations. If data are indeed nonlinear, then appropriate transformations to achieve linearity can be attempted and/or nonlinear models can be applied. If lumping of data is causing the problem, then important biological phenomena are being masked; the researcher should go back and collect more data over shorter time frames. Another biologically reasonable approach would be to employ time-series analyses, but the topic is beyond our scope here.

Outliers and Influential Observations

An outlier among residuals is one that is far greater than the rest in absolute residual value and lies 3–4 standard deviations or further from the mean of the residuals. The outlier is a peculiarity and indicates a data point atypical of the data set. Writers have proposed rules for examining outliers (Draper and Smith 1981:152; Dillon and Goldstein 1984:252–70).

The researcher should carefully examine an outlier but should not remove it until clear justification, either biological or statistical, exists. If one can show that an outlier resulted from observer error, a malfunction of equipment, unusual or otherwise atypical animal behavior (for example, a vehicle caused an animal to move), or other readily apparent factors, then the data point should be removed from the data set. Outliers might exist, however, due to a failure by the researcher to adequately sample from the entire range of activities of the animal. Such failures might also result from an inadequate division of the data into time periods, lumping of sexes or age classes, an improper mixing of habitat types, and related factors. Obvi-

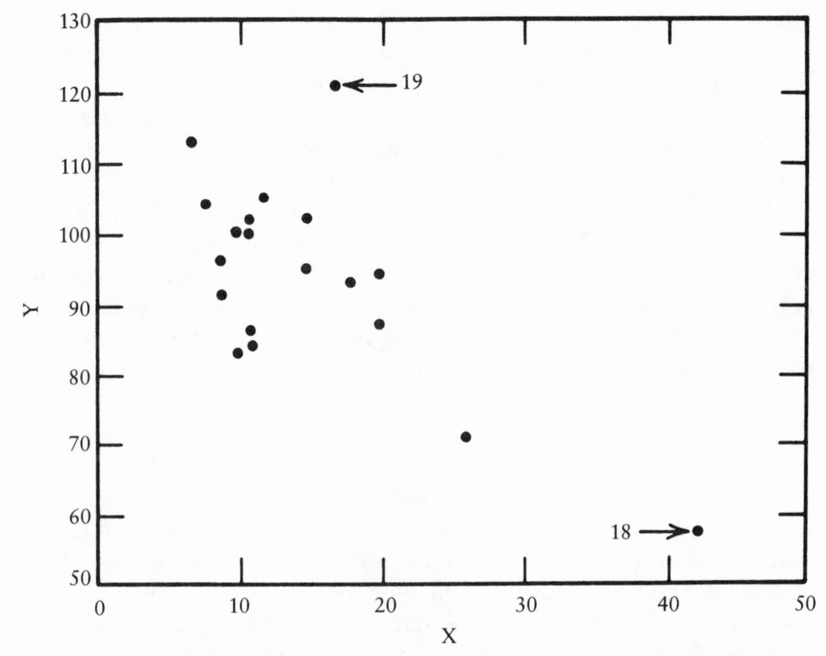

Figure 7.8. A regression with an observation (19) that is not influential and one (18) that may well be. *X* represents the age of a child at first word (in months) and *Y* represents the child's score in an aptitude test. (From N. R. Draper and H. Smith, *Applied Regression Analysis*, Second Edition, Copyright © 1981 by John Wiley & Sons, Inc. Reprinted by permission of John Wiley & Sons, Inc.)

ously, the researcher can anticipate and eliminate or at least greatly reduce most of these problems by carefully planning the research design.

The fact that an observation provides a large outlier does not necessarily mean that it is influential in the final regression model. For example, in figure 7.8, the observation marked 19 will certainly be an outlier for most models fitted through the data, although it is not influential because of its proximity to the swarm of neighboring *X*-values; that is, it cannot greatly effect the estimated regression coefficients. Observation 18, however, may not have a large residual, but it will certainly influence the final model because of its isolation from the other data points.

Polynomial and Nonlinear Regression

The selection of linear models (regression and otherwise) is usually more a matter of statistical convenience than a depiction of biological reality.

Linear relationships, however, are relatively straightforward. A model has little utility if other researchers and managers find it difficult to understand. Thus, data should be analyzed first by linear models unless compelling reasons exist to use nonlinear models; failure of data to approach linearity is a primary reason. Further, analyzing highly nonlinear data by use of linear models, even after appropriate data transformations, is unwise.

There are two basic methods of applying nonlinear relationships in statistical models. The first is to transform variables by various orders of polynomial transformations. Linear models are, specifically, first-order polynomial models. Second-order, or quadratic, models are obtained by squaring a variable (X^2), resulting in a U-shaped relationship. Third-order, or cubic, models are obtained by cubing a variable (X^3), resulting in a curving relationship. Higher-order models are, of course, possible, but they become increasingly difficult to visualize.

If a first-order model is not appropriate, then a second-order might be. Likewise, if a second-order model is not adequate, then a third-order might be. One should not, however, add higher-order terms until after exhausting other possible solutions (such as transformations to linearize data). After all, it is difficult to explain that an animal's response to habitat is based on the cube of foliage volume, the square of snag density, and shrub cover (first-order). If a linear model can achieve the research goals, then one should use it. If, even after transformations, the linear model cannot reach an appropriate level of resolution, however, the researcher should evaluate nonlinear functions.

Meents, Rice, Anderson, and Ohmart (1983) evaluated the influence of first-, second-, and third-order independent variables on predictions of bird abundance using MR. They found that polynomial variables resulted in statistically significant regressions in many cases in which linear variables did not do so (table 7.6). They concluded that while linear relationships were dominant, curvilinear relationships provided better explanations of birds and vegetation.

The most important finding of this study was the impact of nonlinear relationships on the biological interpretation of results. First, responses of birds varied between season, being linear during parts of the year but nonlinear during other parts. This, of course, greatly complicates the use of models, because our season-specific models should be of different statistical forms. Second, nonlinear relationships have important management implications. For example, the density of woodpeckers might be related to snag density in a curvilinear fashion, increasing to a certain point but then decreasing with further increases in snags. More important, woodpeckers may disappear when snags reach some nonzero but low critical

Table 7.6. Number of times a polynomial variable was significant in general regression models

	Linear not significant		Linear significant		
Season	Squared	Cubed	Squared	Cubed	Only linear significant
Winter	11	11	10	7	15
Spring	15	12	15	13	31
Summer	12	4	13	12	25
Fall	12	11	9	11	20
Total	50	38	47	43	91

Source: Meents et al. 1983, table 4.

density. In this example, a linear model would mask important biological relationships and likely would result in faulty management decisions.

Purely nonlinear models exist in a form other than the standard linear least squares model; Draper and Smith (1981, chap. 10) give examples of such nonlinear equations. Nonlinear models are solved through relatively complicated iterative calculations. The form of the equations and the iterative methods used will determine the outcome of the model. Thus, prior biological and statistical experience is necessary to select appropriate starting values. Beyond the scope of this book, nonlinear regression is a difficult, but potentially useful, method of developing regression models. Nonlinear programs are becoming a regular part of most computer software statistical packages; this will certainly lead to increased use (and concomitant misuse) of the techniques.

Determining the Best Model

Possibly the most difficult aspect of regression analysis is the selection of the model that results in the most appropriate representation of the phenomenon under study, that is, the best model. Depending upon the approach, there may be no unique solution to this problem. Solutions involve both quantitative and qualitative interpretations of biology and statistics. Draper and Smith (1981, chap. 6) extensively discussed this topic and addressed all-possible subsets regression, sequential procedures of variable selection (backward, forward, and stepwise selection), and related topics (see also Dillon and Goldstein 1984:234–42). Draper and Smith considered stepwise selection the best variable selection procedure. They concluded that stepwise selection is economical of computer time (as compared with all-possible subsets regression) and also avoids using more predictor variables than necessary while improving the model at every step. As Draper and Smith noted and as we have shown throughout this section, variable selection procedures should not be used purely as a method of variable reduction; the use of these procedures does not negate the importance of evaluating residual errors. Rather, variable selection procedures should be viewed as a means of refining models when the researcher has already determined that appropriate variables are being considered.

Related to the determination of best models is the area of model testing or validation. Model validation is an often-discussed, but seldom applied, area of wildlife-habitat relationships. Validation applies to many types of model development, including multiple regression and discriminant analysis.

Associations between Two Data Sets: Canonical Correlation Analysis

Using multiple regression, we intend to find combinations of the original predictor variables that best explain the variation in the dependent (criterion) measure. In canonical correlation analysis (CCA) we have a similar goal, but here we seek to simultaneously explain variation in two sets of data—typically wildlife species abundances and habitat characteristics. This explanation is achieved by two linear combinations, one for the predictor set and one for the dependent set, whose product-moment correlation is as large as possible. In CCA variates are computed from both sets of variables. A variate is analogous to a component in principal components analysis, except that a variate consists of a maximally correlated predictor and a criterion part. A maximum of M variates can be extracted, where M is the number of variables in the smallest of the two data sets. As in PCA, in CCA the variates are uncorrelated with each other (orthogonal) (Dillon and Goldstein 1984:337–38). It is important to note, however, that under CCA either set of data can be considered as the predictor; both sets are on the same statistical basis (Pimentel 1979:103).

Researchers should use CCA in analyzing several independent and dependent variables simultaneously. It is appropriate when the dependent variables are themselves correlated. Thus, CCA can help clarify complex relationships that reflect the relationship between the two data sets. It is helpful to note that CCA reduces to multiple linear regression when only one dependent variable is available.

Using CCA for descriptive purposes requires no distributional assumptions. In such cases, the dependent and independent variables can be measured at the nominal or ordinal scale. In order to formally test the significance of the relationship between canonical variates, however, the data should meet the requirements of multivariate normality and homogeneity of variance (Dillon and Goldstein 1984:339).

CCA, for example, might describe the relationship between m morphological variables (dependent variables) and p habitat variables (independent or predictor variables) measured over the same set of individuals; this analysis is often termed ecomorphological analysis. If we define X as the matrix corresponding to m columns of morphological data and Y as the p columns of habitat data, then CCA finds linear combinations of the independent variables and of the dependent variables that are maximally correlated. Of the possible linear combinations of the independent and dependent variables, CCA finds that particular pair of variables most highly correlated with each other. This variable pair becomes the first pair of canonical variates. The second pair of variates is the most highly correlated pair out of all possible linear combinations orthogonal to the first variates, and so on, until m pairs (if $m < p$) of canonical variates have been found (Neff and Marcus 1980:132).

Canonical correlation analysis has received limited attention in the wildlife sciences, although other disciplines have given the method a good deal of attention (Smith 1981). CCA is more difficult to interpret than more widely known methods, such as PCA, MR, and DA. Great potential, however, exists for establishing correlational relationships in biological research, given our knowledge of the complex and interactive nature of most biological systems. CCA should provide an effective means for making sense out of what otherwise might be an unwieldy number of bivariate correlations between two sets of variables (Dillon and Goldstein 1984:359).

Pimentel (1979, chap. 7), Neff and Marcus (1980), and Dillon and Goldstein (1984, chap. 9) offered excellent discussions of CCA. Smith (1981) summarized several studies that used CCA to analyze biological data. Folse (1981) used CCA to evaluate relationships between grassland birds, their habitat, and their food supply. Further discussion and examples of CCA with applications to wildlife-habitat analysis can be found in the symposium papers edited by Capen (1981).

Classification: Discriminant Analysis and Its Relatives

Discriminant analysis is widely applied throughout the scientific disciplines, including wildlife science. Discriminant analysis refers to a general group of methods, each of which has slightly different objectives. The overall goal of discriminant analysis, however, is the classification of individuals into specific groups (such as species, habitat types). Researchers can use methods of discriminant analysis to evaluate similarities and differences among sites or individual samples; discriminant analysis thus resembles PCA in its ordination capabilities. Unlike PCA, however, discriminant analysis starts with sets of groups and a sample from each group. Thus, while the goals of PCA and some applications of discriminant analysis are similar, the experimental designs for collecting data differ markedly.

Much confusion exists in the literature, including that specific to wildlife science, concerning specific methods within the general area of discriminant analysis. Discriminant analysis can be viewed as a family of methods; progression between each specific method depends on one's objectives for analysis at each successive step. For example, a valid discriminant analysis requires that a significant difference between groups exists. Such statistical significance is tested by one of several methods, depending upon the number of groups involved. If significant group separation exists, then one might wish to examine the classification ability of the discriminant functions. In addition, information on the degree of similarity between groups and the variables primarily responsible for this distance (separation) will probably be of interest (as it usually is in wildlife studies). Finally, one might wish to produce a prediction equation that assigns a new observation (unknown) to a group.

In this section we will refer to discriminant function analysis (DFA) as the method used to examine relationships between two or more multivariate groups with the goal of classifying observations. Unfortunately, many people refer to the general area of discriminant analysis as DFA. Pimentel (1979:187–90) detailed the many additional specific terms that are used in discriminant analysis; Pimentel's book provides a good source of information for deeper investigations into this area of analysis.

Also related to discriminant analysis is the broad area of multivariate analysis of variance, or MANOVA. We will return to MANOVA later in this chapter. Finally, a term frequently used but seldom defined in ecological literature is *discriminant space; canonical space* is a synonym (Pimentel 1979:243). Discriminant space is formed from Euclidean distances by rotating original variable axes so the angles between any pair of axes have cosines equal to the correlation between the variables. Pimentel (1979:188, 247) referred to this space as one whose total form is greater than the sum of the variables defining the space.

How Discriminant Analysis Works

Dillon and Goldstein (1984, chap. 10) provided an especially lucid description of discriminant analysis that we will summarize here, while incorporating material from Pimentel (1979, chap. 10) and Neff and Marcus (1980). Under discriminant analysis, researchers evaluate one categorical dependent variable and a set of independent variables. Although there is no requirement that these independent variables be continuous in nature, discriminant analysis often performs poorly when independent variables are categorical. The categorical dependent variable is a grouping factor that places each observation into one and only one predefined group. In this way, DFA differs from nonhierarchical cluster classification, which can assign an observation to multiple groups. We might be interested in examining differences among species or study sites based on a series of habitat characteristics. After all individuals are assigned to these groups, we further wish to "discriminate" among the groups on the basis of the value of the predictor variables (such as habitat characteristics), interpreted through observed scores (locations) of observations on the set of independent variables. Discriminant analysis is thus a method for classifying observations into one of two or more mutually exclusive groups and, further, determining the degree of dissimilarity of observations and groups and the specific contribution of each independent variable to this dissimilarity.

Discriminant Functions

Discriminant analysis involves deriving linear combinations of the independent variables that will discriminate between the predefined groups such that misclassification rates will be minimized. Discriminant analysis can be viewed as a "scoring system" that assigns observations in the sample with a score that is a weighted average of those observations' values from the set of independent variables. Once a score has been determined, it can be transformed into an *a posteriori* probability that gives the likelihood of the individual belonging to each of the groups.

Figure 7.9 presents the two-group case of discriminant analysis, showing a scatter diagram of individuals and the projection of the discriminant function in the sample space (defined by just two variables in this case, X_1 and X_2). A discriminant function in this space is the line connecting the two groups. Specifically, the linear discriminant function between two groups is the product of a discriminant coefficient vector and a vector of measurements: that is, $S^{-1}(x_1 - x_2)$, where S^{-1} is the inverse of the pooled sample variance-covariance matrix and the x's are the matrices (p characteristics $\times n$ observations)of observationsin deviations takenfrom each group. The linear discriminant function equation takes a form similar to the multiple regression equation:

$$D = B_0 + B_1X_1 + B_2X_2 + \dots + B_pX_p$$

where the X's are the values of the independent variables, and the B's are the discriminant coefficients derived from the data. For example, by finding the weighted average of independent variables such as canopy volume,

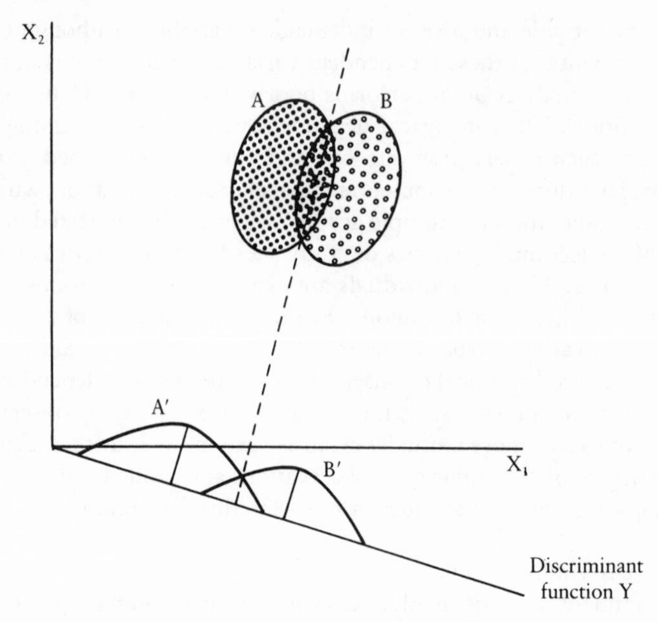

Figure 7.9. Graphical illustration of two-group discriminant analysis. (From W. R. Dillon and M. Goldstein, *Multivariate Analysis: Methods and Applications*, Copyright © 1984 by John Wiley & Sons, Inc. Reprinted by permission of John Wiley & Sons, Inc.)

shrub cover, and litter depth, one can obtain a score that distinguishes, say, habitat used by one rodent species from another. In discriminant analysis, the weights or coefficients are estimated so that they result in the best separation between the groups. The above equation can be used to calculate the discriminant score for each observation.

With only two groups, a single function of the p measurements will account for all differences between groups possible with the data set. For more than two groups, however, one weighted combination might distinguish some of the groups but not others. Unfortunately, with more than two groups, numerous functions may be required to distinguish between groups not well separated by the first discriminant function. In such cases, interpretation of results becomes increasingly complicated as the number of groups increases. Because of this complicating factor, we will discuss the two-group case before turning to multigroup considerations.

Statistical Significance

Discriminant analysis helps to determine whether significant differences exist among groups. The mean value of the discriminant function is commonly called the group centroid. The Euclidean distances between cen-

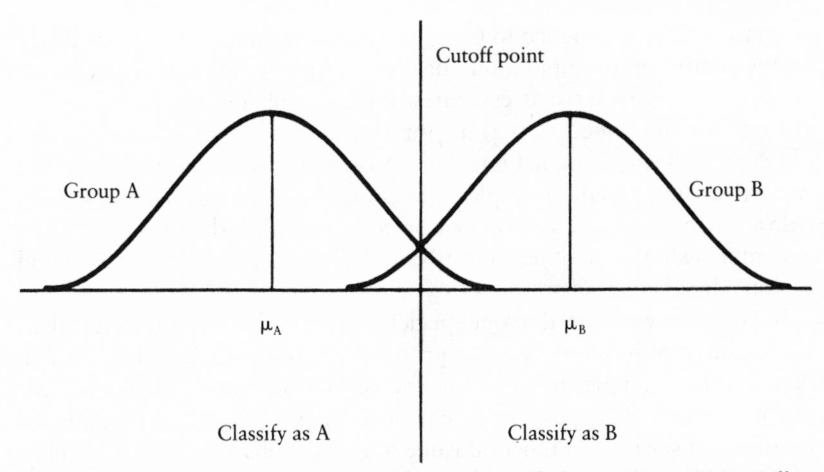

Figure 7.10. Optimal cutting score with equal sample sizes. (From W. R. Dillon and M. Goldstein, *Multivariate Analysis: Methods and Applications*, Copyright © 1984 by John Wiley & Sons, Inc. Reprinted by permission of John Wiley & Sons, Inc.)

troids in the discriminant space have been called Mahalanobis's generalized distances, D^2 (see Dillon and Goldstein 1984:366–67, for computation). This statistic can be used to determine whether the between-group differences in mean discriminant scores are statistically significant. Large values of D^2 indicate that the groups are sufficiently spread in terms of mean separation. In such cases it is likely that new observations could be successfully classified on the basis of the characteristics measured. For the two-group case we calculate Hotelling's two-sample T^2-statistic, the multivariate analogue of the univariate t-statistic (see Dillon and Goldstein 1984:367–69).

Classification

After developing linear discriminant functions, the researcher calculates classification rules for assigning observations to the *a priori* groups. In general, these rules specify the probability that an observation with a discriminant score D belongs to group i. Cutpoints are designated on each discriminant function axis. In the two-group case where there is equal probability of an observation belonging to either group—that is, the prior probabilities, denoted $P(G_i)$, are equal—the cutpoint would be midway between the two groups (the dashed line in fig. 7.10).

The prior probability is an estimate of the likelihood that an observation belongs to a particular group when no additional information about it is available. For example, if 40 percent of the observations used to derive the discriminant function were from Group 1, then the probability of a new observation being assigned to Group 1 is 0.40. If the sample is considered representative of the population, the observed proportions of observations in each group can serve as estimates of prior probabilities. For example, say we have developed a discriminant function for classifying two species of *Peromyscus* based on habitat characteristics measured around trap sites (we will show a specific example of this situation for more than two species below). If we have no *a priori* reason to believe that the probabilities of assignment should be other than 50-50, then the prior probabilities would be considered equal.

If, however, we knew that one species was more abundant than the other, we might want to alter the prior probabilities to reflect this difference in density; for example, to 60-40 if one species accounted for 60 percent of the captures. It is possible to calculate the probability of obtaining a particular discriminant function value of D if the observation is a member of Group 1 or Group 2. To calculate this probability, we assume that the observation belongs to a particular group and then estimate the probability of the observed discriminant score given group membership. These probabilities, called conditional probabilities, are denoted $P(D|G_i)$, where G_i is the ith group.

The conditional probability of D given the group membership suggests the likelihood of the score for members of a particular group. When group membership is unknown, however, an estimate of the likelihood of membership in the various groups is needed, given the information that was used to develop the discriminant function and conditional probabilities. This is called the posterior probability and is denoted $P(G_i|D)$; it is calculated from the conditional probability and prior probability. A case is classified, based on its discriminant score D, in the group for which the posterior probability is the largest; that is, it is assigned to the most likely group based on its discriminant score (Norusis 1985:81–83).

Markedly different sample sizes between groups can substantially influence results of classification analysis. The cutpoint in the case of equal sample sizes is shown in figure 7.10, whereas the effect of unequal sample sizes is depicted in figure 7.11. We see that if Group A is substantially smaller than Group B, then the cutpoint will be closer to the centroid of Group A than it is to the centroid of Group B. With unequal n's a weighted cutpoint can be calculated (Dillon and Goldstein 1984:369–70). If difference in sample size is ignored, however, we see that the unweighted cutpoint results in the optimal classification of Group B but substantially misclassifies members of Group A. Extremely nonnormal and skewed distributions will also adversely influence results of classification analysis, especially if these problems are unequal between groups and/or sample sites and periods. If at all possible, one should maintain equality of sample sizes in

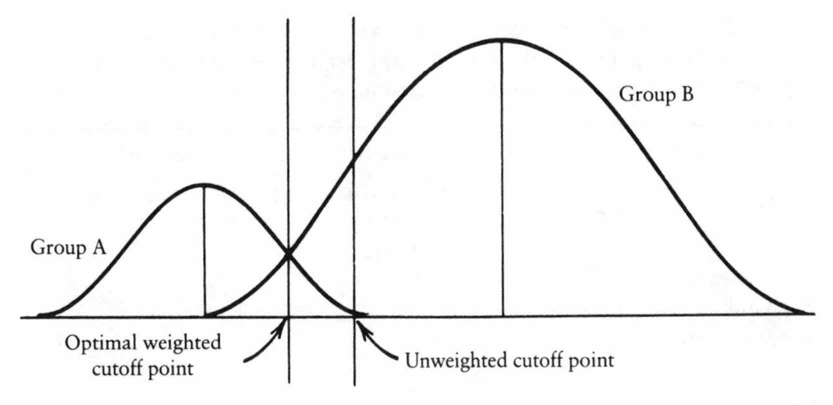

Figure 7.11. Optimal cutting score with unequal sample sizes. (From W. R. Dillon and M. Goldstein, *Multivariate Analysis: Methods and Applications*, Copyright © 1984 by John Wiley & Sons, Inc. Reprinted by permission of John Wiley & Sons, Inc.)

classification analysis, even if this requires subsampling from groups with the largest *n*'s.

Results of classification analysis are usually presented in a simple table that shows the percentage of cases correctly classified for each group. In addition, one can also show the groups that misclassified cases were placed in or closest to. In the three-group case shown in table 7.7, setting the prior probabilities equal for all groups results in a roughly 0.33–0.33–0.33 probability of correct classification. Thus, a classification substantially above 33 percent indicates that the discriminant function was useful in identifying that group (for example, 87.9 percent for MacGillivray's warbler). Species not well separated from each other might show high classifications (misclassifications) for a different species; here, many orange-crowned warblers were misclassified as MacGillivray's warblers (Morrison 1981).

Interpretation

If the researcher finds a significant difference between groups, then discriminant analysis can be used to identify the contribution of each independent variable to the overall discrimination. Dillon and Goldstein (1984:372–75) discussed the approaches available for interpreting discriminant analysis; we summarize several of these here.

Most researchers have interpreted discriminant analysis by the group means and associated univariate *F*-values for each independent variable, and/or the magnitudes of the standardized discriminant coefficients (weights). If, however, the independent variables are intercorrelated, then

Table 7.7. Classification matrix derived from discriminant function program showing actual and predicted species (group) membership for singing male warblers based on habitat use on the deciduous tree and nondeciduous tree sites

	Predicted group membership[a]		
Actual group	Orange-crowned	MacGillivray's	Wilson's
	Deciduous Tree Sites		
Orange-crowned warbler	*13.0*	47.8	39.1
MacGillivray's warbler	9.1	*87.9*	3.0
Wilson's warbler	25.9	14.8	*59.3*
	Nondeciduous Tree Sites		
Orange-crowned warbler	*16.1*	58.1	25.8
MacGillivray's warbler	16.1	*74.2*	9.7
Wilson's warbler	26.1	13.0	*60.9*

Source: Morrison 1981, table 5.
[a]Numbers in italics denote percentage correctly classified.

both approaches can give misleading indications of the importance of each variable. For example, two intercorrelated variables might be splitting the discriminant coefficient between them, with both variables thus appearing to be only marginal contributors to the separation. Alternatively, one coefficient might be inflated while the other is near zero. Many of the earlier publications in the wildlife literature used discriminant coefficients to interpret group separations in this way, ignoring potential bias of correlations between variables.

A more appropriate approach for interpretation uses discriminant loadings. Similar to principal component loadings, discriminant loadings give the simple correlation of an independent variable with a discriminant function. Most of the canned statistical packages will produce discriminant loadings. Because loadings reflect common variance among predictors, they are less subject (but not immune) to instability caused by predictor intercorrelations and thus are more useful in interpretation than are discriminant coefficients.

The partial F-value approach essentially partitions out the variance in the independent variable of interest that is already explained by the other variables. Covariance-controlled partial F-values are calculated by computing for each independent variable a one-way analysis-of-covariance where the covariates are the remaining $p - 1$ variables. In practice, the use of discriminant loadings, we believe, will give the most straightforward and useful interpretation of a discriminant analysis. This is given, of course, assuming adequate sample sizes and data at least approaching normality.

A stepwise selection procedure is commonly used to reduce the set of independent variables that best separate the groups under consideration. Intercorrelations among independent variables, however, make difficult the assessment of the unique effects that individual independent variables have on the dependent variable (group). Multicollinearity will also substantially impact results of discriminant analysis.

Dillon and Goldstein (1984:241) suggest that all variable selection procedures be considered useful approaches for selecting good subsets of variables, but not necessarily the single best set of variables. This sound advice applies to all multivariate methods. Researchers should review the Dillon and Goldstein material (1984:234–42) regarding specific steps to take when using variable selection procedures.

Multiple Discriminant Analysis

Like two-group discriminant analysis, multiple or multigroup discriminant analysis (that is, analysis of more than two groups) seeks to find an axis that maximizes the ratio of between-groups to within-group variability of projections of observations onto this axis. Two or more groups might not be separable by the use of one axis; a three-group illustration of this is shown in figure 7.12. In general, with G groups and p independent variables, there are a total of $\min(p, G - 1)$ possible discriminant axes. Because there will usually be far more independent variables than groups, at most $G - 1$ axes will be possible; the number of statistically significant axes will likely be even less (Dillon and Goldstein 1984:394–95).

With multigroups, the discriminant functions are generated so that the resulting discriminant scores on each new axis are uncorrelated with the scores on any previously generated axis. With only two groups, of course, only one axis exists. The discriminant functions are calculated (extracted) so that the observed variation in the data set appears in decreasing orders of magnitude. The discriminant axes need not be orthogonal; only their discriminant scores (that is, the projections of observations onto the axes) are uncorrelated (Dillon and Goldstein 1984:396–97). With more than two groups, the canonical variate axes will be in the directions of maximum dispersions among group centroids; these directions will seldom be coincident with any of the lines connecting pairs of centroids (these are the discriminant functions) (Neff and Marcus 1980:159).

Interpretation

A major goal of multiple discriminant analysis is, first, to determine if the group centroids are statistically different. Overall significance is determined by assessing the significance of the Wilks's *lambda* statistic. Wilks's *lambda* is simply the ratio of the within-groups sum of squares to the total

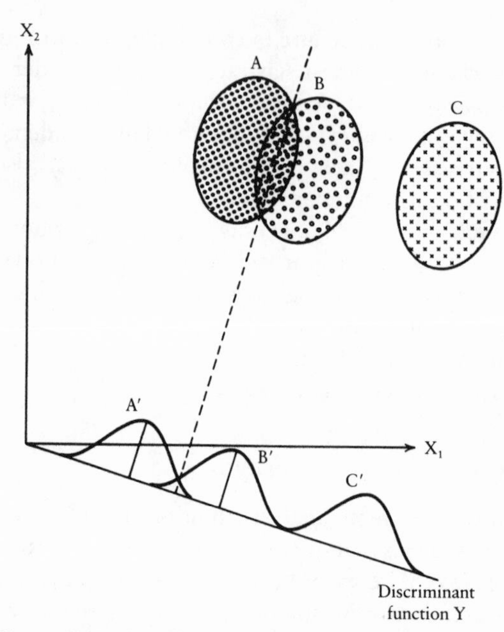

Figure 7.12. Graphical illustration of three groups and a single discriminant function. (From W. R. Dillon and M. Goldstein, *Multivariate Analysis: Methods and Applications*, Copyright © 1984 by John Wiley & Sons, Inc. Reprinted by permission of John Wiley & Sons, Inc.)

(within + between) sum of squares (see Dillon and Goldstein 1984:404–5, 420–22). If Wilks's *lambda* is significant, however, it provides little information on the ability of the discriminant function(s) to classify observations. It is interpreted only to indicate that an overall significant difference among groups exists; that is a very necessary but relatively uninformative step.

Table 7.8 shows a typical presentation of the results of a discriminant analysis. For the "all species" analysis, note that two significant functions were derived. Box's *M* is the test of equality of the variance-covariance matrices among species; here it was significant in all comparisons, meaning that the species did have different variances (in multivariate habitat use). A significant Box's *M* means that the formal test of the null hypothesis of equality of group centroids is technically invalid. As noted earlier in this chapter, multivariate analyses are considered robust to violation of assumptions. In the face of a significant Box's *M*, many researchers qualify their presentation by stating that results should be considered descriptive in nature, not strict tests of a null hypothesis. This is especially appropriate in

Table 7.8. Discriminant function analysis of small mammal habitat use, western Oregon

Group discriminant function	Relative variance (%)	Cumulative variance (%)	Wilks's λ	χ^2	df	P	Box's M (P)
All species							
DF I	75.1	75.1	0.700	93.45	12	< 0.0001	—
DF II	24.9	100.0	0.910	24.77	5	0.0002	< 0.0001
Rodents only (excluding *Zapus*)							
DF I[a]	98.0	98.0	0.716	68.76	6	< 0.0001	< 0.0001
Sorex only							
DF I	100.0	100.0	0.789	9.00	2	0.0111	< 0.0001

Source: Morrison and Anthony 1989, table 4.
[a]DF II accounted for 2 percent of the variance and was not significant at $P = 0.45$.

nonexperimentally controlled wildlife research, where unequal variances are likely a true biological property of the way different species use the habitat. This is given, of course, assuming adequate numbers of samples.

Of special interest in multiple discriminant analysis is determination of the difference among all groups in the analysis. For example, while we first need to know if five species vary, overall, in habitat use, we will also wish to know which species were significantly different from which other species and, further, how far apart the groups actually were. Actual distances between species will help in the interpretation of the ecological relations among all groups analyzed, that is, the animal "community" under study. This can be viewed as the multivariate extension of the univariate case where we first determine overall group difference by ANOVA and then apply a multiple range test to determine the differences.

The most useful statistic for summarizing intergroup differences in the multivariate case is the Mahalanobis distance, D^2. The values of D^2 are usually given in the form of a contingency table that compares each group with all other groups (table 7.9). It is also possible to use this matrix of D^2-values in a cluster analysis to compare the groups (Neff and Marcus 1980:165).

Plots, often used to illustrate the results of discriminant analysis, are especially useful in the multigroup case. Using the statistically significant discriminant functions, the group centroids can be plotted in two- or three-dimensional discriminant space. In practice, the first two or three functions

Table 7.9. Significance ($P < 0.05$) between pairs of species as determined by Mahalanobis's distance for bird habitat use at Blodgett forest, California, during summer (lower left triangle) and winter (upper right triangle)

	HAWO	WHWO	PIWO	HAFL	DUFL	MOCH	CBCH	RBNU	BRCR	GCKI	RCKI
Hairy woodpecker (HAWO)	—		*	*				*			
White-headed woodpecker (WHWO)		—	*								*
Pileated woodpecker (PIWO)	*	*	—			*		*	*	*	*
Hammond's flycatcher (HAFL)			*	—							
Dusky flycatcher (DUFL)	*	*	*	*	—						
Mountain chickadee (MOCH)		*	*	*	*	—	*	*			*
Chestnut-backed chickadee (CBCH)			*		*	*	—		*		*
Red-breasted nuthatch (RBNU)			*		*			—	*	*	*
Brown creeper (BRCR)			*		*	*			—		*
Golden-crowned kinglet (GCKI)			*		*	*				—	*
Ruby-crowned kinglet (RCKI)											—

Source: Adapted from Morrison, With, and Timossi 1986, table 5.

are usually adequate for ecological interpretation of results; most of the variation in the data set is explained by these first few functions.

One can further evaluate groups by plotting ellipses for each group. The ellipse for each group is centered on the group centroid, and the shape of the ellipse for each group depends on which of the two within-group covariance matrices is used. Recall that an assumption of multivariate analysis is equality of the group variance-covariance matrices. If this assumption holds or is simply assumed to hold, then the pooled within-group covariance matrix is used, and the shapes of all of the ellipses is the same (using standardized discriminant coefficients results in circles for each group). If, however, the variance-covariance matrices are not equal, then the separate variance-covariance matrices are used, and the ellipses can take various shapes and orientations.

Figure 7.13 depicts the use of ellipses based on separate variance-covariance matrices to graphically display habitat relationships of seven small mammal species. Although variations in sample size and variance-covariance matrices make absolute, interspecific comparisons based on these ellipses difficult (Carnes and Slade 1982), the ellipses provide useful aids for ecological interpretations of the results of discriminant analysis. Figure 7.13 also graphically depicts the results of D^2, showing statistically significant separation among species in discriminant space.

Researchers can often improve the interpretation of discriminant analysis by rotating the matrix of discriminant coefficients. Group centroids, discriminant loadings, and group ellipses can all be based on these rotated coefficients (Dillon and Goldstein 1984:416).

Numerous additional examples of the use of discriminant analysis can be found in the ecological literature, especially that published since the mid-1970s. Indeed, today it would be unusual not to find at least several papers dealing with multivariate statistics in any issue of most journals. Many of the specific procedures used in the earliest papers have now been called into question, as later researchers have examined problems with sample sizes, transformations applied to variables, measures of the niche derived from multivariate methods, and the like (see Capen 1981; Carnes and Slade 1982; Williams 1983; Verner, Morrison, and Ralph 1986; Morrison et al. 1990). Although many of these problems have disappeared from the most recent papers, a critical problem involving inadequate sample sizes remains.

Logistic Regression Analysis

It is not improper to apply discriminant analysis to evaluation of qualitative (categorical) independent variables. Remember, however, that discrimi-

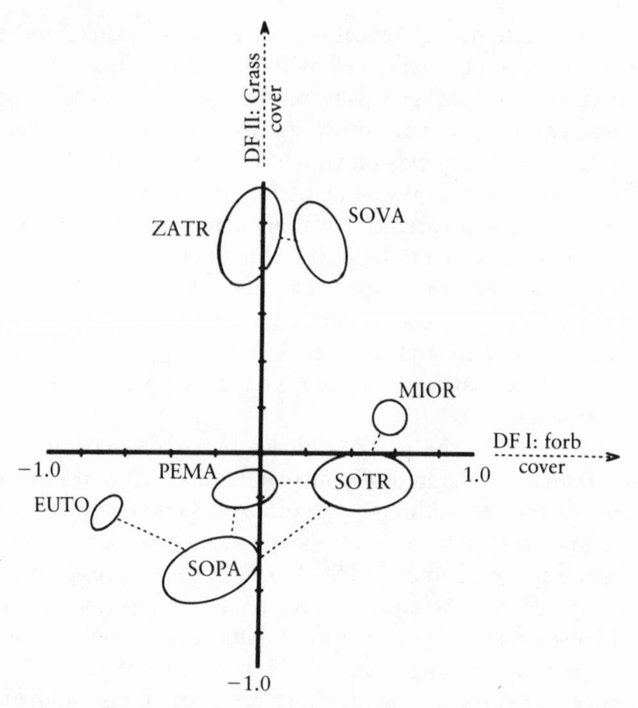

Figure 7.13. Two-dimensional ordination of small mammal habitat use as determined by discriminant analysis. Ellipsoids represent 95 percent confidence intervals of species centroids. Broken lines connect pairs of species not significantly ($P > 0.05$) different in habitat use as determined by pairwise multivariate tests based on the Mahalanobis distance (for example, *Microtus oregoni* is not significantly different from *Sorex trowbridgii* but is significantly different from *Peromyscus maniculatus*). Species codes: PEMA, *Peromyscus maniculatus;* MIOR, *Microtus oregoni;* EUTO, *Eutamias townsendii;* ZATR, *Zapus trinotatus;* SOVA, *Sorex vagrans;* SOPA, *S. pacificus;* SOTR, *S. trowbridgii.* (From Morrison and Anthony 1989, fig. 1.)

nant analysis requires that the variables exhibit multivariate normality with equal covariances; categorical variables will not satisfy these assumptions (Hassler, Sinclair, and Kallio 1986). An alternative method, termed logistic regression, does not require that these assumptions be met. Logistic regression (LR) can be used to analyze independent variables that are true categorical data (such as age, sex, coat color) and those that have been summarized into categories (such as height intervals, size classes of food items). Observations are assigned to one unique category of each variable. Like other multivariate techniques, LR can be used to develop predictive models.

Hosmer and Lemeshow (1989) discussed general applications of LR in their recent text. Press and Wilson (1978) and Hassler, Sinclair, and Kallio (1986) analyzed comparisons of LR and discriminant analysis. Capen, Fenwick, Inkley, and Boynton (1986) and Brennan, Block, and Gutierrez (1986) described specific uses, including the advantages and disadvantages, of LR in wildlife studies. A detailed presentation of LR and other logistic models is beyond the scope of this book. These methods will, however, see increasing use in the future; most of the major statistical software packages now offer LR.

Related Methods

Discriminant analysis is related to various other multivariate methods, especially canonical correlation analysis and MANOVA; Dillon and Goldstein (1984:417–29) detailed these relations. Multivariate analysis of variance embodies a large number of experimental designs that are analogous to univariate factor analyses and other higher-way experimental designs. For example, with MANOVA one could examine the influence of five classes of variables on habitat use of five species of birds while simultaneously considering seasonal and/or temporal interactions. James and Lockerd (1986) used a nested MANOVA model to adjust for differences in study sites that could obscure differences in habitat use between small mammal species. There have not, however, been many applications of factorial or other experimental designs to multivariate data analysis. Dillon and Goldstein (1984:420–29), Harris (1985, chap. 4), and Norusis (1985, chap. 6) provided good discussions of MANOVA.

The Cart before the Horse: Sample Size Analysis

Inadequate sample size is probably the most critical problem found in papers using multivariate methods. Regardless of the refinement of the study design and the care taken in recording data, proper ecological interpretation of multivariate analyses is difficult at best; interpretations based on inadequate samples are wasted efforts. In Chapters 4 and 5 we identified methods of determining proper sample sizes in the univariate case; those methods apply here on a variable-by-variable basis. In multivariate analyses, however, our problems are increased by orders of magnitude given our desire to interpret ecological interactions in many dimensions for many species. Thus, do not get the cart before the horse; any study must be pulled by a strong data set!

Johnson (1981) outlined general guides for determining adequate sample sizes in multivariate studies, noting that more observations were needed when the number of independent variables was large. Many published studies, however, have only slightly more observations than variables, and sometimes even fewer. Johnson thought that an appropriate minimum sample size for multivariate analysis might be twenty observations, plus three to five additional observations for each variable in the analysis. Larger sample sizes do not, however, provide an answer to poorly designed studies. In such cases, larger numbers of samples will not decrease unwanted variability. As Johnson stated, "Calls for larger samples are the 'knee-jerk' reaction when variability is excessive" (1981:56).

In a study of habitat use by birds in Oregon, Morrison (1984) found that a minimum of thirty-five habitat plots was necessary to obtain stable results (stability was determined when means and variances did not change with an increase in numbers of samples). Morrison's review of the wildlife-habitat literature showed, however, that very few studies met the minimum numbers recommended by Johnson. A similar study by Block, With, and Morrison (1987) found that even larger numbers of plots (up to seventy-five or more) were needed in habitat analysis. If adequate sample sizes are not obtained, then one can place little confidence in the ecological interpretations of multivariate analyses.

Thus, the few studies that have evaluated sample size show that one must have somewhere between thirty-five and seventy-five samples per dependent group in order to apply multivariate methods. This wide range of minimum samples should be used as a guide for designing an individual study; that is, realize that *at least* this number of samples will be needed. Remember, also, that these minimum sample sizes apply to each biological period being analyzed, and the appropriate n might vary among periods. Sample size is a study-specific question that must be evaluated on that basis.

Sample size analysis is not difficult; this is a question of proper study design and not the analytical techniques per se. The wide use of multivariate methods to analyze data, however, has not been accompanied by an improvement in study design.

How Well Does It Work? Model Validation

In many ways, this final section on model validation examines the most important aspect of multivariate analysis, especially when the analysis is to be used in management. Simply, validation determines how much confidence we should place in our analyses, whether they are used for predictive or descriptive purposes. Recall the summary of the model-building process given in figure 7.6. If, under "Development," we determine that our analysis met our goals for the study, there remains the critical process of model validation (verification) and maintenance. Maintenance acknowledges that biological systems change over time and that most models are specific to area and time. (See Chapter 6 for an overview of model validation. This section specifically refers to validation of statistical models.)

Even if a model is never applied outside the area for which it was developed, factors can change within the area of concern. Thus, validation must continue over time to ensure a continued suitability for research needs. Unfortunately, few studies have considered validation in the initial design stage.

Draper and Smith (1981:419–20) distinguished two different types of data sets that must be considered in designing a validation procedure. If the model has been developed using observations taken across a long time span, then the data are considered longitudinal. One can test the stability of the estimated coefficients derived from longitudinal data by fitting the model on shorter time spans. If the estimated coefficients show trendlike patterns, then use of the fitted model based on all the data would be unwise. This, of course, relates back to our discussion of sample size and the need to examine results as one proceeds through a study. If model validation indicates that shorter time periods should be used, a researcher likely will not have a sufficient number of samples within each of the shorter periods. The study will, in essence, have been a waste of time.

Data that can be considered as information collected at the same "point" in time is termed cross-sectional. For example, the shorter time intervals noted above can be considered cross-sectional if the biological phenomenon under study remained mostly static during the period. Here again, one should use quantitative measures to justify the time periods used. In ecological studies, researchers most often simply state that data collection was confined to some predefined biological period, such as the "breeding season." As we saw in Chapter 4, however, such descriptions of periods are often much too broad.

Models are validated in two basic ways: the entire data set may be randomly divided into two parts, with one part used to develop the model, and the other, to evaluate the model's suitability when "unknown" or "new" data are presented; and/or data sets are collected from different times and/or locations, and these various sets are then used in validation. Employing the first approach, sometimes called cross-validation, researchers have used different percentages of data for each divided data set, and more complicated statistical procedures (such as jackknife and bootstrap methods) are available (see Dillon and Goldstein 1984:392–93; Meyer et al. 1986; Lanyon 1987). Results of models resulting from "all" data and those from validation sets are generally evaluated in terms of similarity in component loadings and coefficients (magnitude and direction), and classification analysis. Here again, the "success" of the model is based on one's goals

Table 7.10. Residual (observed − predicted) statistics for densities of five bird species on six study plots at the research ranch (RR), 1983

Model	Species	$\bar{x}$	SD
Multiple linear regression	Cassin's sparrow	6.41	44.66
	Botteri's sparrow	−12.70	15.98
	Black-throated sparrow	−140.15*	62.17
	Brown towhee	168.64*	44.15
	Blue grosbeak	5.25	3.67
Principal components regression	Cassin's sparrow	−195.35*	78.01
	Botteri's sparrow	6.40	14.58
	Black-throated sparrow	85.36*	32.34
	Brown towhee	45.08*	16.90
	Blue grosbeak	0.80	5.25

Source: Maurer 1986, table 5 © 1986 The Wildlife Society. Predicted values for RR plots were obtained from statistical models relating density to habitat variables on the Santa Rita Experimental Range, 1982–83.
*95 percent CI for $\bar{x}$ residual does not include 0.

in developing the model in the first place. Further, because there is probably no single "best" model, there is no single best or unique solution in model validation. Thus, validation is a process for determining when and if a statistical model meets some minimal set of adequacy criteria.

Studies of habitat use increasingly emphasize model validation (see Marcot, Raphael, and Berry 1983; Capen et al. 1986; Maurer 1986; and Morrison, Timossi, and With 1987). Maurer (1986) found that initial models developed using both multiple linear regression and principal components regression performed poorly when validated using data from a site different from that used to develop the models (table 7.10). Morrison, Timossi, and With (1987) used two different strategies to validate multiple regression models that were developed to predict bird abundance from habitat data. First, they counted birds on the same set of points during 1983 and 1984 and used models based on 1983 data to predict bird abundance for 1984; this they termed a "same place, different time" validation. Second, they counted birds at a different set of points in 1985 and then used the combined 1983–84 bird-vegetation models to predict the 1985 abundances; this they termed a "different place, different time" validation. They found that their models generally underestimated bird abundance. They concluded that their models failed to adequately predict bird abundance but did indicate presence-absence for most species. The study does illustrate, however, a useful approach to validating models.

"Canned" Statistical Packages

The advent of high-speed computers and, later, readily available statistical programs has led to the ever-increasing use of multivariate analyses. Unfortunately, this has also led to an increase in the absolute number of misuses of these methods. Previously only available on mainframes, many of these statistical packages are now available on microcomputers. The most powerful and widely available statistical packages are the Statistical Analytical System (SAS), the Biomedical Computer Programs (BMDP), the Statistical Package for the Social Sciences (SPSS), and the System for Statistics (SYSTAT). Each of these is well documented and supported; several or all of these, plus various other packages, are available at most universities.

Novice users of these packages should be aware that the default settings for the specific analytical methods often must be adjusted for each application. Further, each package has a detailed set of options and statistics that must be specifically requested. For example, violation of assumptions are seldom identified unless specifically requested (such as Box's M), the default P-values may not be appropriate for your analysis, and/or the variable selection procedure must be specified (for example, stepwise, forward, all subsets, and F-to-enter values for stepwise regression).

Many university computer centers offer short courses in the use of statistical packages, and an increasing number of professors are including such material in graduate-level courses. Further, the user manuals for each of the most popular packages are becoming increasingly user-friendly. This is especially true of SPSS, which offers a general user's manual and both basic and advanced manuals that explain most statistical procedures. Various statistical textbooks also include examples using output from one or more of the major packages; these texts include those by Berenson, Levine, and Goldstein (1983), Tabachnick and Fidell (1983), Afifi and Clark (1984), and Harris (1985).

Literature Cited

Afifi, A. A., and V. Clark. 1984. *Computer-aided multivariate analysis*. Belmont, Calif.: Lifetime Learning Publications.

Ben-Shahar, R., and J. D. Skinner. 1988. Habitat preferences of African ungulates derived by uni- and multivariate analyses. *Ecology* 69:1479–85.

Berenson, M. L., D. M. Levine, and M. Goldstein. 1983. *Intermediate statistical methods and applications: A computer package approach*. Englewood Cliffs, N.J.: Prentice-Hall.

Block, W. M., K. A. With, and M. L. Morrison. 1987. On measuring bird habitat: Influence of observer variability and sample size. *Condor* 89:241–51.

Brennan, L. A., W. M. Block, and R. J. Gutierrez. 1986. The use of multivariate statistics for developing habitat suitability index models. In *Wild-*

life 2000: Modeling habitat relationships of terrestrial vertebrates, ed. J. Verner, M. L. Morrison, and C. J. Ralph, 177–82. Madison: Univ. of Wisconsin Press.

Capen, D. E., 1981. *The use of multivariate statistics in studies of wildlife habitat.* USDA Forest Service General Technical Report RM–87.

Capen, D. E., J. W. Fenwick, D. B. Inkley, and A. C. Boynton. 1986. Multivariate models of songbird habitat in New England forests. In *Wildlife 2000: Modeling habitat relationships of terrestrial vertebrates*, ed. J. Verner, M. L. Morrison, and C. J. Ralph, 171–75. Madison: Univ. of Wisconsin Press.

Carnes, B. A., and N. A. Slade. 1982. Some comments on niche analysis in canonical space. *Ecology* 63:888–93.

Colgan, P. W., ed. 1978. *Quantitative ethology*. New York: John Wiley and Sons.

Collins, S. L. 1983. Geographic variation in habitat structure of the black-throated green warbler (*Dendroica virens*). *Auk* 100:382–89.

Collins, S. L., F. C. James, and P. G. Risser. 1982. Habitat relationships of wood warblers (Parulidae) in north central Minnesota. *Oikos* 39:50–58.

Cooley, W. W., and P. R. Lohnes. 1971. *Multivariate data analysis*. New York: John Wiley and Sons.

Digby, P. G. N., and R. A. Kempton. 1987. *Multivariate analysis of ecological communities*. New York: Chapman and Hall.

Dillon, W. R., and M. Goldstein. 1984. *Multivariate analysis: Methods and applications*. New York: John Wiley and Sons.

Draper, N. R., and H. Smith. 1981. *Applied regression analysis*. 2d ed. New York: John Wiley and Sons.

Folse, L. J., Jr. 1981. Ecological relationships of grassland birds to habitat and food supply in East Africa. In *The use of multivariate statistics in studies of wildlife habitat*, ed. D. E. Capen, 160–66. USDA Forest Service General Technical Report RM–87.

Green, R. H. 1971. A multivariate statistical approach to the Hutchinsonian niche: Bivalve molluscs in central Canada. *Ecology* 52:543–56.

Harris, R. J. 1985. *A primer on multivariate statistics*. 2d ed. Orlando, Fla.: Academic Press.

Hassler, C. C., S. A. Sinclair, and E. Kallio. 1986. Logistic regression: A potentially useful tool for researchers. *Forest Products Journal* 36:16–18.

Hill, M. O. 1974. Correspondence analysis: A neglected multivariate method. *Journal of the Royal Statistical Society*, ser. C, 23:340–54.

Hill, M. O., and H. G. Gauch, Jr. 1980. Detrended correspondence analysis: An improved ordination technique. *Vegetatio* 42:47–58.

Hosmer, D. W., and S. Lemeshow. 1989. *Applied logistic regression*. New York: John Wiley and Sons.

Hutchinson, G. E. 1957. Concluding remarks. *Cold Spring Harbor Symposium on Quantitative Biology* 22:415–27.

James, D. A., and M. J. Lockerd. 1986. Refinement of the Shugart-Patten-Dueser model for analyzing ecological niche patterns. In *Wildlife 2000: Modeling habitat relationships of terrestrial vertebrates*, ed. J. Verner, M. L. Morrison, and C. J. Ralph, 51–56. Madison: Univ. of Wisconsin Press.

Johnson, D. H. 1981. The use and misuse of statistics in wildlife habitat studies. In *The use of multivariate statistics in studies of wildlife habitat*, ed. D. E. Capen, 11–19. USDA Forest Service General Technical Report RM–87.

Karr, J. R., and T. E. Martin. 1981. Random numbers and principal components: Further searches for the unicorn? In *The use of multivariate statistics in studies of wildlife habitat*, ed. D. E. Capen, 20–24. USDA Forest Service General Technical Report RM–87.

Kenkel, N. C., and L. Orloci. 1986. Applying metric and nonmetric multidimensional scaling to ecological studies: Some new results. *Ecology* 67:919–28.

Kruskal, J. B., and M. Wish. 1978. *Multidimensional scaling*. Quantitative applications in the social sciences, no. 11. Beverly Hills, Calif.: Sage Publications.

Lanyon, S. M. 1987. Jackknifing and bootstrapping: Important "new" statistical techniques for ornithologists. *Auk* 104:144–46.

Marascuilo, L. A., and J. R. Levin. 1983. *Multivariate statistics in the social sciences: A researcher's guide*. Monterey, Calif.: Brooks-Cole.

Marcot, B. G., M. G. Raphael, and K. H. Berry. 1983. Monitoring wildlife habitat and validation of wildlife-habitat relationships models. *Transactions of the North American Wildlife and Natural Resources Conference* 48:315–29.

Maurer, B. A. 1986. Predicting habitat quality for grassland birds using density-habitat correlations. *Journal of Wildlife Management* 50:556–66.

Maurer, B. A., L. B. McArthur, and R. C. Whitmore. 1981. Habitat associations of breeding birds in clearcut deciduous forests in West Virginia. In *The use of multivariate statistics in studies of wildlife habitat*, ed. D. E. Capen, 167–72. USDA Forest Service General Technical Report RM–87.

Meents, J. K., J. Rice, B. W. Anderson, and R. D. Ohmart. 1983. Nonlinear relationships between birds and vegetation. *Ecology* 64:1022–27.

Meyer, J. S., C. G. Ingersoll, L. L. McDonald, and M. S. Boyce. 1986. Estimating uncertainty in population growth rates: Jackknife vs. bootstrap techniques. *Ecology* 67:1156–66.

Miles, D. B. 1990. A comparison of three multivariate statistical techniques for the analysis of avian foraging data. *Studies in Avian Biology* 13:295–308.

Morrison, D. F. 1967. *Multivariate statistical methods*. 2d ed. New York: McGraw-Hill.

Morrison, M. L. 1981. The structure of western warbler assemblages: Analysis of foraging behavior and habitat selection in Oregon. *Auk* 98:578–88.

Morrison, M. L. 1984. Influence of sample size on discriminant function analysis of habitat use by birds. *Journal of Field Ornithology* 55:330–35.

Morrison, M. L., and R. G. Anthony. 1989. Habitat use by small mammals on early-growth clear-cuttings in western Oregon. *Canadian Journal of Zoology* 67:805–11.

Morrison, M. L., C. J. Ralph, J. Verner, and J. R. Jehl, Jr., eds. 1990. Avian foraging: Theory, methodology, and applications. *Studies in Avian Biology* no. 13.

Morrison, M. L., I. C. Timossi, and K. A. With. 1987. Development and testing of linear regression models predicting bird-habitat relationships. *Journal of Wildlife Management* 51:247–53.

Morrison, M. L., K. A. With, and I. C. Timossi. 1986. The structure of a forest bird community during winter and summer. *Wilson Bulletin* 98:214–30.

Neff, N. A., and L. F. Marcus. 1980. A survey of multivariate methods for systematics. New York: privately published. Printed by the American Museum of Natural History, New York.

Neter, J., and W. Wasserman. 1974. *Applied linear statistical models*. Homewood, Ill.: Richard D. Irwin.

Noon, B. R. 1986. Summary: Biometric approaches to modeling—the researcher's viewpoint. In *Wildlife 2000: Modeling habitat relationships of terrestrial vertebrates*, ed. J. Verner, M. L. Morrison, and C. J. Ralph, 197–201. Madison: Univ. of Wisconsin Press.

Norusis, M. J. 1985. *SPSS^x advanced statistics guide*. New York: McGraw-Hill.

Pielou, E. C. 1984. *The interpretation of ecological data*. New York: John Wiley and Sons.

Pimentel, R. A. 1979. Morphometrics. Dubuque, Iowa: Kendall/Hunt Publishing Co.

Press, S. J., and S. Wilson. 1978. Choosing between logistic regression and discriminant analysis. *Journal of the American Statistical Association* 73:699–705.

Rexstad, E. A., D. D. Miller, C. H. Flather, E. M. Anderson, J. W. Hupp, and D. R. Anderson. 1988. Questionable multivariate statistical inference in wildlife habitat and community studies. *Journal of Wildlife Management* 52:794–98.

Shugart, H. H., Jr. 1981. An overview of multivariate methods and their application to studies of wildlife habitat. In *The use of multivariate statistics in studies of wildlife habitat*, ed. D. E. Capen, 4–10. USDA Forest Service General Technical Report RM–87.

Smith, K. G. 1981. Canonical correlation analysis and its use in wildlife habitat studies. In *The use of multivariate statistics in studies of wildlife habitat*, ed. D. E. Capen, 80–92. USDA Forest Service General Technical Report RM–87.

Sneath, P. H. A., and R. R. Sokal. 1973. *Numerical taxonomy*. San Francisco: W. H. Freeman.

Tabachnick, B. G., and L. S. Fidell. 1983. *Using multivariate statistics*. New York: Harper and Row.

Tacha, T. C., W. D. Warde, and K. P. Burnham. 1982. Use and interpretation of statistics in wildlife journals. *Wildlife Society Bulletin* 10:355–62.

Verner, J., M. L. Morrison, and C. J. Ralph, eds. 1986. *Wildlife 2000: Modeling habitat relationships of terrestrial vertebrates*. Madison: Univ. of Wisconsin Press.

Wartenberg, D., S. Ferson, and J. F. Rohlf. 1987. Putting things in order: A critique of detrended correspondence analysis. *American Naturalist* 129:434–48.

Wesolowsky, G. O. 1976. *Multiple regression and analysis of variance*. New York: John Wiley and Sons.

Whittaker, R. J. 1987. An application of detrended correspondence analysis and non-metric multidimensional scaling to the identification and analysis of environmental factor complexes and vegetation structures. *Journal of Ecology* 75:363–76.

Williams, B. K. 1983. Some observations on the use of discriminant analysis in ecology. *Ecology* 64:1283–91.

Index